ADAPTIVE COOPERATIVE SYSTEMS

Adaptive and Learning Systems for Signal Processing, Communications, and Control

Editor: Simon Haykin

Werbos / THE ROOTS OF BACKPROPAGATION: From Ordered Derivatives to Neural Networks and Political Forecasting

Krstić, Kanellakopoulos, and Kokotović / NONLINEAR AND ADAPTIVE CONTROL DESIGN

Nikias and Shao / SIGNAL PROCESSING WITH ALPHA-STABLE DISTRIBUTIONS AND APPLICATIONS

Diamantaras and Kung / PRINCIPAL COMPONENT NEURAL NETWORKS: Theory and Applications

Tao and Kokotović / ADAPTIVE CONTROL OF SYSTEMS WITH ACTUATOR AND SENSOR NONLINEARITIES

Tsoukalas and Uhrig / FUZZY AND NEURAL APPROACHES IN ENGINEERING

Hrycej / NEUROCONTROL: Towards an Industrial Control Methodology

Beckerman / ADAPTIVE COOPERATIVE SYSTEMS

ADAPTIVE COOPERATIVE SYSTEMS

Martin Beckerman

Oak Ridge National Laboratory
Oak Ridge, Tennessee

A Wiley-Interscience Publication
JOHN WILEY & SONS, INC.
New York / Chichester / Weinheim / Brisbane / Singapore / Toronto

This text is printed on acid-free paper.

Copyright © 1997 by John Wiley & Sons, Inc.

All rights reserved. Published simultaneously in Canada.

Reproduction or translation of any part of this work beyond that permitted by Section 107 or 108 of the 1976 United States Copyright Act without the permission of the copyright owner is unlawful. Requests for permission or further information should be addressed to the Permissions Department, John Wiley & Sons, Inc., 605 Third Avenue, New York, NY 10158-0012.

Library of Congress Cataloging in Publication Data:
Beckerman, Martin.
    Adaptive cooperative systems / Martin Beckerman.
      p. cm. -- (Adaptive and learning systems for signal processing, communications, and control)
    ISBN 0-471-01287-4 (cloth : alk. paper)
    1. Adaptive control systems. 2. Adaptive signal processing. 3. Neural networks (Computer science) I. Title. II. Series.
TJ217.B37 1997
629.8'36--dc20   96-44719

Printed in the United States of America

10 9 8 7 6 5 4 3 2 1

CONTENTS

Preface   xv

1 Introduction   1

    1.1 Prologue / 1
       1.1.1 Allosteric Proteins / 1
       1.1.2 Mutual Synchronization in Oscillator Populations / 3
       1.1.3 Order-Disorder Transitions in Atomic Lattices / 3
       1.1.4 Markovianess and a Multiplicity of States / 4
       1.1.5 Nonlinear Dynamics / 5

    1.2 Plan of the Book / 6
       1.2.1 Equilibrium Dynamics / 6
       1.2.2 Neural and Image-Processing Domains / 7
       1.2.3 Nonlinear Dynamics Far from Equilibrium / 8
       1.2.4 Rhythms and Synchrony / 9

    1.3 Epigenesis of the Central Nervous System / 11
       1.3.1 Genetic Instruction / 11
       1.3.2 Mechanical, Chemical, and Electrical Interactions / 12

    1.4 Synaptic Plasticity / 15
       1.4.1 Hebbian Synapses / 15
       1.4.2 Experience-Dependent Modifications / 16

    1.5 Neural Assemblies / 17
       1.5.1 Dynamic Adaptability / 17
       1.5.2 Assembly Coding / 18

    1.6 References / 19

2 Thermodynamics, Statistical Mechanics, and the Metropolis Algorithm   23

    2.1 Introduction / 23

    2.2 Key Concepts / 24

2.3   The Measure of Uncertainty / 25

2.4   The Most Probable Distribution / 28
  2.4.1   Configurations and Weight Factors / 28
  2.4.2   Entropic Forms / 30

2.5   The Method of Lagrange Multipliers / 31

2.6   Statistical Mechanics / 33
  2.6.1   Formal Structure of the Theory / 33
  2.6.2   External Parameters / 36

2.7   Thermodynamics / 38
  2.7.1   Equilibrium States / 38
  2.7.2   The Correspondence between Statistical Mechanics and Thermodynamics / 39

2.8   The Ensembles of Statistical Mechanics / 42
  2.8.1   Microcanonical Ensemble / 42
  2.8.2   The Canonical Ensemble (Gibbs Distributions) / 43
  2.8.3   Helmholtz Free Energy / 44
  2.8.4   Energy Fluctuations / 44
  2.8.5   Grand Canonical Ensemble / 45

2.9   The Monte Carlo Method / 46
  2.9.1   Definition / 46
  2.9.2   Study of Reaction Processes / 47
  2.9.3   Problems in Statistical Mechanics / 49

2.10 Markov Chains / 50
  2.10.1   Definitions / 50
  2.10.2   Transition Probabilities / 51
  2.10.3   Convergence to Stationary Distributions / 53

2.11 The Metropolis Algorithm / 54

2.12 Concluding Remarks / 56

2.13 Further Reading / 57

2.14 References / 58

3   Cooperativity in Lattice Systems   61

3.1   Introduction / 61
  3.1.1   Configurations and the Counting Problem / 62
  3.1.2   Objectives of the Chapter / 63

3.2   Key Concepts / 65

3.2.1 Entropic and Energetic Potentials / 65
3.2.2 Phase Transitions / 65
3.2.3 Order Parameters / 68
3.2.4 Critical Exponents and Scaling / 69
3.2.5 Correlation Lengths / 70

3.3 The Ising Model / 70
3.3.1 Ferromagnetism / 70
3.3.2 The Classical Limit and XY-Model / 72
3.3.3 Historical Development of the Ising Model / 73
3.3.4 The Ising Hamiltonian and Partition Function / 74
3.3.5 Thermodynamic Parameters / 75

3.4 The Ising Model in One Dimension / 76

3.5 The Ising Model in Two Dimensions / 80

3.6 Strong Cooperativity and Peierls's Argument / 83

3.7 Mean-Field Theory / 85
3.7.1 The Weiss Molecular Field Equation / 85
3.7.2 The Bragg-Williams (Random Mixing) Method / 86
3.7.3 The Bethe Approximation and Synopsis of Mean-Field Results / 89

3.8 The Lattice Gas Model of Fluids / 91
3.8.1 The Lattice Gas / 92
3.8.2 The van der Waals Equation / 94
3.8.3 The Triple and Critical Points / 95

3.9 The Renormalization Group / 96
3.9.1 Coupled Degrees of Freedom / 96
3.9.2 The Kadanoff Block Spin Construction / 97
3.9.3 The Renormalization Group and Fixed Points / 98
3.9.4 Widom-Kadanoff Scaling / 101

3.10 Summary / 103

3.11 Additional Reading / 105

3.12 References / 105

4 Simulated Annealing   **108**

4.1 Introduction / 108
4.1.1 Energy Landscapes / 108
4.1.2 Multiple Constraints and Selectivity / 110
4.1.3 Searching for Optimal Solutions / 110

4.2 Objectives / 111

4.3 Kinetic Ising Model / 112
   4.3.1 The Master Equation / 113
   4.3.2 Spin Kinetics / 113

4.4 Order-Disorder Transitions in Binary Alloys / 115
   4.4.1 Order-Disorder Transitions / 115
   4.4.2 Particle Kinetics and Binary Alloys / 116

4.5 Spin Glasses / 117
   4.5.1 Introduction / 117
   4.5.2 Annealed and Quenched Random Variables / 119
   4.5.3 Replicas / 119
   4.5.4 The TAP Equations / 121
   4.5.5 The Parisi Order Parameter and Ultrametricity / 122
   4.5.6 Critical Behavior / 123
   4.5.7 The Energy Landscape and Ergodicity Breaking / 125

4.6 Combinatorial Optimization and NP-Completeness / 126
   4.6.1 Combinatorial Optimization / 126
   4.6.2 The World of NP-Completeness / 126

4.7 Optimization by Simulated Annealing / 128
   4.7.1 Introduction / 128
   4.7.2 The Simulated Annealing Algorithm / 129
   4.7.3 The Cooling Schedule and Convergence of the Algorithm / 131

4.8 The Traveling Salesman and Graph-Partitioning Problems / 134
   4.8.1 Traveling Salesman Problems / 134
   4.8.2 Graph Partitioning / 136

4.9 Microcanonical Annealing / 137

4.10 Continuous Simulated Annealing / 140

4.11 Further Reading / 142

4.12 References / 142

5 The Patterning of Neural Connections   146

5.1 The Patterning of Neural Connections in the Central Nervous System / 147
   5.1.1 Laminar Structure of the Mammalian Lateral Geniculate Nucleus / 148
   5.1.2 The Retinotectal Projection in Lower Vertebrates / 150

5.2 Multiple Interactions / 150

5.3   Objectives of the Chapter / 152
5.4   Developmental Models / 153
      5.4.1 Minimalist Marker Models / 155
      5.4.2 Molecular Gradients / 156
5.5   The Multiple Constraint Model / 156
      5.5.1 Position-Independent Affinity / 157
      5.5.2 Fiber-Fiber Repulsion / 158
      5.5.3 Nearest-Neighbor Correlated Activity / 159
      5.5.4 Position-Dependent Affinity / 159
      5.5.5 Multiple Stable States in the Retinotectal Projection / 160
5.6   Morphogenesis of the Lateral Geniculate Nucleus / 162
      5.6.1 Interaction Potentials / 163
      5.6.2 Induction of the Laminar Transition / 166
      5.6.3 Trapping of the Transition by the Blind Spot / 167
5.7   Growth Cone Guidance and Neurite Outgrowth / 168
      5.7.1 Chemoattractants / 169
      5.7.2 Chemorepulsants / 170
      5.7.3 Cell Adhesion Molecules / 171
5.8   Signal Transduction and Integration / 172
      5.8.1 Dynamic Regulation of Cell Surface Receptors / 172
      5.8.2 Intracellular Calcium Signaling / 173
      5.8.3 G-Proteins / 174
5.9   Dynamics / 176
      5.9.1 Reorganization in the Adult Cortex / 176
      5.9.2 Time Course of Developmental Events / 177
5.10 References / 177

6 Markov Random Fields   182

6.1   Definitions and Introductory Remarks / 182
      6.1.1 The Markov P-Process / 182
      6.1.2 Markov Random Fields and Neighborhood Systems / 183
      6.1.3 Gibbs Random Fields and Clique Potentials / 185
      6.1.4 The Gibbs-Markov Equivalence / 186
6.2   Random Fields and Image Processes / 187
      6.2.1 Random Field Models / 187
      6.2.2 Gibbs Distributions and Simulated Annealing / 188
      6.2.3 Potentials and Deterministic Approaches / 189
6.3   The Hammersley-Clifford Theorem for Finite Lattices / 189

6.4 Random Fields on Graphs / 191
    6.4.1 Cooperative Assemblies / 191
    6.4.2 Gibbs-Markov Equivalence for Random Fields on Graphs / 192
    6.4.3 Higher-Order Interactions / 196

6.5 Random Field Models / 196
    6.5.1 Besag's Auto Models / 197
    6.5.2 The Autobinomial Model / 198
    6.5.3 Wide-Sense Markov Processes / 199
    6.5.4 Gauss-Markov Random Fields / 201

6.6 Simultaneous Autoregressive Models / 202
    6.6.1 Simultaneous Autoregressive Random Fields / 202
    6.6.2 Image Reconstruction / 204
    6.6.3 Fourier Computation and Block Circulant Matrices / 206

6.7 The Method of Geman and Geman / 208
    6.7.1 Maximum A posteriori (MAP) Estimation / 209
    6.7.2 Clique Potentials / 211
    6.7.3 The Posterior Distribution / 212
    6.7.4 The Gibbs Sampler / 213

6.8 Maximizer of the Posterior Marginals / 215
    6.8.1 Optimal Bayesian Estimation / 215
    6.8.2 Segmentation and Reconstruction / 216
    6.8.3 The MPM Algorithm / 218

6.9 Iterated Conditional Modes of the Posterior Distribution / 219
    6.9.1 Annealing and Quenching / 219
    6.9.2 Quenching and ICM / 219
    6.9.3 The ICM Algorithm / 223

6.10 Coupled Markov Random Fields / 223
    6.10.1 The Line Process / 223
    6.10.2 Coupled Markov Random Fields and Integration / 226

6.11 Compound Gauss-Markov Random Fields / 230

6.12 Mean-Field Annealing / 232
    6.12.1 Mechanical Models, Graduated Nonconvexity, and Mean-Field Theory / 232
    6.12.2 Mean-Field Annealing / 233
    6.12.3 Convex Functions and Jensen's Inequality / 234
    6.12.4 The Mean-Field Annealing Algorithm / 236

- 6.13 Graduated Nonconvexity / 237
  - 6.13.1 Weak Continuity Constraints / 238
  - 6.13.2 Elimination of the Line Process / 239
  - 6.13.3 Nonconvexity / 241
  - 6.13.4 Convex Approximation / 241
- 6.14 Effective Potentials / 242
- 6.15 Renormalization Group Simulated Annealing / 243
  - 6.15.1 Kadanoff Transformations / 244
  - 6.15.2 The Gidas Algorithm / 245
- 6.16 References / 248

7 The Approach to Equilibrium   253

- 7.1 Nonequilibrium Dynamics / 253
  - 7.1.1 Relaxation and Equilibrium Fluctuations / 253
  - 7.1.2 Brownian Motion / 254
  - 7.1.3 Markov Processes / 255
- 7.2 Outline of the Chapter / 256
- 7.3 The Fokker-Planck Equation / 256
  - 7.3.1 Brownian Motion / 256
  - 7.3.2 Fokker-Planck Equation for Brownian Motion / 257
- 7.4 The Langevin Equation / 259
  - 7.4.1 Formal Solution / 260
  - 7.4.2 The Mean Square Displacement / 261
  - 7.4.3 Properties of the Random Force / 263
- 7.5 Stochastic Differential Equations / 265
  - 7.5.1 The Wiener Process / 265
  - 7.5.2 Global Optimization / 266
- 7.6 Fluctuations and Dissipation / 267
- 7.7 The Regression of Spontaneous Fluctuations / 270
  - 7.7.1 Stationary Processes / 270
  - 7.7.2 The Velocity Correlation Function / 271
- 7.8 The Velocity Increments / 272
  - 7.8.1 The Mean Incremental Change in Velocity / 272
  - 7.8.2 The Mean Square Incremental Change in Velocity / 275
- 7.9 Further Reading / 276
- 7.10 References / 277

8 Synaptic Plasticity   **278**

    8.1   Synaptic Plasticity in the Visual Cortex / 278
        8.1.1   Forms of Synaptic Plasticity / 278
        8.1.2   Use-Dependent Changes in Synaptic Efficiency / 279

    8.2   Models of Synaptic Modification / 280
        8.2.1   Feature Selectivity / 280
        8.2.2   Stability / 281
        8.2.3   Information Processing in the Visual System / 282

    8.3   Outline of the Chapter / 283

    8.4   Dynamics / 284
        8.4.1   Phase Space / 284
        8.4.2   Trajectories in Phase Space / 285
        8.4.3   Conservative and Dissipative Systems / 286

    8.5   Liouville's Theorem / 286
        8.5.1   Hamilton's Equations / 286
        8.5.2   Liouville's Theorem of Volume Conservation / 287
        8.5.3   Initial Conditions / 289

    8.6   BCM Theory: Single-Cell Formulation / 290
        8.6.1   The Linear Integrator / 290
        8.6.2   Rule for Synaptic Modification / 292
        8.6.3   Spatial and Temporal Competition / 294
        8.6.4   Stability of the Fixed Points / 295

    8.7   Cortical Response Properties / 297
        8.7.1   Circular Environment / 297
        8.7.2   Classical Rearing / 297
        8.7.3   Orientation Selectivity and Binocular Interactions / 298
        8.7.4   Receptive Field Properties / 301

    8.8   Mean-Field Network / 303
        8.8.1   Mean-Field Approximation / 303
        8.8.2   The Cortical Network / 306

    8.9   Neurophysiological Basis / 307
        8.9.1   Dendritic Spines / 307
        8.9.2   NMDA and AMPA Receptors / 308
        8.9.3   The LTP/LTD Crossover and the BCM Modification Threshold / 309

    8.10   From Genes to Behavior / 310
        8.10.1   CaM Kinase / 311
        8.10.2   Griffith *Drosophila* Data / 312

      8.10.3 Movement of the BCM Threshold / 313
      8.10.4 Plastic Gates / 315
 8.11 Principal Component Neurons / 315
      8.11.1 Introductory Remarks / 315
      8.11.2 Principal Components / 316
      8.11.3 Principal Components and Constrained Optimization / 317
      8.11.4 Hebbian Learning and Synaptic Constraints / 318
      8.11.5 Oja's Solution / 319
      8.11.6 Linsker's Model / 320
 8.12 Synaptic and Phenomenological Spin Models / 321
      8.12.1 Phenomenological Spin Models / 321
      8.12.2 Synaptic Models in the Common Input Approximation / 322
 8.13 Objective Function Formulation of BCM Theory / 325
      8.13.1 Projection Pursuit / 326
      8.13.2 Objective Function Formulation of BCM Theory / 327
 8.14 References / 329

9 Rhythms and Synchrony   335

 9.1 Biological Rhythms and Synchrony / 335
      9.1.1 Nonlinear Dynamics / 335
      9.1.2 Excitable Membranes / 336
      9.1.3 Population Oscillations / 337
      9.1.4 Neural Rhythms and Synchrony / 338
 9.2 Outline of the Chapter / 340
 9.3 Phase-Coupled Oscillators / 342
      9.3.1 Mutual Synchronization / 342
      9.3.2 Entrainment in a Rotator Model / 343
      9.3.3 Mean-Field Model / 344
      9.3.4 The Random Pinning Model / 346
      9.3.5 Lattice Model / 349
      9.3.6 Frequency Plateaus / 350
 9.4 Clustering of Globally Coupled Phase Oscillators / 352
 9.5 Phase Space / 356
      9.5.1 Linear Stability Analysis / 356
      9.5.2 Nullclines / 359
      9.5.3 Poincaré-Bendixson Theorem / 361
      9.5.4 Hopf Bifurcation / 362

- 9.6 Population Dynamics in the Wilson-Cowan Model / 362
- 9.7 Oscillations and Synchrony in the Visual Cortex and Hippocampus / 367
  - 9.7.1 Mean-Field Model of Cortical Oscillations / 368
  - 9.7.2 Delay Connections and Nearest-Neighbor Interactions / 372
  - 9.7.3 Burst Synchronization / 373
  - 9.7.4 Rhythmic Population Oscillations in the Hippocampus / 374
  - 9.7.5 Feature Integration / 378
- 9.8 Neural Excitability and Oscillations / 381
  - 9.8.1 The Hodgkin-Huxley Equations / 382
  - 9.8.2 The Morris-Lecar Model / 385
  - 9.8.3 Waves and Synchrony in Systems of Relaxation Oscillators / 390
- 9.9 Spindle Waves / 394
  - 9.9.1 Ionic Mechanisms / 395
  - 9.9.2 Network Mechanisms / 399
- 9.10 Calcium Oscillations, Excitable Media, and Cellular Automata / 400
  - 9.10.1 Calcium Oscillations and Spiral Waves / 401
  - 9.10.2 Excitable Media / 401
  - 9.10.3 Cellular Automata / 403
- 9.11 Additional Reading / 406
- 9.12 References / 406

Glossary 413

Index 421

PREFACE

The world about us is composed of a multitude of nonlinear dynamic systems, each containing a number of interacting elements and each operating either near or at equilibrium or far from equilibrium. These systems, of which the human brain is perhaps the most spectacular example, are intrinsically adaptive and cooperative. During the past few years, striking advances in our understanding of these systems have taken place, and their study has become a major thrust in science. We have witnessed a revolution in genetics and molecular biology, the emergence of new experimental techniques in brain research, and advances in our theoretical modeling capabilities made possible by digital computers.

The study of adaptive cooperative systems is a multidisciplinary endeavor, drawing upon findings in physics, mathematics, statistics, computer science, and other fields depending upon the application domain, for inspiration and guidance. This is true in narrowly defined activities such as digital image processing as well as in broad areas of study such as brain research. Thus, we observe a confluence of biological, statistical, and physical reasoning in the simulated annealing algorithms used in conjunction with Markov random fields to carry out image-processing tasks. The brain contains multiple, interacting levels of organization. The goal of computational neuroscience is to integrate information from the multiple levels in order to elucidate the organization principles, explore structure-function relations, and identify the information processing tasks being carried out. A multidisciplinary approach is a central feature of computational neuroscience, which has emerged to join with anatomy, biology, pharmacology, physiology, psychology, biochemistry, and biophysics in contributing to our understanding of brain function.

We are only in the early stages of developing an understanding of dynamic systems such as the brain. Neuroscience is barely one hundred years old and computational neuroscience is far younger. As a result, we are still in the process of establishing a conceptual framework and worldview to guide inquiry into the behavior of adaptive cooperative systems. At this stage in the evolution of our thinking, theoretical models such as Ising ferromagnets and Sherrington-Kirkpatrick spin glasses, communities of coupled nonlinear oscillators, Hebbian synapses, and ion channels linked through Hodgkin-Huxley dynamics are especially valuable in helping us formulate questions to ask. Some of these theoretical constructs describe self-organization in physical systems while others represent processes in biological systems. Although the application domains may differ, the processes being modeled are conceptually linked to one another, sharing a common language of probabilistic reasoning and a dynamics described by systems of coupled ordinary differential equations.

My objectives in writing *Adaptive Cooperative Systems* is to present a unified treatment of self-organizing processes, drawing upon examples from physics, spatial statistics, image processing, and brain science. This will be accomplished by developing the formal structure of the probabilistic reasoning, and then showing how this machinery can be used in each domain to construct algorithms which generate useful near-optimal solutions to problems. The algorithms will be applied to order-disorder transitions in magnetic (spin) systems, the epigenesis of the vertebrate visual system, and image reconstruction tasks. This will be followed by an exploration of the nonequilibrium dynamics of open systems, presented in the context of synaptic plasticity in the visual cortex, and rhythmicity and synchrony in the central nervous system.

Spin systems occupy a special place in the evolution of the theory of adaptive cooperative systems. Several key concepts that will reappear throughout the book originated in studies of these simple model systems. The image-processing domain is interesting not only because it complements research in the visual system, but also because it provides us with variety of annealing and annealing-like algorithms. These algorithms emerge from the combination of an equilibrium dynamics with random field models designed to capture correlations in the data. Synaptic plasticity is a central theme in brain theory. It will be studied first from the viewpoint of equilibrium dynamics and then in greater detail using nonequilibrium dynamics. Rhythms and synchrony, our fourth topic area, permits us to introduce the notion of temporal order while exploring some of the most dramatic examples of adaptivity and cooperativity in nonlinear dynamic systems.

A number of themes will recur throughout our discussions of these topics. For example, we will encounter in various places in the book mutual synchronization and cluster formation, spontaneous symmetry breaking and feature selectivity, integration by labeling and feature binding, order parameters and mean fields, scaling, and renormalization groups. The aforementioned concepts and principles, together with an information processing viewpoint, serve to link the problem domains to one another, and comprise an intellectual heritage shared by researchers exploring self-organizing phenomena in their respective fields of research.

I have included discussions of the biochemical and neurophysiological substrate underlying the phenomena in the brain being modeled, in order to establish context for the theoretical models. The experimental results that I have presented are intended to not only support the models but also highlight future challenges. The revolutionary developments in genetics, molecular biology, and electophysiology have led to an explosive growth in the literature. The experimental literature is so vast that only a few of the many important topics can be mentioned briefly, and it is my hope that my all-too-short overviews will serve as a guidepost to the literature for the interested reader. The chapter on synaptic plasticity and Bienenstock-Cooper-Munro (BCM) theory includes mentions of N-methyl-D-aspartate (NMDA) and α-amino-3-hydroxy-5-methyl-4-isoxazolepropionate (AMPA) receptors, long-term potential (LTP) and long-term depression (LTD), protein kinase networks, and intracellular calcium signaling. The development of the equilibrium theory of how eye-specific afferents segregate themselves in the vertebrate visual system is accompanied by a sketch of the experimental findings on cell-surface molecules and their up- and down-regulation, chemotropic factors, G-proteins, and signal transduction. Finally, many of the recent studies of neural oscillations have been motivated by experimental results acquired using simultaneous multicellular recording techniques. These results are presented in the chapter on rhythms and synchrony and in several other places where appropriate.

It is worth noting that several sources of tension naturally arise in broad endeavors that cross specialty boundaries. First, there is a normal tension between the desire to extract general principles shared by many different systems independent of their domain and the drive to elucidate the essential uniqueness of each (biological) system. Second, there are problems in communication due to differences in domain knowledge and research goals among researchers in physical, biological, engineering, and mathematical disciplines. Third, there are problems associated with the voluminous literature. As is perhaps inevitable in any rapidly expanding field, there are experimental results whose significance are unclear and those that are seemingly inconsistent. There are gaps between what is known, and accepted by all as such, and hard conclusions regarding perception, learning, memory, and so on. The objective of the research producing the theoretical models described in the book is primarily to elucidate mechanisms, i.e., to address "how" questions and their direct consequences. A start towards answering the far more difficult questions of a "what" nature is represented by the information-processing studies. We all recognize that the tensions described above can be highly productive if they take the form of a dialogue, and it is my hope that the book may assist in some small measure in achieving this goal.

Many people contributed in various ways to the success of this project. In particular, I wish to thank my colleagues Mark Bear, Elie Bienenstock, Rama Chellappa, Avis Cohen, Leon Cooper, Scott Fraser, George Gerstein, Charles Gray, Basilis Gidas, Nathan Intrator, Anil Jain, Joe Malpeli, and Harel Shouval for discussions that were of great assistance to me in writing the book. These researchers, as well as many others whom I have not had the opportunity of meeting personally, have laid the groundwork for a dynamics of adaptive cooperative systems. Any error or omission in presenting their contributions are mine alone. My friend and colleague Dean Hartley III read portions of the manuscript and made helpful suggestions for improvement. During the early stages of the project, Simon Haykin and James Anderson provided guidance and suggestions that were greatly appreciated. I also benefitted from the constant support of my editor, George Telecki, and Rose Leo Kish at Wiley. Finally, I would not have written this book without the love and support of my family. I gratefully acknowledge my daughter Elana for her enthusiasm, interest, and excitement for the project, and my wife Barbara for her continued encouragement, forbearance, and editorial assistance during the many hours devoted to the writing of the book.

MARTIN BECKERMAN

Oak Ridge

1

INTRODUCTION

1.1 PROLOGUE

Cooperative processes, or algorithms, are procedures that generate large-scale, or global, effects through sequences of small-scale, or local, operations. These procedures underlie life as we know it. They are encountered in a variety of physical and chemical systems, and are found in biological systems at all levels of organization. Before proceeding further, let us consider a few specific examples.

1.1.1 Allosteric Proteins

Allosteric proteins are molecular agents that control and coordinate biochemical events in cells. These proteins possess two distinct binding sites, one active and one nonactive. When these sites couple to one another through conformational changes, namely alterations in the three-dimensional structure of the molecule, the binding of one ligand will regulate the affinity of the protein for the second. These adaptive cooperative processes are central to cell function.

Allosteric proteins are often made up of a number of identical subunits so that allosteric and multi-unit cooperative effects accompany one another. The symmetric arrangement of many identical subunits enables these allosteric proteins to function in a switchlike manner, rapidly undergoing a transition from the tight (T) inactive conformation to a relaxed (R) active conformation, and vice versa. This global cooperative property assists allosteric molecules in performing their regulatory functions in cells. If an multi-unit allosteric protein (enzyme) is regulated by negative feedback, it can respond rapidly to small changes in inhibitory ligand concentration.

2 INTRODUCTION

Hemoglobin, heme plus a polypeptide chain wrapped around it, is an allosteric protein designed to pick up, deliver, and release oxygen. There are four subunits in hemoglobin, each with its own binding site, and these subunits function cooperatively. Although the tetramer has a lower affinity for oxygen than its separate monomers, the binding of oxygen by one subunit in the tetramer enhances the pickup of oxygen by the others. Conversely, the loss of one oxygen molecule makes it easier to release the others.

The cooperative interactions in hemoglobin generate a sigmoidal dependence of the oxygen binding upon oxygen tension. This type of dose-response curve—a lag, followed by a rapid, almost linear, rise terminating in a saturation plateau—is a characteristic feature of multiple binding sites interacting cooperatively. As a result of the cooperativity among the hemoglobulin subunits, there are two especially stable states. These favored states are those for which the hemoglobin molecules are either fully oxygenated (the R state) or full deoxygenated (the T state) so that, in solution, most hemoglobin molecules are found in one of these states. The dose-response properties of

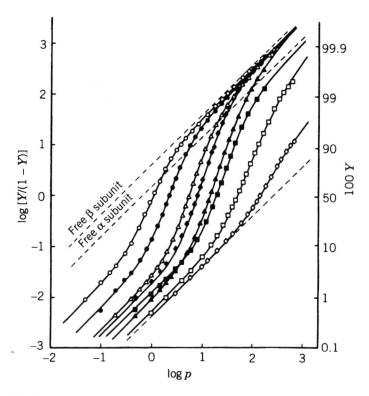

Figure 1.1 Cooperative kinetics in hemoglobin. Oxygen equilibrium curves of hemoglobin solutions are shown for different concentrations of allosteric effectors H^+, Cl^-, CO_2, 2,3-diphosphoglycerate and inositolhexaphosphate. (From Perutz[1]. Reprinted with permission of Cambridge University Press.)

hemoglobin shown in Fig. 1.1 are illustrative of the sigmoidal effect of cooperativity and of the regulation of this activity by a number of different allosteric effectors.

1.1.2 Mutual Synchronization in Oscillator Populations

A second example of cooperativity is the emergence of temporal order in populations of oscillatory units coupled to one another. When isolated, each oscillator moves according to its native frequency. When coupled together, these oscillatory units are able to synchronize their motions provided that the native frequencies are not too dissimilar. This adjustment of rhythmicity in populations of coupled oscillators is called *mutual synchronization*, or *entrainment*. The synchronized rhythmic firing of neurons in central pattern generators, the synchronized firing of pacemaker neurons in the heart, oscillations and waves in the mammalian small intestine, and circadian rhythms are a few representative examples of mutual synchronization.

The coherent chirping of nocturnal crickets heard in many regions of the United States and the fully synchronized flashing of fireflies seen in parts of southeast Asia are two additional examples of mutual synchronization, observed in insect populations. The acoustic signaling of the cricket consists of a sequence of chirps produced by the male rubbing his forewings together. The crickets synchronize their chirpings by responding to the preceding chirp of their neighbors. If a neighbor's chirp precedes his own, a cricket will shorten his chirp and the following interchirp interval. If a neighbor's chirp follows his own, he lengthens his next one or two chirps. Phases may be advanced by up to 160 degrees and retarded by as much as 200 degrees by this mechanism. In the case of fireflies, the males gather at dusk and flash on and off rhythmically to attract females. The firings are initially unsynchronized, but pockets of synchrony develop and expand in time. Eventually an entire population of many thousands of fireflies flashes on and off in unison.

1.1.3 Order-Disorder Transitions in Atomic Lattices

Mutual synchronization in oscillator populations is a temporal analog of phase transition phenomena that produce spatial order in systems such as ferromagnets. In a ferromagnet, spin one-half atoms form a two- or a three-dimensional lattice. The atomic spin elements are coupled to one another through local (Pauli exchange) interactions, and thus the spins are the counterparts of the phase variables in the oscillator populations. At high temperatures the atomic spins are randomly oriented, but as the temperature is lowered below a critical value, they begin to align themselves. As the temperature is decreased further toward zero, the percentage of aligned spins increases, and eventually all spin elements point in the same direction to produce the global effect known as *magnetism*. The transition from a high-temperature disordered arrangement of spins to a low-temperature highly ordered configuration is an example of an *order-disorder phase transition.* Similar order-disorder transitions are encountered in binary alloys and in lattice gases.

It was Arthur Winfree[2] who first noted that dynamic processes occurring far from equilibrium such as the mutual synchronization in oscillator communities are the temporal analogs of order-disorder phase transition phenomena in equilibrium lattice systems. In a ferromagnet the spontaneous magnetization provides us with a measure of the amount of the global order emerging through the sequence of local spin-spin interactions. This order parameter is zero at and above the critical temperature and increases to unity at as the temperature approaches zero. Order parameters, symmetry breaking (for a ferromagnet, the selection of a particular direction in space by the aligned spins) and critical fluctuations (large fluctuations in energy and magnetization in the vicinity of the critical temperature) are a few of the global consequences of cooperativity encountered in order-disorder phase transitions.

Still other examples of order-disorder transitions in physical systems include superconductivity and the coherent emission of laser light. A laser may be viewed as an assembly of coupled (quantum mechanical) oscillators. Laser light in the form of coherent radiation is produced by the cooperative interactions between the oscillating atomic dipoles. In this type of device, an inversion in the population distribution of atomic energy levels is produced by the supplying energy from an external source. At low pumping rates the laser emission is incoherent, namely noisy. However, when the pumping rate exceeds the laser threshold, there is transition to a temporally ordered state in which coherent emission of radiation occurs. As is the case for phase transition phenomena, order parameters, symmetry breaking and critical fluctuations accompany the transition to the synchronous or coherent activity state in this oscillator community.

The efficiency of cooperative strategies in producing global change is strikingly apparent in the folding of a protein. A protein folds through a sequence of small conformational changes. In this cooperative process a protein wraps into a native state corresponding a global minimum in the free energy. The protein is able to find this unique state among many others in an amount of time that is ten orders of magnitude shorter than would be required for an exhaustive search. In this process the cooperativity generates a sequence of local changes in conformation that enables a protein to find its global minimum without a global search.

1.1.4 Markovianess and a Multiplicity of States

Two key concepts appear in our examples of cooperativity. The first is the notion of Markovianess. This concept captures the step-by-step local nature of the interactions in a cooperative system. A system is temporally Markovian if its state at a particular time depends on its state at the immediate preceding time but not on any of its states at earlier times. Similarly a system is spatially Markovian if the states of its constituent elements depend on those of their neighbors but not on the states of units that are spatially remote. These local temporal and spatial properties can be described mathematically using the probabilistic language of Markov chains and processes, and Markov random fields.

Cooperative systems evolve through sequences of small changes from one configuration to another. Large numbers of states may be involved in the evolution of the systems. Among the large numbers of possible states are a much smaller number of favored states. These states are stable against small fluctuations or perturbations of the system. Referring back to our examples, magnetic materials possess multiple stable states. We can simplify our description of the stable states in a ferromagnet by considering the atomic spins to be oriented along a single axis so that each spin either points up or points down. Then there are two favored stable states, those where all spins point up and those where they all point down. If we consider another magnetic material, spin glasses, we find that there are many equivalent global minima plus a multitude of local, suboptimal minima. This multiplicity of local and global minima is encountered in many situations, including our example of proteins undergoing folding. Our second key idea is that of an energy landscape that contains one or more deep minina and perhaps many suboptimal minima.

Thus a cooperative process must contain a mechanism for avoiding the plethora of suboptimal local minina while efficiently searching for a good deep minima. The resulting procedures are typically stochastic, containing a strategy for either avoiding or climbing out of shallow minima. The overall approach is to make a large-grained exploration of the space of possible states early in a manner insensitive to small details and then take small-grained, fine steps later in the process to settle into a deep near-optimal valley. In this equilibrium dynamics the deep valleys once selected are entered into permanently. In many instances the valley selected will depend on the initial starting point in the procedure. In some cases, such as the simple ferromagnet, the particular valley selected will depend on random fluctuations or small asymmetries in the environment.

1.1.5 Nonlinear Dynamics

Cooperative systems, whether in equilibrium or far from equilibrium, tend to exhibit a number of stable (steady) state behaviors. An example of this multiplicity in a system far from equilibrium is illustrated in Fig. 1.2. The membrane potentials of thalamic neurons corresponding to four different steady states are plotted in this figure. Two of the steady states are resting states, and two are oscillatory. Unlike the equilibrium dynamics described above, these behaviors are use-selectable in an additional, different way. Neurons can modify their firing patterns in response to internal and external signals. That is, they can switch from one stable state to another.

This adaptive capability is a distinguishing feature of a nonlinear dynamic system. A linear system will either oscillate or not, and it will do one or the

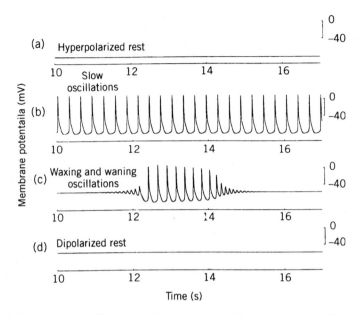

Figure 1.2. Stable states in thalamic relay reurons: (a) Hyperpolarized resting state; (b) slow oscillations at about 3.5 Hz; (c) waxing and waning oscillations in the range 4 to 8 Hz; (d) depolarized resting state. (From Destexhe, Babloyantz, and Sejnowski[3]. Reprinted with permission of the Biophysical Society.)

other over its entire parameter range. Nonlinear systems such as model neurons exhibit different response characteristics as their physical parameters are varied. This capability, plus the existence of multiple steady states, endows these systems with dynamic adaptive abilities.

1.2 PLAN OF THE BOOK

In *Adaptive Cooperative Systems*, two intertwined domains, neural systems and visualp information processing, will provide the focus for our exploration of the dynamics of cooperativity at or near equilibrium and also far from equilibrium.

1.2.1 Equilibrium Dynamics

Our first task will be to develop the mathematical machinery that allows us to construct sequences of states that converge to near-optimal minima. The prototypic algorithm for generating such sequences of states, simulated annealing, was first introduced by Kirkpatrick, Gelatt, and Vecchi,[4] and independently by Cerny.[5] It is built on a procedure for generating Markov chains known as the *Metropolis algorithm*. We will present mathematical theory behind these algorithms and their siblings in Chapters 2 and 4.

The up-down picture of atomic spin ordering invoked in our discussion of ferromagnetism is known as the *Ising model*. The mathematical modeling of magnetic materials such as ferromagnets and spin glasses provides us with a significant portion of our vocabulary of cooperativity. Chapter 3 and the first part of Chapter 4 are devoted to a discussion of cooperativity in lattice systems. We will examine the Ising model in several forms in Chapter 3. One of the most important of these is the mean field approximation. Another is the renormalization group method. We will encounter one or the other of these constructs in all of the later chapters devoted to the exploration of the neural and image-processing domains.

The simulated annealing algorithm was inspired by the formal similarity between the problem of finding the ground (lowest energy) states of magnetic materials such as spin glasses and problems in optimization in the engineering domain. Spin glasses will be discussed at the start of Chapter 4. The remainder of Chapter 4 is devoted to a discussion of the simulated annealing algorithm. Further insight into the dynamics generated by simulated annealing algorithm can be gained using a dynamic theory first applied to studies of Brownian motion. Known as Langevin dynamics, this theory uses the notions of a Markov process, and evolution in small steps, to study the response properties of a system to small perturbations that drive it away from equilibrium. We will explore this dynamics in Chapter 7.

1.2.2 Neural and Image-Processing Domains

The mathematical abstraction of the Ising model known as a *Markov random field* provides us with a way of relating global and local properties of a cooperative system. One of our first tasks in Chapter 6 will be to establish this link. We will then study ways of generating cooperative, or self-organizing, processes that find useful low-energy states. This will be accomplished by merging the dynamics of simulated annealing with the cooperativity expressed by a Markov random field.

Images of natural scenes are highly organized spatial structures. In the rest of Chapter 6, we will use Markov random field models to describe in statistical terms the structure and correlations present in natural images. In the marriage of simulated annealing with Markov random fields studied first by Geman and Geman,[6] we express the statistical properties of the natural images in terms of cooperative interactions among pixel elements. The images are endowed with an artificial equilibrium dynamics that evolves the lattice system through a series of configurations to a near-optimal low-energy state. Depending on the task being addressed, the optimal states are those for which noise, blur, and other artifacts have been removed (image reconstruction) and/or where pixels belonging to the same entity have been identified (segregation and segmentation). We will examine these processes using the method of Geman and Geman, and also using several approximations and alternatives to their approach.

Two ideas emerging from use of Markov random fields as image models are of special interest from the viewpoint of neural architecture and information processing. First, we have the notion of the cooperative integration of information from different sensor modalities. In this cooperative process information from one sensor modality is fed to another and assists in the processing of information in that modality. The second concept is that a visual system may encode attributes of the physical world using constraints. These constraints are of a statistical character, and they serve at least in the early visual processing stages of interest to us in place of specific information about each unique entity.

We will begin our exploration of the neural architecture and information processing in Chapter 5. The goal in Chapter 5 is to explore how topographical mappings develop and how terminals from different eyes segregate into distinct eye-specific layers and patches. In these order-disorder processes, precise patterns of neural connectivity develop through sequences of cooperative interactions from initial less precise patterns. Because of its experimental accessibility and ease of experimental manipulation, the vertebrate visual system has been a favored area of study. We will examine developmental and adult physiological activities in several chapters. In Chapter 5 we will use simulated annealing to explore how multiple global minima endow the developing visual system in lower vertebrates with adaptive capabilities and how local anomalies in the mammalian visual system give rise to global effects. We will study the former using a differential adhesion approach developed by Fraser and Perkel,[7] and the latter using the method of Lee and Malpeli.[8]

1.2.3 Nonlinear Dynamics Far from Equilibrium

Information is processed in a series of stages in the mammalian visual system. From an initial preprocessing stage in the retina information is fed to the lateral geniculate nucleus (LGN) located in the thalamus and from there to the visual cortices. A cell's receptive field is that region of the sensory periphery, namely the area on the retina, whose illumination influences the cell's activity. The receptive fields of retinal ganglion cells and geniculate cells have a concentric, center-surround structure. In contrast to this symmetric arrangement, cells in the primary visual cortex receiving input from the LGN can become orientation selective, responding maximally to short bars of light, of a particular orientation.

The neural circuitry in the primary visual cortex of mammals is profoundly influenced by visual experience during a several week period shortly after birth. This has been demonstrated many times by manipulating the visual environment and observing the changes in the receptive fields. The striking demonstrations of the adaptive capabilities of the visual system during early postnatal life, plus evidence for the formation of ocular dominance columns and iso-orientation patches, have led to the development of a number of theoretical models. These models are based on independent experimental evidence that the locus of the changes observed in the visual system is the synapse. Specifically, the adaptive changes are the result of modifications in the efficiency of synaptic transmission, and accordingly this adaptive cooperative process is termed *synaptic plasticity*.

Our primary objective in Chapter 8 is to study the phenomenon of synaptic plasticity using the dynamic theory of Bienenstock, Cooper, and Munro (BCM).[9] We will explore three aspects of the theory. First, we will study the emergence and changes in orientation selectivity and ocular dominance in different visual environments. Second, we will establish a connection between the mathematical theory and the neurophysiological and biochemical substrate that supports it. Third, we will examine the information-processing functions carried out by BCM neurons.

A number of models of synaptic modification have been developed during the past few years. A key element in all models describing the changes in synaptic efficiency is a nonlinear modification or learning rule. These synaptic models differ from one another in their choice of learning rule, network connectivity, model of the input environment, and dynamic regime. As a result of these differences, each model provides a unique set of insights into the possible information-processing activities carried out in the various layers of the visual system. Our second goal in Chapter 8 is to examine some of these insights. In doing so, we will encounter symmetry-breaking phenomena and simple models based on interactions between continuous-valued spinlike elements.

1.2.4 Rhythms and Synchrony

Neurons in the central nervous system exhibit a number of prominent forms of rhythmic and synchronous activity. Some of these forms are of low frequency and large amplitude, implying that large populations of cells are participating. Others, of high frequency and low amplitude, involve smaller populations of neurons.

A number of different cellular, network, and combined cellular-network mechanisms can generate collective behavior in neural circuits. Some neurons are endowed with intrinsic electrophysiological properties that support rhythmic firing. These cells serve as pacemakers. Others do not, but they still have intrinsic properties that promote rhythmicity when driven by pacemakers. Cells that are not intrinsically oscillatory may become so as a consequence of networks properties. A key observation, due to Wilson and Cowan,[10] is that synchronized oscillations can arise in populations of excitatory and inhibitory neurons reciprocally coupled to one another. Global oscillations, local clusters of synchronized oscillators, and traveling waves can be produced, depending on the parameter regime of the oscillators and the nature of the couplings between units. Our primary objective in Chapter 9 is to identify some of the mechanisms for generating rhythmic behavior in the brain.

We will start by deriving the entrainment condition for communities of coupled oscillators in a dynamic regime where each behaves as a simple rotator. In systems of coupled rotators, a phase-pulling mechanism synchronizes their motions. We will explore the temporal order that emerges when these oscillators are coupled globally to one another through their mean field and also when they are coupled locally through nearest-neighbor interactions. We will follow the studies of mutual synchronization in oscillator communities with an examination of the network mechanisms for generating rhythms and synchrony in populations of excitatory and inhibitory neurons. Again we will explore these processes in networks of globally coupled and also nearest-neighbor-coupled units.

Neurons and neural ensembles exhibit a rich variety of responses over the allowable ranges of their physiological parameters. A number of mathematical tools have been developed that allow us to characterize the stable states in the various parameter regimes in these nonlinear dynamic systems. One of the most important of these is the phase plane portrait. In the phase plane portrait we have a picture of the trajectories representing the sequence of states generated by the dynamics. Phase plane and linear stability analyses can tell us that the sequence of states converges to an equilibrium state, or that oscillatory motion will occur, or that the steady states of the system in a particular parameter regime are unstable. The demonstration that the phase plane trajectories of BCM neurons converge to equilibrium states is one of the main results presented by Bienenstock, Cooper, and Munro. These analytic techniques also forms the basis for much of our exploration of collective oscillations. For these reasons we discuss nonlinear dynamics in a preliminary fashion at the beginning of Chapter 8, and again with an overview of phase plane methods in Chapter 9.

In Chapter 9 we begin with communities of abstract oscillatory units. We next consider populations of structureless excitatory and inhibitory neurons reciprocally coupled to one another. We then add another layer of detail by studying how ionic currents self-organize in a membrane to produce a variety of stable firing states. We begin our exploration of ion channel cooperativity with a brief review of the Hodgkin-Huxley[11] equations. We will follow this preamble with an examination of the phase plane characteristics of the Morris-Lecar[12] model of membrane excitability and oscillations. Network considerations will be introduced by means of a model of relaxation oscillators coupled to one another through nearest neighbor interactions. In this model, due to Somers and Kopell[13] we observe differences between the global effects produced by phase pulling among rotators and those resulting from another mechanism, fast threshold modulation, among relaxation oscillators. We will next look at spindle waves in the thalamocortical system. We will start with the cellular properties that support spindling and then turn to the network mechanisms.

A number of topics are not discussed in the book, as a result of our focus on the neural and image-processing domains. The regulatory capabilities of allosteric proteins arising from the cooperative linkage of active and nonactive sites was recognized first by Monod et al.[14] The theoretical model of concerted transitions between T and R conformations described in Section 1.1 was introduced by Monod et al.[15] shortly thereafter. The rich variety of response properties and mechanisms exhibited by allosteric proteins are described in a monograph by Perutz.[1] This topic will not be discussed further here. Protein folding too will not be discussed further in this book. Results of recent investigations of protein folding have been reported by Dill et al.[16] and Šali, et al.[17] Also cooperativity in lasers will not be explored further in the book. An overview of cooperativity in lasers has been given by Haken.[18]

1.3 EPIGENESIS OF THE CENTRAL NERVOUS SYSTEM

The stochastic procedures that generate an equilibrium dynamics are formulated in terms of energies and temperatures. The former, containing the cooperative interactions and constraints on the process, evaluates the configurations of the system under study, while the latter functions as a control parameter. Procedures based on simulated annealing are presented in Chapters 2 to 6 together with a number of alternative algorithms possessing faster convergence properties. The key element in all cases is the identification of an appropriate energy function for the system under study. In this section we will attempt to delineate the nature of the procedures in the developing mammalian central nervous system.

1.3.1 Genetic Instruction

The development of the mammalian central nervous system, with its laminar structure, its division into anatomically and functionally distinct areas, and its precise synaptic connections, is one of the most remarkable and dramatic processes in nature. Cortical neurogenesis begins with the production of neurons and glia by cells lining the ventricular surface of the neural tube. The newly born neurons multiply, grow, differentiate, segregate, or otherwise sort themselves out, migrate, and aggregate. During this period supporting structures — glial guides and macromolecular matrices, the extracellular matrix, and basal lamina — are erected.

Cell growth and circuit formation follow arrival of the neurons at their cortical destinations. The growing cells within each of the six layers develop their morphologically and electrophysiologically distinct axonal and dendritic structures, and form initial, circuit-forming, synaptic connections. This neurite outgrowth, in many ways paralleling the earlier cell migrations, involves pathfinding, navigation, docking, and connecting. The initial contacts are then refined. Naturally occurring cell death, that is, massive reductions in cell number, follows initial propagation of axons to postsynaptic targets, and in some cases subsequent to initiation of afferent synaptic input. Axons retract inappropriate connections and establish new ones. At this stage of development characteristic features of the different neocortical areas of the various species such as somatosensory barrels and ocular dominance patches emerge, topographically registered and aligned projections from the sensory periphery are established, and receptive fields which respond in a sophisticated manner to input signals are formed.

A combination of explicit instruction and cooperative programming involving multiple interactions guides the development of the vertebrate nervous system. In many instances several strategies are pursued, often in concert, in order to achieve a particular goal. The procedures for self-organization are stochastic in character and produce the exquisite global order through sequences of simple, local cell-cell, cell-substratum, and substratum-substratum interactions performed repetitively.

The combination of explicit instruction and cooperative procedure is highly efficient. The numbers are revealing. There are approximately 10^{11} neurons and perhaps as many as 10^{15} distinct synaptic connections in the central nervous system. Assuming a typical code sequence length of about 7000 base pairs, we find that there fewer than 10^5 gene sequences in human DNA. This remarkable high degree of instructional efficiency can only be achieved by pursuing a strategy of self-organization.

Supporting evidence for the efficiency of this strategy is provided by the many examples of phylogenetic conservation among the molecular agents participating in the cooperative processes. The neural cell adhesion molecule (NCAM) found in vertebrate nervous systems is homologous in structure and function to the neural cell adhesion molecule fasciclin II expressed in insects.[19] The netrins, chemotropic factors identified in embryonic chick brain, have a high degree of sequence similarity to the UNC-6 protein found in the nematode.[20] Further evidence for efficient genetic coding is provided by comparing gene sequences in humans to those of other vertebrates. We find that the degree of similarity is high between humans and other mammals, even though there has been an explosive evolutionary growth of the neocortex. For the chimpanzee the degree of similarity is 98.5%, for the mouse it is 80%, for the cow 75%. Considerable similarity exists as well between gene sequences of humans and lower vertebrates. For the frog the number is 50%, while for fish the value is still 40%.

1.3.2 Mechanical, Chemical, and Electrical Interactions

The developmental procedures incorporate mechanical, chemical, and electrical interactions operating under genetic control in various combinations. Signals transmitted by molecular agents such as NCAM and the netrins mentioned in the preceding paragraph serve a variety of functions. These chemical signals promote cell sorting, migration and aggregation, provide instructions for cell differentiation, and guide neurite outgrowth. For instance,[21] sequences of cell-cell interactions between the floor plate, formed at the ventral midline of the neural tube, and the notocord help establish the polarity, symmetry, and cell type in the vertebrate spinal cord. Additional interactions then guide motor axons to their muscle targets.

Instruction through cooperative interactions is not restricted to the developing neuromuscular system. It plays a well-known role in the immunological system and contributes to the development of other parts of the central nervous systems. We find that in the vertebrate retina, as elsewhere, cell differentiation occurs in an orderly temporal fashion. Retinal ganglion cells form first, then cones and horizontal cells are born, and last, amacrine cells, rods, bipolar cells, and Muller glial cells are created. The modulation of this orderly process by cellular interactions can be demonstrated through *in vitro* dissociation of retinal cells. Germinal cell differentiation is induced by this dissociation into single cells. The resulting retinal cell differentiation is mediated by environmental cues, with earlier formed cells instructing later ones.[22]

Molecular cues in the local microenvironment provide the signals that promote and guide neurite outgrowth during embryogenesis. Growth cones are a motile sensory structure located at the tip of the advancing axon; they explore, interact with, interpret, and respond to signaling molecules in the local environment. These structures were discovered by Ramon y Cajal,[23] and confirmed in the first tissue culture experiments by Harrison.[24] Growth cones extend filopodia, which radiate in all directions to probe their microenvironment. Filopodia are dynamic structures and are able to rapidly change position. In many instances they are capable of differentially adhering to favored surfaces. Growth cones are able to guide the axon to their appropriate target zones without instruction from the soma. The trajectories they produce do not correspond to a random search but instead follow a fairly direct course even when their starting point is perturbed. In some cases misrouted fibers will meander, but the trajectories are still not random.

Local mechanical, chemical, and electrical interactions acting in concert form and refine the patterning of neural connectivity in the mammalian visual system. The wiring is frequently viewed as occurring in two stages. In the first stage, axons navigate to and establish initial synaptic connections. These first patterns of connectivity are diffuse. In many cortical areas, including the visual cortices, neurons are initially overproduced in order to ensure adequate synaptic input. Electrical cell-cell interactions include both spontaneous and experience-dependent interactions. In the second stage, occurring during late development, cooperative and competitive electrical activity-driven processes promote the retraction of inappropriate connections, the formation of new axonal and dendritic structures and synaptic connections, and the appearance of eye-specific domains and orientation selective structures and receptive fields. These synaptic modification processes last through the remainder of gestation, continue to early postnatal development, and are maintained to some extent through adult life.

Electrical activity-driven mechanisms contribute to the refinement of initial sets of axonal projections, and may even provide growth cone guidance

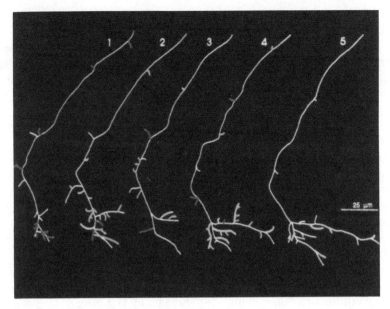

Figure 1.3. Stochastic remodeling of terminal arbors: A three-dimensional reconstruction of a relatively simple arbor taken at hourly intervals. (From O'Rourke et al.[26]. Reprinted with permission of Cell Press.)

through interactions with transient neural populations. For example, in studying the process by which axon collaterals from the LGN first establish connections with target neurons in layer 4 of areas 17 and 18, we find[25] that local dynamic cellular interactions take place between the axon collaterals and subplate neurons along the entire path traversed by the developing LGN axons. That is, the neural substrate through which the axon collaterals pass is not a passive medium. Instead, the axons and neurons undergo necessary local interactions with one another. Ablation of the subplate neurons leads to a failure of the LGN axons to select and invade the appropriate target area.

Evidence for the stochastic character of the interactions is provided by studies of the remodeling of terminal arbors in the primary visual pathway in lower vertebrates such as frogs and fish. In contrast to the visual pathways in mammals, the retina and optic tectum continue to grow throughout the larval stage in frogs and through both development and adult life in fish. The initial contacts are rather orderly, exhibiting considerable dorsoventral topography and, somewhat later, nasotemporal topography. During growth there is an orderly migration of terminal arbors over the tectal surface that maintains the global topography. Locally the situation is quite dynamic. As illustrated in Fig. 1.3, short spikes rapidly extend and retract over the tectal surface probing for optimal locations for placing a stable axon branch. Several hundred of these extension-retraction cycles may occur in a single day.[26] The stochastic nature of the interactions is apparent in this system.

1.4 SYNAPTIC PLASTICITY

The visual system in lower vertebrates is highly adaptive. This point is dramatically illustrated in experiments by Constantine-Paton and Law[27] where a third eye rudiment was implanted in the frog *Rana pipiens*. The results, shown in Fig. 1.4, is a mostly functional three-eyed frog. Perhaps the most interesting aspect of these experiments is the finding that in three-eyed frogs there is an eye-specific segregation of axon terminals into ocular dominance stripes similar to those found in the mammalian primary cortex. As is the case for the mammalian visual system, electrical activity is crucial for the proper formation of these patterns.[28]

1.4.1 Hebbian Synapses

Spatially and temporally correlated patterns of electrical activity produce the refined set of connections found in the central nervous system. The rules for synaptic modification introduced to explain the formation of ocular dominance

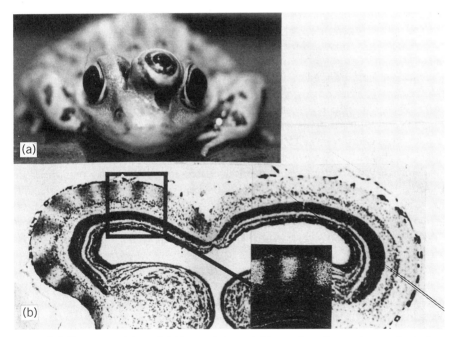

Figure 1.4. Three-eyed frog: (a) *Rana pipiens* eight months after metamorphosis; two complete retinal projections have innervated a single tectal lobe. (b) Eye-specific segregation bands that appear in the optic tectum as a result of the dual innervation. (From Constantine-Paton and Law[27]. Reprinted with permission of the American Association for the Advancement of Science.)

domains and orientation selective receptive fields and structures in the visual cortices are thought to underlie learning and memory as well.

The basic principle governing synaptic modification was enunciated some time ago by Hebb.[29] Hebb's rule states: "When an axon of cell A is near enough to excite a cell B and repeatedly or persistently takes part in firing it, some growth process or metabolic change takes place in one or both cells such that A's efficiency, as one of the cells firing B, is increased." Recognizing that a rule is needed to govern synaptic decrease, as well, Stent[30] added the following, "When a presynaptic axon of cell A repeatedly and persistently fails to excite the postsynaptic cell B while cell B is firing under the influence of other presynaptic axons, metabolic change takes place in one or both cells such that A's efficiency, as one of the cells firing B, is decreased."

According to the resulting two-part covariance hypothesis,[31,32] changes in the levels of pre- and postsynaptic activity determine not only the amplitude but also the sign of the changes in synaptic efficiency. Synapses whose activation is strongly correlated with the firing of postsynaptic cells are strengthened, while synapses that are silent during postsynaptic firing suffer reductions in efficiency. Similarly synapses whose activation is not accompanied by postsynaptic firing exhibit reductions in efficiency. We may generalize Hebb's rule slightly by designating a Hebbian synapse as one in which use-dependent, spatiotemporally correlated activity induces changes in the synaptic state.

Evidence has been acquired supporting the idea that the refinement of the initial set of neural connections is guided by cooperative and competitive electrical-activity-dependent interactions. Blocking experiments, in which electrical activity is selectively prevented, and classical rearing studies, in which early visual experience is manipulated, provide data supporting these notions. In addition hippocampal brain slice preparations have been used to study long-term potentiation (LTP) and long-term depression (LTD), two forms of synaptic plasticity. These investigations and other on LTP and LTD in the visual cortex provide information on electrical-activity-induced synaptic state changes that is consistent with Hebbian mechanisms. Finally, the molecular substrate responsible for the observed changes in synaptic efficiency has been probed, with the result that cooperativity-supporting molecular agents such as the N-methyl-D-aspartate (NMDA) receptor have been identified.

1.4.2 Experience-Dependent Modifications

In more detail, a Na^+ channel blocker, tetrodotoxin (TTX), is used in a blocking experiment to prevent action potentials in a particular pathway. Both spontaneous and experience-dependent activities have been studied using TTX in the developing visual pathways of vertebrates with clear results. Features such as eye-specific patches in the optic tectum of three-eyed frogs, and eye-specific layers in the lateral geniculate nucleus, and ocular dominance columns in layer 4 of the visual cortex of mammals do not develop if electrical activity is prevented. In the second type of experiment mentioned in the above paragraph, visual experience is manipulated. We find that under a classical rearing condition such as monocular deprivation, the absence of proper binocular activity produces both a loss of orientation selectivity and an inhibition of ocular dominance segregation. These data, and others on lid suture, strabismus, reverse suture, and binocular deprivation, demonstrate the profound influence of visual experience on the still developing patterns of connectivity, and they support the covariance hypothesis that correlated firings lead to synaptic stabilization—cells that fire together wire together.

Additional evidence for synaptic modification according to a Hebbian covariance mechanism has been found in hippocampal brain slice measurements. Hippocampal LTP is a long-lasting increase in synaptic strength resulting from a brief period of synaptic stimulation. In associative LTP a pairing of strong and weak input activity with postsynaptic depolarization produces a potentiation of both synapses. To induce associative LTP in the Schaffer collateral/commissural inputs to hippocampal CA1 pyramidal cells, one can either supply a brief high-frequency stimulation of the Schaffer collaterals or provide a low-frequency stimulation paired with a depolarization of the postsynaptic membrane. In associative LTP in CA1, cooperative effects are present in which the magnitude and probability of induction depend monotonically upon the number of synchronized synaptic inputs. This form of LTP is endowed with Hebbian characteristics. A similar form of LTP has been seen in the superficial layers of the visual cortex. Its associativity and cooperativity properties are a consequence of the coactivation of NMDA and non-NMDA receptors. The experimental observations on LTP have been summarized by Kirkwood and Bear,[33] and the data on LTP have been discussed in conjunction with the experimental findings on the reverse process LTD by Bear and Malenka.[34]

1.5 NEURAL ASSEMBLIES

The concept of a Hebbian synapse is accompanied by a second basic tenet of our thinking about the central nervous system. Since the time of Sherrington[35] and Hebb,[29] neurons are thought to act in concert by forming and reforming assemblies. A neural assembly may be formally defined[36] as a collection of neural elements that cooperate with one another. This means that members of an assembly interact with one another. With the introduction of multiple extracellular microelectrode recording techniques, we are able to study neural assemblies in operation by detecting their synchronous activity.

1.5.1 Dynamic Adaptability

The analyses of multielectrode recording data are revealing. We find that neuronal assembly organization is dynamic with modulations of the effective connectivity, or cooperativity, occurring over time scales ranging from milliseconds to seconds. The changes in assembly organization are stimulus-initiated and context-dependent. The adaptive couplings may extend across multiple cortical regions at a number of different frequencies and operate across several time scales. These properties have been demonstrated in experiments by Gochin et al.[34] in the rat, by Ahissar et al.[35] and Vaadia et al.[36] in the awake monkey, and by Bressler et al.,[37] also in the monkey. They have been demonstrated by Nicolelis et al.[38,39] in experiments in the awake rat as well.

In the experiments by Gochin et al.,[34] we observe rapid changes in functional connectivity in the rat dorsal cochlear nucleus. These modulations of the effective connectivity take place during the presence of tonic bursts. The functional coupling of cells through temporally correlated activity is stimulus-initiated and behavioral-context-dependent. In the experiments of Nicolelis et al.,[38,39] we observe the dynamics of assembly operation in networks of cortical, thalamic, and brainstem neurons in the somatosensory system. By examining the stimulus response of single neurons in the ventral posterior medial thalamus, we find that the receptive fields are large and overlapping. The spatial locations of the receptive fields are not fixed but instead shift over the first 35 ms of poststimulus time. The emerging picture of the representation of the face in the awake rat is that of a dynamic and distributed coding of spatiotemporal information by assemblies of neurons linked to one another by networks of feedforward and feedback connections.

There is a convergence of experimental and theoretical findings regarding neural assemblies. Their operation has been observed experimentally and charted in a number of mammals. Their formation and reformation has been studied theoretically and several underlying cellular and network mechanisms have been identified. There is a phylogenetic convergence too. Many of the same strategies found in the perceptual systems of mammals are seen in lower organisms. For instance, in lower organisms such as *Tritonia*, the task performed by a circuit can change in response to alterations in environmental conditions. As shown by Getting,[40] the *Tritonia* escape swim circuit functionally reconfigures itself when changing from a reflexive withdrawal mode to a pattern generator mode.

1.5.2 Assembly Coding

The various types of oscillations observed in electroencephalogram recordings may be associated with different global dynamic behavioral states. We observe high-amplitude, low-frequency oscillations during sleep and anesthesia. However, as shown by Steriade et al.[41] and by Munk et al.,[42] during arousal these temporal patterns are replaced by high-frequency, low-amplitude oscillations. This may be a general finding. Low-frequency oscillations may characterize states of drowsiness and inattention, while high-frequency oscillations accompany states of high attention and arousal.

The high-frequency patterns of synchronous activity among neural assemblies may serve a visual information-processing role. In assembly coding, temporal clusters form and reform in response to input signals. Individual neurons belong at various times to different assemblies and can rapidly associate into a functional group, while at the same time disassociating from a different functional group. The experimental evidence of Vaadia et al.[36] is of a rapid coalescence of cells into functional groups and of their dissociation from competing groups. They find no evidence for an elevation in the mean firing rate in these processes, consistent with another tenet of assembly coding that temporal coherence alone signifies membership in a temporal cluster.

The possible role of assembly coding in visual information processing was promoted by the discovery by Gray et al.,[43,44] Eckhorn et al.,[45] of high-frequency synchronous activity in the mammalian visual system. These experimental studies and others that followed established that clusters of synchronously discharging cells form within one or more columns located in a particular cortical area and within columns located in different cortical regions and hemispheres. These data support a hypothesis advanced by Malsburg,[46] that states that neurons that encode attributes that belong together are integrated, or bound, together by their synchronous firing through cooperative processes mediated by the network connectivity. Under this hypothesis high-frequency synchronous activity in the visual cortex is used for feature binding or temporal tagging analogous to the cooperative labeling in Markov random field models. In Chapters 8 and 9 we will examine the experimental findings in greater detail, and we will evolve the central tenets of Hebb and Sherrington into a mathematical theory using the machinery of nonlinear dynamics.

1.6 REFERENCES

1. Perutz, M. F. (1990). *Mechanisms of Cooperativity and Allosteric Regulation in Proteins*. Cambridge: Cambridge University Press.

2. Winfree, A. T. (1967). Biological rhythms and the behavior of populations of coupled oscillators. J. Theor. Biol., **16**, 15–42.

3. Destexhe, A., Babloyantz, A., and Sejnowski, T. J. (1993). Ionic mechanisms for intrinsic slow oscillations in thalamic relay reurons. Biophys. J., **65**, 1538–1552.

4. Kirkpatrick, S, Gelatt, C. D., and Vecchi, M. P. (1983). Optimization by simulated annealing. Science, **220**, 671–680.

5. Cerny, V. (1985). Thermodynamical approach to the traveling salesman problem: An efficient simulation algorithm. J. Opt. Th. Appl., **45**, 41–51.

6. Geman, S., and Geman D. (1984). Stochastic relaxation, Gibbs distributions., and the Bayesian restoration of images. IEEE Trans. Pattern Anal. Machine Intell., **PAMI-6**, 721–741.

7. Fraser, S. E., and Perkel, D. H. (1990). Competitive and positional cues in the patterning of nerve connections. J. Neurobiol., **21**, 51–72.

8. Lee, D., and Malpeli, J. G. (1994). Global form and singularity: Modeling the blind spot's role in lateral geniculate morphogenesis. Science, **263**, 1292–1294.
9. Bienenstock, E. L., Cooper, L. N., and Munro, P. W. (1982). Theory for the development of neuron selectivity: Orientation specificity and binocular interaction in visual cortex. J. Neurosci., **2**, 32–48.
10. Wilson, H. R., and Cowan, J. D. (1972). Excitatory and inhibitory interactions in localized populations of model neurons. Biophys. J., **12**, 1–24.
11. Hodgkin, A. L., and Huxley, A. F. (1952). A quantitative description of membrane current and its application to conduction and excitation in nerve. J. Physiol., **117**, 500–544.
12. Morris, C., and Lecar, H. (1981). Voltage oscillations in the barnacle giant muscle fiber. Biophys. J., **35**, 193–213.
13. Somers, D., and Kopell, N. (1993). Rapid synchronization through fast threshold modulation. Biol. Cybern., **68**, 393–407.
14. Monod, J., Changeux, J.-P., and Jacob, F. (1963). Allosteric proteins and cellular control systems, J. Mol. Biol., **6**, 306–329.
15. Monod, J., Wyman, J., and Changeux, J.-P. (1965). On the nature of allosteric transitions: A plausible model. J. Mol. Biol., **12**, 88–118.
16. Dill, K. A., Fiebig, K. M., and Chan, H. S. (1993). Cooperativity in protein-folding kinetics. Proc. Nat. Acad. Sci. USA, **90**, 1942–1947.
17. Šali, A., Shakhnovich, E., and Karplus, M. (1994). How does a protein fold?, Nature, **369**, 248–251.
18. Haken, H. (1975). Cooperative phenomena in systems far from thermal equilibrium and in nonphysical systems. Rev. Mod. Phys., **47**, 67–121.
19. Harrelson, A. L., and Goodman, C. S. (1988). Growth cone guidance in insects: Fasciclin II is a member of the immunoglobulin superfamily. Science, **242**, 700–707.
20. Serafini, T., Kennedy, T. E., Galko, M. J., Mizayan, C., Jessell, T. M., and Tessier-Lavigne, M. (1994). The netrins define a family of axon outgrowth-promoting proteins homologous to c. elegans UNC-6. Cell, **78**, 409–424.
21. Jessell, T. M., and Dodd, J. (1992). Floor-plate-derived signals and the control of neural cell pattern in vertebrates. Harvey Lectures, **86**, 87–128.
22. Harris, W. A., and Holt, C. E. (1990). Early events in the embryogenesis of the vertebrate visual system: Cellular determination and pathfinding. Ann. Rev. Neurosci., **13**, 155–169.
23. Ramon y Cahal, S. (1890). Sur l'origine et les ramifications des fibres nerveuses de la moelle embryonaire. Anat. Anz., **5**, 609–613.
24. Harrison, R. G. (1910). The outgrowth of the nerve fiber as a mode of protoplasmic movement. J. Exp. Zool., **17**, 521–544.
25. Ghosh, A., and Shatz, C. J. (1993). A role for subplate neurons in the patterning of connections from thalamus to neocortex. Development, **117**, 1031–1047.
26. O'Rourke, N. A., Cline, H. T., and Fraser, S. E. (1994). Rapid remodeling of retinal arbors in the tectum with and without blockade of synaptic transmission. Neuron, **12**, 921–934.
27. Constantine-Paton, M., and Law, M. I. (1978). Eye-specific termination bands in tecta of three-eyed frogs. Science, **202**, 639–641.

28. Cline, H. T., Debski, E. A., and Constantine-Paton, M. (1987). *N*-methyl-D-aspartate receptor antagonist desegregates eye-specific stripes. Proc. Nat. Acad. Sci. USA, **84**, 4342–4345.
29. Hebb, D. O. (1949). *Organization of Behavior*. New York: Wiley.
30. Stent, G. S. (1973). A physiological mechanism for Hebb's postulate of learning. Proc. Nat. Acad. Sci. USA, **70**, 997–1001.
31. Sejnowski, T. J. (1977). Storing covariance with nonlinearly interacting neurons. J. Math. Biol., **4**, 303–321.
32. Rauschecker, J. P., and Singer, W. (1979). Changes in the circuitry of the kitten visual cortex are gated by postsynaptic activity. Nature, **280**, 58–60.
33. Kirkwood, A., and Bear, M. F. (1994). Hebbian synapses in visual cortex. J. Neurosci., **14**, 1634–1645.
34. Bear, M. F., and Malenka, R. C. (1994). Synaptic plasticity: LTP and LTD. Curr. Opin. Neurobiol., **4**, 389–300.
35. Sherrington, C. (1941). *Man on His Nature. The Gifford Lectures, Edinburgh 1937–38*. Cambridge: Cambridge University Press.
36. Gerstein, G. L., and Turner, M. R. (1990). Neural assemblies as building blocks of cortical computation. In E. L. Schwartz (ed.), *Computational Neuroscience*. Cambridge: MIT Press.
83. Gochin, P. M., Gerstein, G. L., and Kaltenbach, J. A. (1990). Dynamic temporal properties of effective connections in rat dorsal cochlear nucleus. Brain Res., **510**, 195–202.
35. Ahissar, E., Vaadia, E., Ahissar, M., Bergman, H., Arieli, A., and Abeles, M. (1992). Dependence of cortical plasticity on correlated activity of single neurons and on behavioral context. Science, **257**, 1412–1415.
36. Vaadia, E., Haalman, I., Abeles, M., Bergman, H., Prut, Y., Slovin, H., and Aertsen, A. (1995). Dynamics of neuronal interactions in monkey cortex in relation to behavioural events. Nature, **373**, 515–518.
37. Bressler, S. L., Coppola, R., and Nakamura, R. (1992). Episodic multiregional cortical coherence at multiple frequencies during visual task performance. Nature, **366**, 153–156.
38. Nicolelis, M. A. L., Lin, R. C. S., Woodward, D. J., and Chapin, J. K. (1993). Dynamic and distributed properties of many-neuron ensembles in the ventral posterior medial thalamus of awake rats. Proc. Nat. Acad. Sci. USA, **81**, 4586–4590.
39. Nicolelis, M. A. L., Baccala, L. A., Lin, R. C. S., and Chapin, J. K. (1995). Sensorimotor encoding by synchronous neural ensemble activity at multiple levels of the somatosensory system. Science, **268**, 1353–1358.
40. Getting, P. (1989). Emerging principles governing the operation of neural networks. Ann. Rev. Neurosci., **12**, 185–204.
41. Steriade, M., Amzica, F., and Contreras, D. (1996). Synchronization of fast (30–40 Hz) spontaneous cortical rhythms during brain activation. J. Neurosci., **16**, 392–417.
42. Munk, M. H., Roelfsema, P. R., König, P., Engel, A. K., and Singer, W. (1996). Role of reticular activation in the modulation of intracortical synchronization. Science, **272**, 271–274.

43. Gray, C. M., König, P., Engel, A. K., and Singer, W. (1989). Oscillatory responses in cat visual cortex exhibit inter-columnar synchronization which reflects global stimulus properties. Nature, **338**, 334–337.
44. Gray, C. M., and Singer, W. (1989). Stimulus-specific neuronal oscillations in orientation columns of cat visual cortex. Proc. Nat. Acad. Sci. USA, **86**, 1698–1702.
45. Eckhorn, R., Bauer, R., Jordan, W., Brosch, M., Kruse, W., Munk, M., and Reitbock, H. J. (1988). Coherent oscillations: A mechanism for feature linking in the visual cortex? Biol. Cybern, **60**, 121–130.
46. Malsburg, C. von der (1981). The correlation theory of brain function. Internal report 81–2. Max-Planck-Institute for Biophysical Chemistry, Gottingen.

2

THERMODYNAMICS, STATISTICAL MECHANICS, AND THE METROPOLIS ALGORITHM

2.1 INTRODUCTION

In this chapter we will establish the formal structure and accompanying technical vocabulary for studying adaptive cooperative systems. The language we will use is probabilistic, or stochastic, in character. The adoption of a probabilistic language is based upon developments that began nearly three hundred years ago and have coalesced during the last fifty years with a number of seminal contributions to information theory, probabilitic inferencing, and statistical mechanics. Many of the concepts underlying the study of cooperative systems first appeared in classical statistical mechanics and thermodynamics. We will examine these concepts in the thermodynamic setting that gave them meaning, and we will delineate the correspondence between the probabilistic, universal formalism, and thermodynamics.

The subject we know as thermodynamics evolved rapidly in the period of time following the overthrow of the caloric theory, in which heat was regarded as a fluid substance, and the return to a mechanistic theory, where heat was correctly identified with molecular activity. Daniel Bernoulli[1] had developed a modern billiard-ball model in his kinetic theory of gases in 1738, but this mechanistic view had been rejected by those favoring a caloric approach. The return to a mechanical model had its beginnings with Carnot's (still caloric) paper[2] in 1824. It culminated with Max Planck's quantum theory[3,4] in

1900–1901, which answers problems posed by radiant heat phenomena, and with the formulations of statistical mechanics by Boltzmann[5] and Gibbs[6] relating the macroscopic thermodynamic observations to microscopic, stochastic processes.

The first and second laws of thermodynamics were formulated during the decades following the appearance of Carnot's paper. The first law is a statement of the conservation of energy, expressed in a thermodynamic setting in terms of internal energy, work done by the system, and heat absorbed by it. The second law asserts the existence of a nondecreasing parameter of state, the entropy. The term "entropy" had been coined from the Greek word for transformation by Clausius[7] in a paper read to the Züricher naturforschende Gesellschaft, in 1865, and published several times in 1865 and 1867. According to Clausius, the increase in entropy of a system that is not thermally isolated is associated with the differential amount of heat absorbed.

In 1948 Shannon[8] presented a generalization of the thermodynamic concept of entropy to that of a universal measure of uncertainty. In 1956 Jaynes[9] showed that this measure can be used as a starting point for statistical inferencing, in general, and for generating probability distributions including those of the statistical mechanics of Boltzmann and Gibbs, in particular. A paper by Elsasser[10] is but one of a number of noteworthy antecendents for these steps in the evolution of stochastic reasoning. Another convergent line of reasoning, that of Bayesian inferencing, will be sketched shortly.

2.2 KEY CONCEPTS

Several key concepts will be introduced in Chapter 2. We will start in Section 2.3 by deriving an information-theoretic measure of missing information and uncertainty, known as entropy. We will find in Section 2.4 that, in the special case where the probabilities are known from measurements of the relative frequencies of a set of events, the maximum entropy distribution is the most probable one. We will proceed to use this measure to deduce the maximally noncommittal form of the probability distribution for the vast majority of experimental instances, where the available data are limited. The approach used by us is known as the *method of Lagrange's undetermined multipliers*. It is described in Section 2.5 and is followed in Section 2.6 by a presentation of the formal structure of our inferencing theory.

The goal in Sections 2.7 and 2.8 is to elucidate the correspondences between our inferencing theory, classical statistical mechanics, and thermodynamics. These correspondences endow the concepts and formalism associated with statistical mechanics and thermodynamics with a universality far beyond the confines of the phenomena for which they were originally invented/discovered. The uniform probability distribution, and the broad class of exponential forms that includes the Gibbs distributions, emerge as maximum entropy distributions subject to simple moment constraints. These distributions will be used in the later chapters on simulated annealing and Markov random fields. In the case of the Gibbs distribution, an important correspondence can be established between the information-theoretic and thermodynamic entropies. We will examine this identification in Section 2.7, and then examine the microcanonical and canonical (Gibbs) distributions in greater detail in Section 2.8.

An important step in the evolution of the theory is the Metropolis algorithm for simulating a Gibbs distribution. The main result of Section 2.10 is the convergence theorem for Markov chains, which states that the probabilities for transitions between states will approach a unique stationary distribution in the limit of large chains provided that a stationary probability distribution exists. The Metropolis sampling algorithm (Section 2.11) is designed to generate a Markov chain in a manner that satisfies the requisite conditions for the convergence theorem to hold. We present the Metropolis algorithm following the establishment of the necessary terminology, and properties of Markov chains, and precede both with an introduction (Section 2.9) to the Monte Carlo method, of which the Metropolis alogorithm is an example.

2.3 THE MEASURE OF UNCERTAINTY

Suppose that we have a variable X, which can assume the values $x_1, x_2, \ldots, x_r$. We will denote the corresponding probabilities for these values as $p_i, i = 1, \ldots, r$. We now introduce a function S, which we call the *entropy* of the probability distribution. We will regard this entropy function as representing a measure of the uncertainty associated with probability distribution. Alternatively, we look at this function as representing the amount of missing information which, if provided, would allow us to infer with complete certainty the value for the variable X. Our goal is to determine the form of the entropy function based solely upon consistency conditions with respect to continuity, monotonicity, and composition. In more detail, we require the following:

1. The uncertainty measure S is a continuous function of the probabilities associated with each outcome. That is, we have

$$S = (p_1, p_2, \ldots, p_r) \tag{2.1}$$

2. If the probabilities are all equal to one another, then the uncertainty

$$A(r) = S\left(\frac{1}{r}, \frac{1}{r}, \ldots, \frac{1}{r}\right) \tag{2.2}$$

is a monotonic increasing function of r. That is, when the possible outcomes are equally likely, the uncertainty increases as the number of possibilities grows.

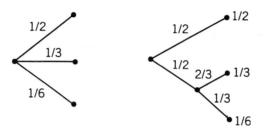

Figure 2.1. Decomposing a choice: Two alternative decompositions of set of probabilities.

3. The uncertainty of a distribution does not depend on the way we group probabilities. If a choice can be decomposed into two successive choices, the original uncertainty measure should equal the weighted sum of the component uncertainties. This is a consistency requirement that can be best illustrated with a simple example. Let us consider the trio of probabilities $p_1 = 1/2$; $p_2 = 1/3$, and $p_3 = 1/6$, and let us decompose these in two different ways as shown in Fig. 2.1. In the one step partitioning shown on the left, we have $S(1/2, 1/3, 1/6)$. Then, according to our composition requirement, we should have

$$S\left(\frac{1}{2}, \frac{1}{3}, \frac{1}{6}\right) = S\left(\frac{1}{2}, \frac{1}{2}\right) + \left(\frac{1}{2}\right) S\left(\frac{2}{3}, \frac{1}{3}\right)$$

The first term on the right-hand side takes into account the initial partition into two branches each assigned probability 1/2. The second term describes the second branching weighted by the probability of choosing the antecedent branch. Since we end up with the same probabilistic decomposition at the end, the uncertainty associated with the one-step partitioning must equal that of the more elaborate decomposition.

More generally, let us place each of the r probabilities into one of s distinct groups, with probabilities p_1 to p_{m1} placed into the first group, probabilities p_{m1+1} to p_{m1+m2} put into the second group, and so on. The probability for observing one of the elements in the first group is $q_1 = (p_1 + p_2 + \cdots + p_{m1})$, the total probability for observing an element of the second group is $q_2 = (p_{m1+1} + \cdots + p_{m1+m2})$ and so on. In terms of these groupings, our composition rule assumes the general form

$$\begin{aligned}S(p_1, p_2, \ldots, p_r) = {} & S(q_1, q_2, \ldots, q_s) \\ & + q_1 S\left(\frac{p_1}{q_1}, \ldots, \frac{p_{m_1}}{q_1}\right) \\ & + q_2 S\left(\frac{p_{m_1+1}}{q_2}, \ldots, \frac{p_{m_1+m_2}}{q_2}\right) + \cdots\end{aligned} \quad (2.3)$$

In this expression, the weight factors, q_1, q_2, etc., take into acccount that the uncertainties due to the additional groupings are encountered with probabilities q_i.

To deduce the form of our uncertainty measure S we consider a special case that allows us to simplify the mathematics. We now proceed as follows: We represent the probabilities as ratios of integers m_i, relying on the continuity condition to handle nonrational cases. Thus we write

$$p_i = \frac{m_i}{M}, \quad M = \sum_{i=1}^{n} m_i \tag{2.4}$$

Next we subdivide each p_i to the point where each grouping is generated by events with probability $1/M$. Within each group there are m_i elements. These too are equally likely, and for this particular decomposition we have

$$S\left(\frac{1}{M}, \ldots, \frac{1}{M}\right) = S(p_1, \ldots, p_n)$$
$$+ p_1 S\left(\frac{1}{m_1}, \ldots, \frac{1}{m_1}\right) + p_2 S\left(\frac{1}{m_2}, \ldots, \frac{1}{m_2}\right) + \cdots$$

We may rewrite this expression using our definition of $A(r)$:

$$A(M) = S(p_1, p_2, \ldots, p_n) + \sum_i p_i A(m_i) \tag{2.5}$$

Let us further simplify the situation and assume that the m_i are each equal to an integer m. Our equation then becomes

$$A(M) = S\left(\frac{1}{n}, \ldots, \frac{1}{n}\right) + A(m) \sum_i \left(\frac{1}{n}\right) = A(n) + A(m)$$

or

$$A(nm) = A(n) + A(m) \tag{2.6}$$

This additivity rule can be easily generalized. If we consider a case where we successively partition the elements into equally likely groups of equally likely elements, we obtain a relation of the form

$$A(n^s) = sA(n) \tag{2.7}$$

The solution to this equation can be immediately given. It is

$$A(n) = K \ln n \tag{2.8}$$

where K is an arbitrary positive constant. We now substitute this value for $A(n)$ back into the more general expression, and solve for $S(p_1,\ldots,p_n)$:

$$S(p_1, p_2, \ldots, p_n) = A(M) - \sum_i p_i A(m_i)$$

$$= \sum_i p_i [A(M) - A(m_i)]$$

$$= -K \sum_i p_i \ln\left(\frac{m_i}{M}\right)$$

and therefore we obtain the fundamental result that

$$S(p_1, p_2, \ldots, p_n) = -K \sum_i p_i \ln p_i \qquad (2.9)$$

Equation (2.9) is the result we set out to obtain. It gives our measure of uncertainty as to the value of the variable X. Its form was derived from our three consistency conditions. Its uniqueness has not been demonstrated by us, but Shannon[8] has shown that this entropic form is in fact unique with respect to the continuity, monotonicity, and composition requirements. In the next section we will establish an important connection between this entropic form and probabilities interpreted as relative frequencies.

2.4 THE MOST PROBABLE DISTRIBUTION

2.4.1 Configurations and Weight Factors

Let us now consider the situation in which we have N elements, and we place each element into one of M bins. The elements are labeled by the bins into which they are placed, and there are n_1 elements in the first bin, n_2 elements in the second bin, and so on. The elements are otherwise indistinguishable, and the quantities $p_i = n_i/N$, represent the frequencies, or probabilities, of finding an element belonging to a particular bin. There are many instances where situations of this type arise. Two examples are (1) systems of N molecules distributed among $i = 1, \ldots, M$ quantum states and (2) messages containing N symbols chosen from an alphabet of M letters, the ith letter occurring n_i times. A distribution of the N elements among the M cells constitiutes a particular configuration of the system. We are interested in determining the number of ways a specific configuration can be realized. The answer is given by the multinomial coefficient)

$$W(\{n_1,\ldots,n_M\}) = \frac{N!}{n_1! n_2! \ldots n_M!} \qquad (2.10)$$

Here are two examples to clarify the meaning of what is being counted. Suppose that we have $N = 2$, and $M = 2$. That is, we have two elements and two bins. There are three possible configurations: We may place both objects in bin one, we may put one element in each of the two bins, or we may insert both objects in the second bin. These cases are symbolized as (1) $n_1 = 2$, $n_2 = 0$; (2) $n_1 = 1$, $n_2 = 1$; and (3) $n_1 = 0, n_2 = 2$. The polynomial coefficients, or weights, calculated by means of Eq. (2.10) are 1, 2, and 1, respectively. In our second case we may place the first element in either the first bin or the second bin. The location of the second element is determined by this placement, and therefore the weight $W = 2$. In the first and third configurations, there are no alternatives; the placement is completely fixed, and thus $W = 1$.

A more elaborate example is this: Suppose that we again have $M = 2$ but now we take $n_1 = 2$ and $n_2 = 4$, so $N = 6$. There are 15 configurations. Let us examine the configurations. There are five configurations having the first of the two elements of cell one occurring first. In obvious notation, these are

$$122221, 122212, 122122, 121222, 112222$$

There are four additional configurations with the first "1" occurring second:

$$212221, 212212, 212122, 211222$$

three more configurations with the initial "1" third:

$$221221, 221212, 221122$$

two with the first "1" fourth:

$$222121, 222112$$

and one with it "1" fifth:

$$222211$$

In this enumeration we observe that the order of the M groups is important but not the ordering within a given cell.

Let us now apply Stirling's formula

$$N! \cong \sqrt{2\pi N} \left(\frac{N}{e}\right)^N = \sqrt{2\pi N}\, e^{-N} N^N$$

to each term in the weight and then take the logarithm of that term, assuming that the n_i and N are large:

$$\ln N! = \frac{1}{2}\ln(2\pi) + \frac{1}{2}\ln N - N + N \ln N$$

or

$$\ln N! \cong N \ln N, \quad N \to \infty$$

We then find that

$$\ln W(\{n_i\}) = N \ln N - \sum_{i=1}^{M} n_i \ln n_i$$

and finally

$$\frac{K}{N} \ln W(\{n_i\}) = -K \sum_{i=1}^{M} \frac{n_i}{N} \ln \frac{n_i}{N} = -K \sum_{i=1}^{M} p_i \ln p_i \qquad (2.11)$$

We therefore find that maximizing the entropy corresponds to maximizing the multiplicity or weight factor. That is, maximizing the entropy is equivalent to determining the set $\{n_i\}$, which can be realized in the greatest number of ways. We may summarize this section by noting, as had Planck, that in those instances where the probabilities can be interpreted as a frequency distribution, the maximum entropy distribution is the most probable one.

2.4.2 Entropic Forms

The ln W expression

$$S = K \ln W \qquad (2.12)$$

is treated as essential in some formulations of statisical mechanics. The relationship was first noted by Boltzmann, and the formula is carved on Boltzmann's memorial in Vienna. The relationship between S and $\ln W$ is represented as a proportionality by Boltzmann in his *Lectures on Gas Theory*. The full equality, with the constant K identified as Boltzmann's constant, was first stated by Planck in his *Wärmestahlung*. When written in inverse form

$$W = e^{S/K} \qquad (2.13)$$

this relationship stresses the rapid increase in the number of states with increasing entropy. In arriving at this relationship, Planck's starting point was Boltzmann's[11] concept of heat motion as molecular chaos. That is, in modern terms, as a stochastic process in which the velocities of the molecules in the gas are not correlated. As noted by Planck (*Wärmestrahlung*, p. 117),[4] this statistical principle underlies the concept of entropy and the related concept of temperature. In questioning the nature of entropy, Planck noted its dependence in some manner on the as yet unspecified probability of that state. The neutralization of temperature differences is connected to an increase in entropy, and it must be that a uniform distribution of elements in a chaotic state is more probable than any other distribution. Planck next makes the observation that the concept of entropy and the second law of thermodynamics are universal as are the rules of probability. Entropy is closely connected with probability, and in fact "the entropy of a physical system in a definite state depends solely on the probability of that state" (*Wärmestrahlung*, p.118).

The $p \ln p$ form first appeared in Boltzmann's[12] study of Maxwell's[13] distribution law for the velocities of the molecules of a gas. Boltzmann's analysis, known as the *H-theorem*, described the time evolution of this velocity distribution. Maxwell, Boltzmann, and Planck established the relationship between a macroscopic quantity, entropy, and probability distribution of states, with nature favoring states of maximum entropy. Their efforts were accompanied by Gibb's development of the canonical and grand canonical ensembles.

With Shannon's result we have taken the first step toward establishing a broad context for the entropic form and the accompanying probability distributions. An important second step will be taken in the next sections, where we derive the distributions of classical statistical mechanics by maximizing entropy subject to constraints on expectations. Upon doing so, we will have shown that probabilistic forms such as the Gibbs distributions have a generality beyond that considered by the originators of statistical mechanics.

2.5 THE METHOD OF LAGRANGE MULTIPLIERS

In most situations the data available are too limited to permit an unambiguous determination of the probabilities for the various states of a system. The problem of how to proceed when this occurs is a long-standing one. The first step in answering the question of how to make reasonable inferences, or construct hypotheses, from limited information was the identification of our measure of uncertainty. We now proceed with the next step in which the limited data are represented by constraints. Our goal in this step is to determine the form of the probabilities when the entropy is maximal subject to the constraints. The mathematical technique we will employ to obtain a formal solution to this problem is known as *Lagrange's method of undetermined multipliers*. We will briefly outline this method in this section and then continue with our main objective in the following section.

Suppose that we wish to find the extremum of a function $f(x, y)$. We know from elementary calculus that the extremum can be identified by differentiation; specifically

$$df(x, y) = \frac{\partial f(x, y)}{\partial x} dx + \frac{\partial f(x, y)}{\partial y} dy = 0$$

For arbitrary independent variations in the x- and y-directions, the extremum will occur when

$$\frac{\partial f(x, y)}{\partial x} = 0 = \frac{\partial f(x, y)}{\partial y}$$

Now, further suppose that we still wish to determine the maximum or minimum of the function f. However, we only want to find the extremum along a curve

$$g(x, y) = C$$

where C is a constant, rather than finding the extremum over the entire x-y plane. This additional restriction imposes a constraint on the allowed variations in the x- and y-directions; that is, both x and y can no longer be independently varied but, instead, are subject to the condition that

$$dg(x, y) = \frac{\partial g(x, y)}{\partial x} dx + \frac{\partial g(x, y)}{\partial y} dy = 0$$

Combining the two conditions on the variations produces the requirement for an extremum that

$$\frac{\partial f(x, y)/\partial x}{\partial g(x, y)/\partial x} = \frac{\partial f(x, y)/\partial y}{\partial g(x, y)/\partial y}$$

If λ denotes the common ratio, then the above expression is equivalent to the pair of conditions

$$\frac{\partial f(x, y)}{\partial x} - \lambda \frac{\partial g(x, y)}{\partial x} = 0$$

$$\frac{\partial f(x, y)}{\partial y} - \lambda \frac{\partial g(x, y)}{\partial y} = 0$$

These two expressions can be obtained by defining a new function

$$h(x, y) = f(x, y) - \lambda g(x, y)$$

and then finding the extremum, assuming that both x and y coordinates can be varied independently. The parameter λ is known as a *Lagrange multiplier*.

These results can be generalized easily to n independent variables, $x_1, \ldots, x_n$. In this case the requirements for an extremum are given as

$$\frac{\partial f(x_1, \ldots, x_n, y)}{\partial x_i} - \lambda \frac{\partial g(x_1, \ldots, x_n, y)}{\partial x_i} = 0$$

and the generating function assumes the form

$$h(x_1, x_2, \ldots, x_n) = f(x_1, x_2, \ldots, x_n) - \lambda g(x_1, x_2, \ldots, x_n)$$

These results can be further extended to instances where we have $k = 1, \ldots, m$, $m < n$, constraint equations. The restriction on the indices is given to remind us that we must have fewer constraint equations than independent variables. In this most general situation we introduce m Lagrange multipliers, one for each constraint equation of the form

$$g_k(x_1, x_2, \ldots, x_n) = C_k$$

The conditions for an extremum become

$$\frac{\partial f(x_1, \ldots, x_n, y)}{\partial x_i} - \sum_k \lambda_k \frac{\partial g_k(x_1, \ldots, x_n, y)}{\partial x_i} = 0$$

and the generating function becomes

$$h(x_1, x_2, \ldots, x_n) = f(x_1, x_2, \ldots, x_n) - \sum_k \lambda_k g_g(x_1, x_2, \ldots, x_n)$$

We now apply this procedure for finding the extremum of a function subject to constraints given by the last three expressions to derive the form of the maximally noncommittal probability distributions.

2.6 STATISTICAL MECHANICS

2.6.1 Formal Structure of the Theory

Let us consider a physical quantity x, which can occur in $i = 1, 2, \ldots, n$ discrete states x_i, and let p_i denote the corresponding probabilities. Our starting point for the main derivations of the theory is our expression for the entropy given by Eq. (2.9). The probabilities are normalized to unity,

$$\sum_{i=1}^{n} p_i = 1 \qquad (2.14)$$

and the information available from the data are represented as $k = 1, 2, \ldots, m$ expectation values for the measurable quantities f_k,

$$\sum_{i=1}^{n} p_i f_{ik} = \langle f_k \rangle \qquad (2.15)$$

The goal is to assign values for the probabilities, namely to describe statistically the set of allowed states for the quantities x. A solution to this problem is provided by the maximum entropy principle. This principle asserts that *we should make inferences using probability distributions that maximize the entropy subject to the given constraints.* In other words, we should construct probability distributions that are consistent with data and are otherwise maximally noncommittal with respect to information that is not available.

As shown in Section 2.4, our formal solution is obtained by finding the extremum of the generating function

$$h(p_1, p_2, \ldots, p_n) = \sum_i p_i \ln p_i + \lambda^1 \sum_i p_i + \sum_k \lambda_k \sum_i f_{ik} p_i \qquad (2.16)$$

There are three terms on the right-hand side of Eq. (2.16). The first term is the entropic function we wish to maximize. The second term is the normalization constraint, and the third term is the data constraint. We must then solve the equation

$$\delta \sum_i (p_i \ln p_i + \lambda^1 p_i + \sum_k \lambda_k f_{ik} p_i) = 0 \qquad (2.17)$$

where

$$\delta \equiv \delta p_i \frac{\partial}{\partial p_i} \qquad (2.17')$$

The result of carrying out the differentiation with respect to the probabilities yields the formal solution to the problem we set out to solve. Our solution is

$$p_i = e^{-\lambda_0 - \sum_k \lambda_k f_{ik}} \qquad (2.18)$$

where $\lambda_0 = 1 + \lambda$.[1] We may recast Eq. (2.18) into a more familiar form by making use of the normalization condition, Eq. (2.14):

$$p_i = \frac{1}{Z} e^{-\sum_k \lambda_k f_{ik}} \qquad (2.19)$$

In Eq. (2.19),

$$Z = e^{\lambda_0} = \sum_i e^{-\sum_k \lambda_k f_{ik}} \qquad (2.20)$$

is the normalizing constant for the probabilities, and is called the *partition function*. (It is traditional to denote this normalization constant by the symbol Z, which is taken from the German term, *Zustandssumme*). The exponential probability distributions given by Eq. (2.19), with the normalization condition Eq. (2.20), maximize the entropy while taking into account constraints on expectations.

The partition function contains the known information about the system under study. The expectation values, entropy, and variances can be obtained formally from the partition function through differentiation of its logarithm with respect to the Lagrange multipliers. To derive the expression for the expectation values, we first observe that

$$\frac{\partial \ln Z}{\partial \lambda_k} = \frac{\partial \ln Z}{\partial Z}\frac{\partial Z}{\partial \lambda_k} = \frac{1}{Z}\frac{\partial Z}{\partial \lambda_k}$$

We next observe the effect of differentiation with respect to the Lagrange multipliers:

$$\frac{\partial Z}{\partial \lambda_k} = \frac{\partial}{\partial \lambda_k}\left(\sum_i e^{-\Sigma_l \lambda_l f_{il}}\right) = -\sum_i f_{ik} e^{-\Sigma_l \lambda_l f_{il}}$$

and therefore

$$\frac{1}{Z}\frac{\partial Z}{\partial \lambda_k} = -\sum_i f_{ik} P_i = -\langle f_k \rangle$$

which yields the desired result

$$\langle f_k \rangle = -\frac{\partial \ln Z}{\partial \lambda_k} \qquad (2.21)$$

We derive the relationship for the entropy by inserting the expression for the probabilities back into Eq. (2.9). Upon doing so, we find that

$$S = K \ln Z + K \sum_k \lambda_k \langle f_k \rangle \qquad (2.22)$$

The third quantity of interest is the variance in the f_k's. To derive this relationship, we again use the interchangeablity of differentiation with respect to the Lagrange multipliers and evaluation of the moments of f_k, and then we exploit the relationship between the moments and the partial derivatives with respect to the Lagrange multipliers of the logarithm of the partition function:

$$\langle f_k^2 \rangle = \sum_i f_{ik}^2 P_i = \frac{1}{Z}\sum_i \frac{\partial^2}{\partial \lambda_k^2} e^{-\Sigma_l f_{il} \lambda_l}$$

$$= \frac{1}{Z}\frac{\partial^2 Z}{\partial \lambda_k^2} = \frac{1}{Z}\frac{\partial}{\partial \lambda_k}(-Z\langle f_k \rangle)$$

$$= -\frac{\langle f_k \rangle}{Z}\frac{\partial Z}{\partial \lambda_k} - \frac{\partial \langle f_k \rangle}{\partial \lambda_k}$$

$$= \langle f_k \rangle^2 + \frac{\partial^2 \ln Z}{\partial \lambda_k^2}$$

and therefore we have variances

$$\sigma_k^2 = \langle f_k^2 \rangle - \langle f_k \rangle^2 = \frac{\partial^2 \ln Z}{\partial \lambda_k^2} \tag{2.23}$$

Equations (2.19) to (2.23) constitute our solution to the inferencing problem we set out to solve. Let us summarize how we obtained this answer. We started by deducing a unique measure of uncertainty for probabilistic inferencing. In the special case where a probability distribution can be found by counting the relative frequencies for a set of possible outcomes, maximizing our measure of uncertainty, the entropy, yields the commonsense solution that we pick the most probable distribution. We then set out to define the formal character of the probability distributions that we may use for inferencing, or hypothesizing. We require that these probabilities be consistent with the available, limited data, and be otherwise maximally noncommital with respect to missing information. The resulting parsimonious probability distributions are given by Eqs. (2.19) and (2.20). The mean and variance in the data, and the entropy, can be calculated from the logarithm of the partition function by means of Eqs. (2.21) to (2.23). Although we did not discuss the entropy expression in any detail, Eq. (2.22) may be recognized as a Legendre transformation between the entropy, treated as a function of the expectations, $\langle f_k \rangle$, and the logarithm of the partition function, regarded as a function of the Lagrange multipliers λ_k.

The formal solution we have presented contains as special cases the three major probability distributions of classical statistical mechanics. The first of these is the microcanonical ensemble, or uniform distribution. The second and third of these distributions are the canonical and grand canonical ensembles, often referred to as *Gibbs distributions*. The identification with the canonical ensemble of statistical mechanics occurs when we restrict the expectations to a single quantity — the energy, identified with the internal energy of a system at thermal equilibrium. In the case of the grand canonical ensemble, we add a second equilibrium quantity — the mole number. Before we examine this identification in more detail, we need to establish a set of general relationships dealing with external parameters. This is done to allow for a probabilistic interpretation of work and heat.

2.6.2 External Parameters

Let us now consider the case where the probabilities depend on an external parameter α. The partition function may then be regarded as a function of the Lagrange multipliers and the external parameter. We can exhibit this dependence by writing

$$Z = Z(\lambda_1, \ldots, \lambda_m, \alpha) = \sum_i e^{-\sum_k f_{ik}(\alpha)\lambda_k} \tag{2.24}$$

We are interested in assessing the effects of variations in the value for this parameter on the partition function. We begin by noting that

$$\frac{\partial \ln Z}{\partial \alpha} = \frac{1}{Z}\frac{\partial Z}{\partial \alpha}$$

We then evaluate the partial derivative of the partition function with respect to the external parameter and find that

$$\frac{\partial Z}{\partial \alpha} = \sum_i \sum_k \left(-\lambda_k \frac{\partial f_{ik}(\alpha)}{\partial \alpha}\right) e^{-\Sigma_l f_{il}(\alpha)\lambda_l}$$

$$= -Z \sum_k \lambda_k \sum_i p_i \frac{\partial f_{ik}(\alpha)}{\partial \alpha}$$

and therefore we see that

$$\frac{\partial \ln Z}{\partial \alpha} = -\sum_k \lambda_k \left\langle \frac{\partial f_k(\alpha)}{\partial \alpha} \right\rangle$$

Now let us define the quantity $\langle df_k \rangle$ to be

$$\langle df_k \rangle \equiv \left\langle \frac{\partial f_k(\alpha)}{\partial \alpha} \right\rangle d\alpha$$

By definition,

$$d \ln Z = \frac{\partial \ln Z}{\partial \alpha} d\alpha + \sum_k \frac{\partial \ln Z}{\partial \lambda_k} d\lambda_k$$

Upon noting that

$$\frac{\partial \ln Z}{\partial \alpha} d\alpha = -\sum_k \lambda_k \langle df_k \rangle$$

We find that

$$d \ln Z = -\sum_k (\lambda_k \langle df_k \rangle + \langle f_k \rangle d\lambda_k) \qquad (2.25)$$

where we have used the previous expression to rewrite the first term and Eq. (2.21) to recast the second term. We will use this expression in Section 2.7.2 where the volume serves as the external parameter of interest.

2.7 THERMODYNAMICS

2.7.1 Equilibrium States

Equilibrium states of thermodynamic systems may be characterized by their simplicity and stability. These states are simple in the sense that they are completely described by a small number of macroscopic variables called *parameters of state*. These parameters of state have the property that they are independent of the past history of the system, being defined solely by their instantaneous values. Mathematically speaking, a quantity is a parameter of state if the integrated changes in a system produced by a small increment in that quantity depend on the initial and final states alone and not on the details of the path taken in going from the initial to final state. The thermodynamic entropy is a parameter of state. The internal energy of the system is another state variable as are the temperature and the mole numbers of the component species.

Equilibrium states are stable in the sense that once formed, they do not easily change in time. They are states of maximum entropy for a given value of the total energy. They are also states of minimum (free) energy for a given value of the entropy. In thermodynamic systems stable states are usually not static but, instead, are maintained dynamically. Suppose that we have a system containing a number of internal walls, which serve to partition that system into cells. The parameters of state for the system, such as the total energy E and the particle number N, are taken as fixed. The value for a given parameter of state is composed of the sum of the individual values for each of the cells. These individual values may vary from cell to cell, depending on the placement of the internal walls. The set of values for these variables, such as the set of energies $\{\varepsilon_i\}$, where i ranges from 1 to M, for M cells, serves to define a configuration of the system. The equilibrium state is the collection of configurations into which the system will evolve once the internal constraints are removed. As expected, the equilibrium configurations tend to be fairly homogeneous in their distributions of the values of these quantities among the cells. These equilibrium configurations are stable against fluctuations in energy and the other state variables.

In cooperative systems nearest-neighbor, interparticle interactions produce correlations between constituents. Equilibrium states in cooperative systems tend to be disordered at high temperatures, and highly ordered at low temperatures. As the temperature is lowered in these systems, a few elements become ordered. This ordering is propagated throughout the system by means of long-range correlations, resulting in an rapid cooperative ordering of the remaining elements in the assembly at the transition, or critical, temperature. Examples of properties emerging through these collective actions include spontaneous magnetization, arising from Pauli exchange interactions between neighboring spin-1/2 atoms in solid lattices, and superconductivity, in which weak interactions between conduction electrons at sufficiently low temperature leads to highly correlated motion.

In the next section we will identify the statistical mechanical expectation value of the energy, $\langle E \rangle$, with the thermodynamic internal energy. The energy function is frequently referred to as the *hamiltonian*, and is then designated by the symbol H. For systems of noninteracting particles (ideal gases), the hamiltonian consists of a sum of the individual kinetic and potential energy terms of the constituent molecules. For the cooperative systems of interest to us, the hamiltonian will consist of a sum of interaction energies involving pairs of neighboring elements.

2.7.2 The Correspondence between Statistical Mechanics and Thermodynamics

In the last section we derived a set of relations pertaining to external parameters. Typical of these parameters are volume and external magnetic fields. These thermodynamic variables are related to various forms of work such as mechanical (volume) and magnetic (external fields). Work (and heat) differ from quantities such as internal energy and entropy in a fundamental way — they are not parameters of state. That is, heat and work are not history-independent properties of state. Small, differential changes in work done by a system, and heat added to a system, are termed *inexact differentials* to reflect this difference in character. It is customary to indicate the inexact character of such differentials by means of a "slashed" d, or by some other symbolic change. We will not do so, having noted their character.

Let $E_i = E_i(V)$ represent a set of energy levels of a system of particles, and let these energy levels depend on an external parameter, the volume of the system, denoted by V. Furthermore suppose that we only know the mean energy $\langle E \rangle$ (or equivalently, the total energy of the system):

$$\sum_i E_i(V) p_i(E_i) = \langle E \rangle \tag{2.26}$$

As usual, the probabilities are normalized to unity

$$\sum_i p_i(E_i) = 1 \tag{2.27}$$

Applying our solution, Eqs. (2.19) and (2.20), to this problem yields the probability distribution

$$p_i(E_i) = \frac{1}{Z} e^{-\lambda_1 E_i} \tag{2.28}$$

and corresponding partition function

$$Z = \sum_i e^{-\lambda_1 E_i} \tag{2.29}$$

The entropy, partition function and energy are related to one another through Eq. (2.22):

$$S = K(\ln Z + \lambda_1 \langle E \rangle) \tag{2.30}$$

The last relationship, Eq. (2.30), can be expressed in differential form as

$$dS = K(d \ln Z + \lambda_1 d\langle E \rangle + \langle E \rangle d\lambda_1)$$

We can now use our evaluation of the variation in partition function due to fluctuations in an external parameter given by Eq. (2.25). For the case considered here, this expression becomes

$$d \ln Z = -(\lambda_1 \langle dE \rangle + \langle E \rangle d\lambda_1)$$

Combining the two differential relationships gives

$$dS = K\lambda_1(d\langle E \rangle - \langle dE \rangle) \tag{2.31}$$

Let us now define an inexact differential dQ as the difference between $d\langle E \rangle$ and $\langle dE \rangle$:

$$dQ \equiv d\langle E \rangle - \langle dE \rangle \tag{2.32}$$

where in anticipation of our identification of this difference with a change in heat we are using its standard symbolic designation dQ. Inserting Eq. (2.32) into Eq. (2.31) produces the expression

$$dS = K\lambda_1 dQ \tag{2.33}$$

This is an interesting result insofar as dQ is an inexact differential, while dS is an exact one. This implies that $K\lambda_1$ is an integrating factor for dQ, converting a path-dependent quantity into a path-independent one. We can now set the positive constant K equal to the Boltzmann constant k_B. We further identify the Lagrange multiplier with the inverse product of the Boltzmann constant and the temperature. That is,

$$K = k_B \tag{2.34}$$

and

$$\lambda_1 = \frac{1}{k_B T} \equiv \beta \tag{2.35}$$

Thus we have established a thermodynamic correspondence for the variables and parameters appearing in the probabilistic formalism. In establishing this correspondence, we identify $\langle E \rangle$ with the internal energy U characteristic of a system at thermal equilibrium with a heat bath. The quantity dQ represents a differential increment of heat, with

$$dS = \frac{dQ}{T} \qquad (2.36)$$

The second law of thermodynamics consists of several statements. The first part of the second law asserts that for any equilibrium state of a thermodynamic closed system, it is possible to define a thermodynamic entropy. In other words, entropy is a parameter of state. The second part of the second law of thermodynamics states that in any process in which a thermally isolated (adiabatic) system undergoes a transition from one state to another, the entropy will increase. If the system is not thermally isolated, then the third part of the law asserts that the entropy change associated with a differential amount of heat absorbed is given by Eq. (2.36). Note that in the first part of the second law, the entropy was only defined for equilibrium states. If a and b are two equilibrium states of a thermodynamic system, we may express Eq. (2.36) through Clausius' integral

$$S_b - S_a = \int_a^b \frac{dQ}{T} \qquad (2.37)$$

where a and b are equilibrium states. The numerator dQ is the heat exchanged in a reversible process, that is, through a continuous sequence of equilibrium states, which is otherwise arbitrary, and T is an integrating factor of the linear differential.

We now define another inexact differential dW as

$$\langle dE \rangle = \sum_i p_i \frac{\partial E_i(V)}{\partial V} dV = -PdV = -dW \qquad (2.38)$$

where

$$P \equiv -\sum_i p_i \frac{\partial E_i(V)}{\partial V} \qquad (2.39)$$

is the pressure. We recognize W as the work and the expression

$$d\langle E \rangle = dQ - dW \qquad (2.40)$$

as the first law of thermodynamics. In Eq. (2.40), dQ denotes the heat absorbed by the system and dW represents the work done by the system.

As the energy or, alternatively, the temperature of a physical system is systematically lowered, the system evolves toward a lowest energy state known as the *ground-state*. One or several states may be characterized by this energy. If there is more than one state of lowest energy, the system is said to have a *degenerate ground-state*. During the approach to the ground-state, the number of available states decreases. Since the system will settle into one of a few states, the entropy will tend toward minimal (zerolike) value in the low-energy limit. This limiting behavior is sometimes referred to as the third law of thermodynamics.

2.8 THE ENSEMBLES OF STATISTICAL MECHANICS

2.8.1 Microcanonical Ensemble

The simplest possible situation is one in which there is no information available other than the normalization condition on the probabilities. Then a direct application of Eqs. (2.19) and (2.20) yields the result that all probabilities are equally likely. Their values are equal to the reciprocal of the number of states Ω of the system:

$$p_i = e^{-\lambda_0} = \frac{1}{\Omega} \tag{2.41}$$

The entropy in this case is simply the logarithm of the number of states:

$$S = -k_B \sum_i p_i \ln p_i = -k_B \sum_i \frac{1}{\Omega} \ln \frac{1}{\Omega}$$

or

$$S = k_B \ln \Omega \tag{2.42}$$

This formulation is known as the *microcanonical ensemble*. Suppose that we have an isolated system in thermal equilibrium with known internal energy E, volume V, and number of particles N. In a microcanonical ensemble, states characterized by these E, V, N values have probability given by the above expression, while states whose internal energy, volume or number of particles differ from the specified values have zero probability. Yet another way of viewing the microcanonical ensemble is to suppose, for instance, that the energy is known to lie within a small, well-defined range. We may then define

$$p_i = \begin{cases} \Omega^{-1}, & E' \leq E \leq E' + \delta E' \\ 0 & \text{elsewhere} \end{cases} \tag{2.43}$$

where Ω is the number of states lying within the specified energy range.

The uniform distribution of the microcanonical ensemble deserves further comment. This is an important result in itself, for its clearly provides support for an inferencing principle first enunciated by Jacob Bernoulli[14] in 1713. Known as the *principle of insufficient reason*, it states that: If we are considering two hypothesis, or propositions, and there is insufficient evidence to favor one over the other, then we should assign equal probabilities to the propositions. This principle was elaborated on further by Keynes,[15] who renamed it the *principle of indifference*. The issue of the assignment and combination of probabilities is intimately connected to the general subject of inverse probabilities and the measurement of uncertainty. The early work on inverse probabilities by Bernoulli was elaborated upon by Bayes[16] and Laplace,[17] and the subject was reestablished in modern times by Jeffreys.[18] The status of probabilistic reasoning as the method of choice for making inferences was greatly advanced by the seminal contribution to the subject by Cox.[19,20] We will encounter and use Bayesian inferencing later in this work.

2.8.2 The Canonical Ensemble (Gibbs Distributions)

Let us now consider the case of a small physical system in thermal contact with a larger system acting as a heat reservoir. In this frequently encountered situation the total internal energy can fluctuate about its mean value, while V and N remain fixed. An example of this type of system is that of an assembly of atoms in a crystal lattice with the solid acting as the heat reservoir. We have already encountered this ensemble in Section 2.7. There is a correspondence between the quantities defined formally from the maximum entropy principle and those of thermodynamics, for the probabilities given by the canonical and grand canonical ensembles. The canonical ensemble is defined through Eqs. (2.26) to (2.29):

$$p_i = e^{-\beta E_i} = e^{-E_i/k_B T} \tag{2.44}$$

This probability distribution is called a Gibbs distribution, and the exponential factor, $\exp(-\beta E)$, is known as the Boltzmann factor. The partition function can be written as a sum over the states or as a sum over energy:

$$Z = \sum_i e^{-\beta E_i} = \sum_E \Omega(E) e^{-\beta E} \tag{2.45}$$

In the second type of sum, $\Omega(E)$ is the number of states in a narrow energy bin characterized by the value E as in Eq. (2.43), and the sum is over all such energy bins. In applications of the canonical ensemble in statistical mechanics, the number of states is a rapidly increasing function of energy. Consequently the product of the number of states and the Boltzmann factor is sharply peaked about the mean energy.

2.8.3 Helmholtz Free Energy

We now define the Helmholtz free energy A in terms of the logarithm of the partition function for the canonical distribution as

$$Z = e^{-\beta A} \qquad (2.46)$$

The probabilities for the canonical distribution then take the form

$$p_i = e^{-\beta(E_i - A)}$$

and the entropy becomes

$$S = -k_B \sum_i p_i \ln p_i = k_B \beta (\langle E \rangle - A)$$

or

$$A = \langle E \rangle - TS \qquad (2.47)$$

or, equivalently,

$$\ln \Omega = \ln Z + \beta \langle E \rangle \qquad (2.48)$$

In Chapter 4 we will find that the competing effects of the energetic and entropic terms in Eqs. (2.45) and (2.47), and the changes in their balance as the temperature is changed, inspire the simulated annealing algorithm.

2.8.4 Energy Fluctuations

Let us now examine the energy fluctuations in the canonical ensemble. From Eq. (2.23) we have the relation

$$\sigma_E^2 = \frac{\partial^2 \ln Z}{\partial \beta^2}$$

and from Eq. (2.21),

$$\frac{\partial \ln Z}{\partial \beta} = -\langle E \rangle$$

Upon combining these two expressions, we see that

$$\sigma_E^2 = -\frac{\partial \langle E \rangle}{\partial \beta} \qquad (2.49)$$

The heat capacity $C_{N,V}$ is defined as

$$C_{N,V} \equiv \left(\frac{dQ}{dT}\right)_{N,V} \tag{2.50}$$

where the subscripts signify that there are two types of heat capacity, corresponding to the cases where either N or V are held constant during the thermal variations. Since no work is done, the changes in mean energy and heat are identical, and upon noting that $\partial \beta / \partial T = -1/k_B T^2$, we find that

$$\sigma_E^2 = k_B T^2 C_{N,V} \tag{2.51}$$

This important expression tells us that the magnitude of the energy fluctuations is directly proportional to the heat capacity of the system. We will see in the following chapters that the heat capacity (or in molar form the specific heat) increases rapidly in the vicinity of a phase transition and that its variations with temperature helps characterize the nature of the phase transition.

2.8.5 Grand Canonical Ensemble

The above formalism can be easily extended to multicomponent systems where the species numbers $N_1, N_2, \ldots$, fluctuate about their mean values, $\langle N_1 \rangle, \langle N_2 \rangle, \ldots$. In these grand canonical ensembles we have, in addition to the normalism and energy constraints, the requirements that the probabilities satisfy the conditions

$$\sum_i N_{im} p_i(E_i, N_{im}) = \langle N_m \rangle \tag{2.52}$$

The grand partition function Z_G takes the form

$$Z_G = \sum_i e^{-\beta E_i - \beta \sum_m \mu_m N_{im}} \tag{2.53}$$

In the above expression the Lagrange multilpier for the various species of elements have been written in a standard form of products of β and the chemical potentials μ_m.

The grand canonical ensemble has been widely used in quantum statistical mechanics. The hamiltonians for assemblies of noninteracting elements are usually expressed as a sums of kinetic and potential terms, one pair for each element. The kinetic terms are functions of the "single-particle" momenta, and the potential terms are function of the "single-particle" coordinates. In using the grand partition function, the total particle numbers are expressed in terms of occupation numbers, and the appropriate Fermi-Dirac and Bose-Einstein statistics are imposed. The logarithm of the grand partition function takes the form of a product of single-particle terms, and it can be evaluated by elementary means. Examples of systems that can be described in this manner are electromagnetic fields in thermal equilibrium with an enclosure of fixed volume and temperature (photon gases), conduction electrons in metals, and normal (vibrational) modes of low-temperature solids (phonon gases).

In the last two sections we found that the uniform and Gibbs distributions of classical statistical mechanics are maximum entropy distributions. These distributions are consistent with the normalization and known data, and are maximally noncommittal with respect to missing information. The Gibbs distribution emerges when a constraint on the total energy is imposed. In the special case of the Gibbs distribution, if we identify $\langle E \rangle$ with the internal energy of a system at thermal equilibrium with a heat bath, and the Lagrange multiplier for the energy with the inverse of the product of the thermodynamic temperature and the Boltzmann constant, we are able to define expressions that are formally equivalent to those of thermodynamics.

In the next chapter we will endow the mathematical formalism with physical content and meaning. We will construct and use a number of nearest-neighbor, model hamiltonians to study the behavior of simple cooperative systems. The most important of these is the Ising model originally introduced to describe the formation of ferromagnetic domains. We will examine the lattice gas model used to study density fluctuations and liquid-gas phase transitions, and in Chapter 4 we will study models of binary substitutional alloys and models of mixtures of ferromagnetic and antiferromagnetic materials known as *spin-glasses*. Before we begin these explorations, we will find it useful to derive one more general result. Specifically, we will develop an algorithm for simulating Gibbs distributions. This will be done in the next three sections.

2.9 THE MONTE CARLO METHOD

2.9.1 Definition

In many situations the partition function cannot be evaluated analytically. Difficulties in analytic treatments are already apparent in the simple Ising model to be discussed in Chapter 3. The calculation of the thermodynamic quantities of interest for the one-dimensional Ising model is straightforward, but it was nearly 20 years before a solution to the two-dimensional case was presented by Onsager;[21] an analytic treatment of the three-dimensional Ising model does not yet exist. In the remainder of this chapter we introduce the Monte Carlo sampling method as a means for obtaining precise results for complex systems of interacting elements. Our goal is to describe the Metropolis algorithm and demonstrate its convergence to an equilibrium Gibbs distribution. In doing so, we will examine some properties of Markov chains. We will continue the evolution of this approach with the simulated annealing method, to be presented in Chapter 4.

The Monte Carlo method is a device for studying stochastic models of mathematical or physical processes. The Monte Carlo method was introduced in a paper by Ulam and von Neumann[22] in 1947 as a means of solving integral and differential equations. The Monte Carlo method was applied to the study of high-energy nuclear collisions by Goldberger[23] shortly thereafter. The basic idea in the Ulam-von Neumann abstract is that in certain circumstances it is easier to obtain a numerical solution to an equation by finding a stochastic process whose distributions or parameters satisfy the equation under study, and then computing the resulting statistics, than it is to attempt to solve the equation by direct means. The emphasis in their presentation was on the use of a random sampling method for the solution of a deterministic mathematical problem.

2.9.2 Study of Reaction Processes

In the Goldberger[23] study a physical reaction process was simulated. The process under study involved multiple collisions of nucleons with one another, acting under both deterministic and stochastic influences. In the Goldberger simulation the individual collisions were followed on a step-by-step basis. Both classes of influences were precisely imitated. In more detail, interactions of high-energy nucleons with complex nuclei were treated as a sequence of two-body interactions between the incident particles and the nucleons comprising the target nucleus. The interactions of the struck nucleons with the remaining target nucleons were also treated as a sequence of two-body collisions. The entire process, called an *intranuclear cascade*, was treated by Goldberger in a two-dimensional model of the nucleus. The step-by-step sequence of collisions was repeated for sufficient numbers of particles to generate reliable statistical estimates of the physical quantities of interest such as the angular distributions of the emitted particles.

If the bombarding energies are sufficiently low, the intranuclear cascade stage does not occur. Instead, the target and projectile fuse to form an equilibrated compound nucleus, with atomic charge and mass number equal to the sum of those of the reactants. These compound nuclei are produced in highly excited states, and subsequently decay to the ground-state by statistically emitting particles and gamma rays. If the system is sufficiently massive, fission becomes possible. Typical quantities of interest are the kinetic energy spectra of the different species of recoiling heavy reaction product found at a given angle (the so-called double differential cross sections), and the angular distributions of the reaction products. Displayed in Figs. 2.2 and 2.3 are representative results of measurements by Eyal et al.[24] of energy spectra and angular distributions for reactions of oxygen isotopes with carbon nuclei at low bombarding energies. This type of reaction takes place in stellar interiors and was studied for that reason.

Calculations of the quantities of interest for the oxygen-carbon reactions were performed by means of the Monte Carlo method using a stochastic, evaporation model of the deexcitation process. These calculations are represen-

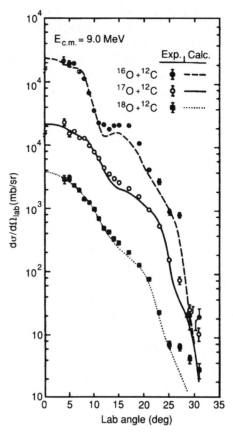

Figure 2.2. Angular distributions: Monte Carlo simulation of the stochastic deexcitation (evaporation) of nuclei formed through a fusion process. Three different reactions are compared, showing experimental (points) and calculated (smooth curves) normalized yields for evaporation residues. (From Eyal et al.[24] Reprinted with permission of the American Physical Society.)

tative of the method. Starting with the initially produced compound nucleus, emission probabilities for six particle species — protons, neutrons, deuterons, tritons, helium-3 nuclei and alpha particles — were computed. These probabilities were compared to a uniformly distributed random number to determine which particle type would be emitted. A candidate kinetic energy for the selected particle was chosen next by drawing another uniformly distributed random number. If the probability for emitting a particle with the proposed energy was greater than the random number, the candidate energy was accepted; otherwise, it was rejected, and another candidate energy was chosen. This two-step process was repeated until an energy was accepted. The center-of-mass emission angle was then randomly selected and the two-step procedure was applied to this variable. The entire calculation was repeated for the nucleus

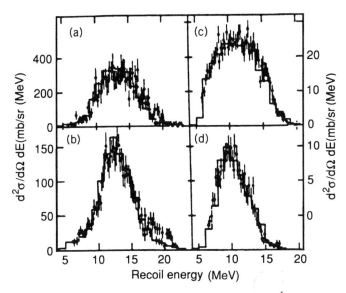

Figure 2.3. Energy spectra. Comparisons of experimental (points) and calculated (histograms) of the double differential cross sections for formation of evaporation residues in the reaction 22.48 MeV ^{16}O on ^{12}C. Panels (a) to (d) show results at laboratory angles of 6, 10, 17, and 21 degrees, respectively. (From Eyal et al.[24] Reprinted with permission of the American Physical Society.)

until particle emission was no longer possible. Evaporation chains was calculated repeatedly, and their results stored, until sufficient statistics were collected. Results of Monte Carlo calculations obtained by Eyal et al.[24] are compared to the data in Figs. 2.2 and 2.3. We observe in these plots that the model predicts rather distinct angular distributions and energy spectra of these heavy reaction products, reflecting the differing recoil kinematics of the various species of emitted particle, in good agreement with the measured values.

2.9.3 Problems in Statistical Mechanics

The Monte Carlo method was extended by Metropolis et al.[25] in 1953 to problems in statistical mechanics. These authors wished to determine properties of substances, such as liquids or dense gases, composed of multiple interacting elements. If we try to solve this type of problem in a direct manner, we must include every possible configuration of the N-molecule system in the evaluation of the partition function. This is not a terribly practical approach. Alternatively, we may use the Monte Carlo method to carry out the summation (integration) by choosing a configuration at random, and then weight the configuration by the appropriate Boltzmann factor. However, most of the contribution to the partition function comes from a small region in configuration space. We would like to avoid randomly sampling large numbers of unlikely states of the system but rather sample states with a frequency in accordance with the Boltzmann factor, and then weight uniformly. In their paper Metropolis et al. present a procedure for carrying out this type of importance sampling.

In the approach of Metropolis et al,[25] hard sphere molecules of a model liquid were arranged in a regular lattice. A molecule was selected at random. The particle was given a small displacement in accordance with a uniformly drawn random number. The energy change ΔE produced by the translation was then calculated. If the energy change was negative, the move was allowed. If the energy change was positive, another random number was chosen, and compared to $\exp(-\beta\Delta E)$. If the random number was less than this quantity, the move was permitted; otherwise, the molecule was left at its former position. In either case another particle was selected next, and the procedure was repeated. The quantities of interest, such as virial coefficients, were then estimated by taking averages over many configurations of the system.

2.10 MARKOV CHAINS

2.10.1 Definitions

Let us consider a system consisting of an array of N elements, $s_1, s_2, \ldots, s_N$, called *spins*, as depicted in Fig. 2.4. Each element s_i, in the array may occur in one of two states—up or down. A configuration of this spin system is an enumeration, or listing, $\{s_1, s_2, \ldots, s_N\}$, of the status (up or down) of each spin element in the array. The elements of the array do not have to be binary-valued. For example, in Chapter 6 we will treat digital images as two-dimensional array of grey-valued pixel elements. In these cases a configuration is a raster-by-raster listing of the pixel grey values.

We can consider each unique configuration of the system as a point in a configuration space. A random variable X is an assignment of a numerical label to each configuration of the system, that is, an association of each point in the configuration space with a number. We designate this assignment by the statement $X = x_i$. More formally, a random variable is a mapping from the configuration space to some label domain. Random variables are quantities that are of physical interest for which we will define probability distributions. For the case of the array of spin elements, a typical random variables is the total energy of the configuration. Another random variable is the magnetization of the system. There may be more than one configuration corresponding

Figure 2.4. Schematic of an array of spins. Up and down orientations of the spin elements correspond to spin values of $+1$ and -1, respectively.

to a particular value of a random variable. We may group together all points in the configuration space for which $X = x_i$. This collection of points is referred to as the state $X = x_i$, and we will designate the probability for this state as $p(X = x_i)$.

We may extend these definitions to two or more random variables. Suppose that X and Y are two random variables defined over the same configuration space. The probability of the state, jointly specified by the statements $X = x_i$ and $Y = y_j$, is denoted by the expression $p(X = x_i, Y = y_j)$. In addition to the joint probability, we may define conditional probabilities; that is, the probability that the random variable Y assumes the value y_j, given that the random variable X, has been observed to assume the value x_i. This second type of probability is designated as $p(Y = y_j | X = x_i)$. These two expressions are related to one another:

$$p(Y = y_j | X = x_i) = \frac{p(X = x_i, Y = y_j)}{p(X = x_i)} \tag{2.54}$$

2.10.2 Transition Probabilities

With these preliminaries completed, we are ready to introduce the idea of a Markov chain. Let us consider a sequence of states $x_1, x_2, \ldots, x_n$ at discrete times $t = 1, 2, \ldots, n$. We are interested in the probability that given that the system is in state x_i at some instant in time, the system will undergo a transition to state x_j at the next instant in time. These conditional probabilities are termed *transition probabilities*. If the transition probabilities are independent of the time index, then we have *stationary transition probabilities*. We may write designate these transition probabilities using an abbreviated notation as

$$p_{ij} = p(X_t = x_j | X_{t-1} = x_i) \tag{2.55}$$

These transition probabilities may be generalized to cases where there are transitions from one state to another in some fixed number of steps. These transitions can take place through many different paths through configuration space. Let us define n-step transition probabilities $p_{ij}^{(n)}$ as the sum over all possible paths connecting the endpoint states j and i. If we put

$$p_{ij}^{(n)} = p(X_t = x_j | X_{t-n} = x_i) \tag{2.56}$$

then we have one-step transition probabilities

$$p_{ij}^{(1)} = p_{ij} \tag{2.57}$$

two-step transition probabilities

$$p_{ij}^{(2)} = \sum_k p_{ik} p_{kj} \tag{2.58}$$

where we sum over all intermediate states k through which the system may pass. By induction,

$$p_{ij}^{(n+1)} = \sum_k p_{ik} p_{kj}^{(n)} \qquad (2.59)$$

Let us now consider the probability for observing a particular sequence of states $x_1, x_2, \ldots, x_n$. This probability may be written as a conditional probability of the general form

$$p(X_t = x_t | X_{t-1} = x_{t-1}, X_{t-2} = x_{t-2}, \ldots, X_1 = x_1)$$

A sequence of states $x_1, x_2, \ldots, x_n$ forms a Markov chain if the probability that the system is in a particular state x_t at time t depends exclusively on the probability for the system to be in state x_{t-1} at time $t-1$. If the sequence of states forms a Markov chain, we do not have to explicitly consider the probabilities for any earlier time steps. That is, the states form a Markov chain provided that

$$p(X_t = x_t | X_{t-1} = x_{t-1}, X_{t-2} = x_{t-2}, \ldots, X_1 = x_1) = p(X_t = x_t | X_{t-1} = x_{t-1}) \qquad (2.60)$$

In words, a sequence of states forms a Markov chain if the probabilities for observing the various sequences of states depend entirely on the one-step transition probabilities p_{ij} and the initial probability distribution.

We now observe that for a Markov chain the (joint) probability for a particular path through configuration space from state j to state k is of the form of a product of one-step transition probabilities

$$p_{jj_1} p_{j_1 j_2}, \ldots, p_{j_{n-1} k}$$

with each transition probability linking a state with its predecessor. The transition probabilities can be arranged in a matrix P:

$$P = \begin{pmatrix} p_{11} & p_{12} & \cdots & p_{1n} \\ p_{21} & p_{22} & \cdots & \cdot \\ \cdot & \cdot & \cdots & \cdot \\ \cdot & \cdot & \cdots & \cdot \\ p_{n1} & \cdot & \cdots & p_{nn} \end{pmatrix} \qquad (2.61)$$

The elements of this matrix are all nonnegative, and the rows sum to unity; that is,

$$\sum_j p_{ij} = 1 \qquad (2.62)$$

Matrices of this type are termed *stochastic*.

2.10.3 Convergence to Stationary Distributions

Before introducing our convergence theorem, we must define a few terms. The first of these is the notion of an irreducible chain. A collection of states is said to be *closed* if none of the states outside of the collection can be reached from any of the states inside that collection. A Markov chain is termed *irreducible* if there exists no closed collection other than the collection of all states. In other words, a Markov chain is irreducible if and only if every state can be reached from every other state. In these chains, interior absorbing states or barriers are not present, where by an *absorbing state* we mean a state of a system which if reached is never exited (i.e., a single state forming a closed set is an absorbing state).

Let w_i denote the normalized, positive-definite probability that the system is in state i. A probability distribution $\{w_i\}$ is termed *stationary* if

$$w_j = \sum_i w_i p_{ij} \qquad (2.63)$$

We then have an important convergence theorem for irreducible and non-periodic (aperiodic) Markov chains.[26] It states that starting from an arbitrary initial state or distribution, the n-step transition probabilities will approach a unique stationary distribution provided that a stationary distribution exists. That is,

$$\lim_{n \to \infty} p_{ij}^{(n)} = w_j \qquad (2.64)$$

with w_j given by Eq. (2.63), and influences of the initial state disappear. If the stationary distribution does not exist, then

$$\lim_{n \to \infty} p_{ij}^{(n)} \to 0 \qquad (2.65)$$

Let us now introduce the relation

$$w_i p_{ij} = w_j p_{ji} \qquad (2.66)$$

This expression may be recognized from first-order reaction kinetics as a statement of the principle of detailed balance, or of microscopic reversibility. If the transition probabilities satisfy the detailed balance relation, then the transition probabilities satisfy Eq. (2.63):

$$\sum_i w_i p_{ij} = \sum_i w_i \left(\frac{w_j}{w_i} p_{ji}\right) = w_j \sum_i p_{ji} = w_j$$

where we have employed the row normalization condition in the last step.

We recall from Section 2.9.3 that one of the goals of Metropolis et al. was develop a way (1) to avoid explicit evaluation of the partition function and (2) to efficiently perform the Monte Carlo sampling by picking configurations with a frequency proportional to the relative importance of that configuration in determining the properties of the system. To achieve these goals, transition probabilities are defined in such a manner that the detailed balance condition is obeyed, with the Gibbs distribution as the stationary distribution.

2.11 THE METROPOLIS ALGORITHM

Let us consider an arbitrary Markov chain with a priori transition probabilities τ_{ij} satisfying the following three conditions:

$$\tau_{ij} \geq 0, \quad \sum_j \tau_{ij} = 1, \quad \tau_{ij} = \tau_{ji} \tag{2.67}$$

The first condition is one of positivity. The second condition is the normalization requirement, that is, the elements of each row must sum to unity; the last requirement is that of symmetry. We then define transition probabilities in terms of the symmetric τ_{ij}'s and probability distribution ratios π_j/π_i as follows:

$$p_{ij} = \begin{cases} \frac{\tau_{ij}\pi_j}{\pi_i}, & \frac{\pi_j}{\pi_i} < 1 \\ \tau_{ij}, & \frac{\pi_j}{\pi_i} \geq 1 \end{cases} \tag{2.68}$$

If $i = j$, we define the transition probabilities as

$$p_{ii} = \tau_{ii} + \sum_{j, \pi_j/\pi_i < 1} \tau_{ij}\left(1 - \frac{\pi_j}{\pi_i}\right) \tag{2.69}$$

This last condition, Eq. (2.69), is needed to ensure that the probabilities sum to unity. We choose the probability distribution to be the Gibbs distribution;

that is, we take

$$\pi_i(E_i) = \frac{1}{Z} e^{-\beta E_i} \quad (2.70)$$

and the probability distribution ratios take the simple form

$$\frac{\pi_j(E_j)}{\pi_i(E_i)} = e^{-\beta \Delta E_{ij}} \quad (2.71)$$

with energy difference

$$\Delta E_{ij} = E_j - E_i \quad (2.72)$$

We immediately observe that by taking ratios of probabilities the dependence on the partition function has been removed. By construction, the transition probabilities p_{ij}, are all greater than or equal to zero, are normalized to unity, and most important, obey the detailed balance relation

$$\pi_i p_{ij} = \pi_j p_{ji}$$

To check that the detailed balance condition is obeyed, let us first assume that the energy change resulting from a transition from state i to state j is negative. Then we have

$$\pi_i p_{ij} = \pi_i \tau_{ij} = e^{-\beta E_i} \tau_{ij}$$

and

$$\pi_j p_{ji} = e^{-\beta E_j} \tau_{ji} e^{-\beta(E_i - E_j)} = e^{-\beta E_i} \tau_{ij} = \pi_i p_{ij}$$

so the condition for detailed balance is satisfied. Now, let us suppose that the energy change is positive. Then we have

$$\pi_i p_{ij} = e^{-\beta E_i} \tau_{ij} e^{-\beta(E_j - E_i)} = e^{-\beta E_j} \tau_{ij}$$

and

$$\pi_j p_{ji} = e^{-\beta E_j} \tau_{ji} = e^{-\beta E_j} \tau_{ij} = \pi_i p_{ij}$$

Therefore the requirement of detailed balance is obeyed for this case as well. Consequently the n-step transition probabilities approach a stationary distribution $w_j = \pi_j$. From the convergence theorem we know that the limiting stationary distribution reached by the n-step transition probabilities is unique, and therefore the stationary distribution generated by the Metropolis algorithm must be the Gibbs distribution.

The Metropolis sampling procedure is as follows: Suppose that at time t the state X_t takes the value x_t. Then select a state Y_t at random from the target distribution in a manner which satisfies the symmetry condition

$$p(Y_t = x_j | X_t = x_i) = p(Y_t = x_i | X_t = x_j) \qquad (2.73)$$

Calculate the energy difference

$$\Delta E = E(Y_t) - E(X_t) \qquad (2.74)$$

If the energy difference is negative, then the transition leads to a state with lower energy, and this transition is allowed. That is, $X_{t+1} = Y_t$. If, on the other hand, the energy difference is positive, then the proposed transition produces an increase in the energy of the system. In this case select a random number ξ in the range 0 to 1. If $\xi < \exp(-\beta \Delta E)$, then the transition is allowed. Otherwise, the transition is not allowed. To summarize, for the case where the transition leads to a decrease in energy ($\Delta E < 0$):

$$X_{t+1} = Y_t \qquad (2.75)$$

In the case where the transition leads to an increase in energy ($\Delta E > 0$), we have

$$X_{t+1} = \begin{cases} Y_t, & \text{prob} = e^{-\beta \Delta E} \\ X_t, & \text{prob} = 1 - e^{-\beta \Delta E} \end{cases} \qquad (2.76)$$

This distribution is sampled with probability proportional to the Boltzmann factor. The procedure generates a Markov chain of transition probabilities that converge to a unique, stable Gibbs distribution.

2.12 CONCLUDING REMARKS

The Metropolis algorithm is representative of the Monte Carlo simulation algorithms that will be employed the next few chapters. We observed that there are the two essential ingredients. These are (1) we must sample in a way that satisfies the symmetry requirement, Eq. (2.67) or (2.73), and (2) we must choose a form for the transition probabilities that obeys detailed balance. There are alternative choices for the sampling procedure, and for the transition ratios, depending on the kinetics of the process being simulated. We will examine several examples in the next chapter, and still others in the following chapters.

2.13 FURTHER READING

Shannon's work on information theory can be found in the book by Shannon and Weaver,[27] as well as in the original articles. There are also a number of excellent texts on information theory, such as the one by Gallager.[28] Discussions of the relationship between information theory and thermodynamics can be found in a short paper by Rothstein[29] and in a longer one by Tribus.[30] A full exposition of classical thermodynamics has been presented by Callen.[31] Another excellent treatment of thermodynamics can be found in the monograph by Pippard.[32] A sophisticated presentation of statistical mechanics, from an information theory viewpoint, has been given by Katz.[33] The book by Chandler[34] is also highly recommended.

The question we addressed in Chapter 2 was: Given a system with a set of possible states, and some limited information expressed as constraints on the probability distribution, how do we best estimate the unknown probability distribution? The answer we gave was to choose the distribution that maximizes entropy subject to the constraints — it is the distribution that is maximally noncommittal with respect to missing information. A generalization of this principle to cases where there is additional information in the form of a prior distribution is called the *principle of minimum cross entropy*. The type of problem addressed by this principle is as follows: Suppose that we have a probability distribution q. New data are acquired, and we wish to update the probability distribution in a way that is consistent with the data, while remaining as close as possible in some sense to the previous distribution. The answer provided by this principle is to find the probability distribution p that is consistent with the constraints and minimizes the cross-entropy $S(p, q)$:

$$S(p, q) = K \sum_i p_i \ln\left(\frac{p_i}{q_i}\right)$$

We may note that when the prior distribution is the uniform distribution, the cross-entropy reduces to the Shannon-Jaynes form. The name "cross-entropy" was coined by Good,[35] who also called this measure the "expected weight of evidence."[36] Another name for $S(p, q)$, which was introduced by Kullback,[37] is the "directed divergence." When applied to continuous distributions, Jaynes[38] refers to this form as an invariant measure. Self-consistency arguments in favor of entropic measures of this type have been presented in a number of papers. Noteworthy among these are the works by Shore and Johnson,[39] van Campenhout and Cover,[40] and Tikochinsky, Tisby, and Levine.[41]

Markov chains are discussed in the classic text on probability theory by Feller.[26] They are also discussed in depth in the monograph by Chung.[42] The early work on the Monte Carlo method is described in detail by Hammersley and Handscomb.[43]

2.14 REFERENCES

1. Bernoulli, D. (1738). On the properties and motions of elastic fluids, especially air (English trans. of "De affectionibus atque motibus fluidorum elacticorum, praecipue autum aëris," *Hydrodynamica, sive de vivibus et motibus fluidorum commentarii*, Sectio Decima, Argentorati: Sumptibus Johannes Reinholdi Dulseckeri, 200–204). In S. G. Brush (ed.), J. P. Berryman (trans.), *Kinetic Theory*, vol 1. Oxford: Pergamon Press, 1965, pp. 57–65.
2. Carnot, S. (1824). Reflections on the motive power of fire, and on machines fitted to develop that power (English trans. of "Réflexions sur la puissance motrice du feu et sur les machines propres à dvelopper cette puissance," Paris: Bachelier). In E. Mendoza (ed.), *Reflections on Motive Power of Fire and Other Papers*. New York: Dover, 1960, pp. 3–22.
3. Planck, M. (1901). Über das Gesetz der Energieverteilung im Normalspektrum. Ann. d. Phys., **4**, 553–563.
4. Planck, M. (1906, 2nd. ed. 1913). *The Theory of Heat Radiation* (English trans. of *Vorlesungen über die Theorie der Wärmestrahlung*, Leipzig, M. Masius, trans., Philadelphia, Blackiston's, 1914). Reprint. New York: Dover, 1959, 1991.
5. Boltzmann, L. (1896, Part I; 1898, Part II). *Lectures on Gas Theory* (English trans. of *Vorlesungen über Gastheorie*, Leipzig, J. A. Barth, S. G. Brush, trans.). Berkeley: University of California Press, 1964.
6. Gibbs, J. W. (1902). *Elementary Principles in Statistical Mechanics*. New Haven: Yale University Press.
7. Clausius, R. (1867). On different forms of the fundamental equations of heat and their convenience for application (English trans. of "Über verschiedene für die Anwendung bequeme Formen der Hauptgleichungen der mechanischen Wärmetheorie," *Anhandlungen über die mechanische Wärmetheorie*, vol. II, Viewig, Braunschweig, 1–44). In R. B. Lindsay (trans.), *The Second Law of Thermodynamics*. Stroudsburg: Dowden, Hutchinson and Ross, 1976, pp. 162–193.
8. Shannon, C. E. (1948). A mathematical theory of communication. Bell System Tech. J., **27**, 379–423, 623–656.
9. Jaynes, E. T. (1957). Information theory and statistical mechanics. Phys. Rev., **106**, 620–630; **108**, 171–190.
10. Elsasser, W. M. (1937). On quantum measurements and the role of the uncertainty relations in statistical mechanics. Phys. Rev., **52**, 987–999.
11. Boltzmann, L. (1877). Über die Beziehung zwischen dem zweiten Hauptsatze der mechanischen Wärmetheorie und der Wahrscheinlichkeitsrechnung, respective den Sätzen über das Wärmegleichgewicht. Wein. Ber., **76**, 373–399.
12. Boltzmann, L. (1872). Weitere Studien über das Wärmegleichgewicht unter Gasmolekulen. Wein. Ber., **66**, 275.
13. Maxwell, J. C. (1860). Illustrations of the dynamical theory of gases: I. On the motions and collisions of perfectly elastic spheres; II. On the process of diffusion of two or more kinds of moving particles among one another, III. On the collision of perfectly elastic bodies of any form. Phil. Mag., **19**, 19–32; **20**, 21–37.
14. Bernoulli, J. (1713). *Ars Conjectandi*. Basil: Thurnisiorum.
15. Keynes, J. M. (1921). *A Treatise on Probability*. London: Macmillan.

16. Bayes, T. (1764). As essay toward solving a problem in the doctrine of chance. Phil. Trans. R. Soc. Lond., **53**, 370–418.
17. Laplace, P. S. (1812, 3rd. ed. 1820). *Théorie analytique des probabilits*. Paris: Courcier.
18. Cox, R. T. (1946). Probability, frequency and reasonable expectation. Am. J. Phys., **14**, 1–13.
19. Cox, R. T. (1961). *The Algebra of Probable Inference*. Baltimore: Johns Hopkins Press.
20. Jeffries, H. (1939). *Theory of Probability*. Oxford: Oxford University Press.
21. Onsager, L. (1944). Crystal statistics I: A two-dimensional model with an order-disorder transition. Phys. Rev., **65**, 117–149.
22. Ulam, S. M., and von Neumann, J. (1947). On combination of stochastic and deterministic processes. Bull. Am. Math. Soc., **53**, 1120.
23. Goldberger, M. L. (1948). The interaction of high energy neutrons and heavy nuclei. Phys. Rev., **74**, 1269–1277.
24. Eyal, Y., Beckerman, M., Chechik, R., Fraenkel, Z., and Stocker, H. (1976). Nuclear size and boundary effects on the fusion barrier of oxygen with carbon. Phys. Rev., **C13**, 1527–1535.
25. Metropolis, N., Rosenbluth, A. W., Rosenbluth, M. N., Teller, A. H. (1953) and Teller, E. (1953). Equation of state calculations by fast computing machines. J. Chem. Phys., **21**, 1087–1092.
26. Feller, W. (1950; 2nd ed., 1957). *An Introduction to Probability Theory and Its Applications*. New York: Wiley.
27. Shannon, C. E., and Weaver, W. (1949). *The Mathematical Theory of Communication*. Urbana: University of Illinois Press.
28. Gallager, R. G. (1968). *Information Theory and Reliable Communication*. New York: Wiley.
29. Rothstein, J. (1952). Information and thermodynamics. Phys. Rev., **85**, 185.
30. Tribus, M. (1961). Information theory as the basis for thermostatics and thermodynamics. J. Appl. Mech., **83**, 1–8.
31. Callen, H. B. (1960). *Thermodynamics*. New York: Wiley.
32. Pippard, A. B. (1957). *The Elements of Classical Thermodynamics*. Cambridge: Cambridge University Press.
33. Katz, A. (1967). *Principles of Statistical Mechanics*. San Francisco: W. H. Freeman.
34. Chandler, D. (1987). *Introduction to Modern Statistical Mechanics*. New York: Oxford University Press.
35. Good, I. J. (1964). Maximum entropy for hypothesis formulation, especially for multidimensional contingency tables. Annals. Math. Statist., **34**, 911–934.
36. Good, I. J. (1950). *Probability and the Weighting of Evidence*. London: Charles Griffin.
37. Kullback, S. (1959). *Information Theory and Statistics*. New York: Wiley.
38. Jaynes, E. T. (1963). Information theory and statistical mechanics. In K. W. Ford (ed.), *1962 Brandeis Summer Institute Lectures in Theoretical Physics*. New York: W. A. Benjamin, pp. 181–218.

39. Shore, J. E., and Johnson, R. W. 1980. Axiomatic derivation of the principle of maximum entropy and the principle of minimum cross entropy. IEEE Trans. Inform. Theory, **IT-26**, 26–37.
40. Van Campenhout, J. M., and Cover, T. M. (1981). Maximum entropy and conditional probability. IEEE Trans. Informat. Theory, **IT-27**, 483–489.
41. Tikochinsky, Y., Tishby, N. Z., and Levine, R. D. (1984). Consistent inference of probabilities for reproducible experiments. Phys. Rev. Lett., **52**, 1357–1360.
42. Chung, K. L. (1960; 2nd ed., 1967). *Markov Chains with Stationary Transition Probabilities.* Berlin: Springer-Verlag.
43. Hammersley, J. M., and Handscomb, D. C. (1964). *Monte Carlo Methods.* London: Methuen.

3

COOPERATIVITY IN LATTICE SYSTEMS

3.1 INTRODUCTION

The defining characteristic of a cooperative assembly is the existence of interactions among its constituents. The states of a given element in a cooperative system are influences by those of its neighbors in a fundamental way, and the overall assembly cannot be regarded as a collection of independent entities. There are an impressive number of different kinds of cooperative assemblies in nature. A few examples of cooperative physical systems, the subject of this chapter, are liquids, ferromagnets, and superconductors. The interactions responsible for the cooperativity in these systems are the intermolecular forces, exchange interactions, and pairing forces, respectively. If the elements of a cooperative system are arranged spatially in a regular manner, then the assemblage is said to form a lattice.

In treating interactions among elements of lattice systems two classes of approximations have been developed. In the first group of methods we consider only interactions between elements at nearby lattice sites. In these approaches we neglect interactions between elements and their more distant neighbors. In the second class of procedures we include interactions between elements that are not necessarily nearest neighbors, but only in an approximate fashion in terms of an average, or mean, field. In these treatments we ignore effects of fluctuations in the interactions.

3.1.1 Configurations and the Counting Problem

In order to appreciate the difficulties encountered in treating cooperative systems we consider the case of a system composed of two types of elements undergoing nearest neighbor interactions. The interaction energy of our model system in a particular configuration consists of a sum of contributions, one from each pair of nearest neighbors. Suppose that we have N_i elements of type 1 and N_2 elements of type 2. These elements are distributed among $N = N_1 + N_2$ lattice sites in such a way that there is one and only one element at each site. Let N_{11}, N_{12}, and N_{22} denote the number of 1-1, 1-2, and 2-2 nearest-neighbor pairs, respectively, and let ε_{11}, ε_{12}, and ε_{22} denote the corresponding pair interaction energies. The total interaction energy of the given arrangement can be expressed as

$$E_{tot} = N_{11}\varepsilon_{11} + N_{12}\varepsilon_{12} + N_{22}\varepsilon_{22} \tag{3.1}$$

The five quantities N_1, N_2, N_{11}, N_{12}, and N_{22} representing the total number of each kind of element, and each form of element pairing, are interrelated. Suppose that we have a regular lattice in which each site has v nearest neighbors. (The quantity v is known as the coordination number.) If we picture each site of type 1 with a link connecting it to its nearest neighbors, then each 1-1 pair will be connected by two links and each 1-2 pair will be joined by one link. A similar construction will hold for the 2-2 pairs, and by this means we infer the relations

$$2N_{11} + N_{12} = vN_1$$
$$N_{12} + 2N_{22} = vN_2 \tag{3.2}$$

with

$$N_1 + N_2 = N \tag{3.3}$$

The number of independent variables is thereby reduced from five to two. For a specified value for N_1, we may choose N_{11} or N_{12} as the second free quantity. In the former case we find that

$$E_{tot}(N, N_1, N_{11}) = \left(\frac{1}{2}\right)vN\varepsilon_{22} + vN_1(\varepsilon_{12} - \varepsilon_{22}) + N_{11}(\varepsilon_{11} + \varepsilon_{12} - 2\varepsilon_{12}) \tag{3.4}$$

while in the latter instance we obtain

$$E_{tot}(N, N_1, N_{12}) = \left(\frac{1}{2}\right)vN\varepsilon_{22} + \left(\frac{1}{2}\right)vN_1(\varepsilon_{11} - \varepsilon_{22}) + N_{12}\left[\varepsilon_{12} - \left(\frac{1}{2}\right)(\varepsilon_{11} + \varepsilon_{22})\right] \tag{3.5}$$

The partition function, representing a sum over all possible configurations of the lattice system, can be expressed as a sum over our two selected independent variables. Choosing N_1 and N_{12} as our variables allows us to write

$$Z = \sum_{N_1, N_{12}} \Omega(N_1, N_{12}) e^{-\beta E(N_1, N_{12})} \tag{3.6}$$

The density of states $\Omega(N_1, N_{12})$ appearing in this formulation of the problem is difficult to evaluate. This difficulty is known as the *counting problem*. Finding ways to evaluate the partition function is a central goal in the field. At least four distinct lines of attack have been developed. The first class of methods are the analytic treatments known as the *transfer matrix* and *combinatorial methods*. These have been used to obtain exact solutions to the one- and two-dimensional Ising systems. A second, important line of attack, which has provided many valuable insights into cooperative phenomena, are the *mean-field theories* mentioned earlier. This class of techniques includes the Weiss molecular field, Landau's theory, the Bethe approximation, and the van der Waals picture. A third, fairly recent development is the emergence of powerful renormalization group techniques. These have been used to examine critical phenomena in a variety of systems. Fourth, we have the Monte Carlo methods introduced in Sections 2.9–2.11. These methods allow us to bypass the necesssity of evaluating the partition function.

3.1.2 Objectives of the Chapter

In this chapter we will apply the formalism developed in Chapter 2 to the study of cooperative phenomena. In doing so, we will approximate the complex interactions taking place in a system to those involving only nearest neighbors. The systems we will examine have many coupled degrees of freedom. A prevailing attitude in the discipline is that for such systems the detailed character of the interactions generating cooperativity is less important than the presence or absence of the interactions themselves.[1] This view is supported by the many striking similarities in cooperative behavior that has been observed in radically different systems. We will adopt this viewpoint, applying in succession the transfer matrix method, mean-field theory, and renormalization group tenchiques to the nearest-neighbor interactions appearing in the Ising model of a ferromagnet. The literature on the subject is impressive, and we cannot hope to present all, or even most, of the major findings in a single chapter. Given these limitations, our objectives are, as follows:

We will begin our discussion in Section 3.2 with an overview of some of the key concepts such as order parameters, correlations lengths, broken symmetry, critical exponents, and scaling. We will then introduce (Section 3.3) the Ising model of a single ferromagnetic domain. The transfer matrix method will be used in Section 3.4 to deduce the thermodynamic properties of a one-dimensional Ising chain. The mathematics involved in the Onsager solution for the two-dimensional Ising lattice are far more complex than that needed to solve the one-dimensional problem. In Section 3.5 we will simply state the results and then make some comparisons to the analogous findings for the one-dimensional chain. We will see that the one-dimensional chain does not undergo phase transitions, while the two-dimensional lattice does exhibit this consequence of cooperativity. We will present the Peierls-Griffiths[2] argument concerning the reason for this difference in behavior in Section 3.6.

In the next three sections we will explore lattice cooperativity in ferromagnets and fluids from the point of view of mean-field theories. The Weiss molecular field and random mixing (Bragg-Williams) methods represent zeroth-order approximations to an exact treatment of ferromagnets and binary alloys, while the Bethe approximation is a first-order method. We will outline these approaches in Section 3.7 and examine phase transitions in fluids near the triple and critical points in Section 3.8. The lattice gas model is isomorphic to the Ising lattice. We will present this model of a fluid and establish a relationship between the van der Waals picture and the Weiss molecular field.

In addition to binary alloys and lattice gases, biological membranes can self-organize into superlattices. Although we will not discuss these models, it is worthwhile to note that Ising-like models describing the allosteric and cooperative properties of biological membranes have been developed.[3] These membrane models can exhibit both sigmodal (graded response) and rapid, on–off kinetics. In constructing these models, a biological membrane is considered to be organized into translationally symmetric lattices constructed out of pseudoequivalent molecular components (protomers), each of which can exist in one of two states.

We have already noted that there are striking similarities between phase transitions in different cooperative systems. These similarities can be brought out by focusing on the behavior of the thermodynamic parameters for these systems in the vicinity of the transition, or critical, temperature T_c. In this regime the thermal response can be summarized by critical indexes that describe the nature of the divergences in the thermodynamic parameters. One of the key ideas that has emerged from comparisons of critical indexes is that of *scaling laws*. These laws, first proposed by Widom[4,5] and by Kadanoff,[6] relate critical indexes to one another. The theoretical foundation of scaling is contained in an important tool for exploration of critical phenomena called a *renormalization group*. This construct, introduced by Wilson,[7,8] permits the study of problems in which fluctuations persist over many scales of length, and provides useful insights into the emergence of phase transitions. In Section 3.9 we will reexamine the one- and two-dimensional Ising lattices using Wilson's renormalization group (RG) approach. We will describe the scaling antecendents, derive the one-dimensional RG solution, and then contrast these results with those for two dimensions in which fixed points appear.

3.2 KEY CONCEPTS

3.2.1 Entropic and Energetic Potentials

Perhaps the most remarkable feature of the stable equilibrium states of thermodynamic systems is the property known as *order*. At high temperatures the equilibrium configurations of a thermodynamic system tend to be homogeneous and disordered. The situation changes as the temperature is lowered. At low temperatures the stable arrangements of the constituent elements of a system become highly ordered. In Chapter 2 we made a first attempt to characterize the stable, equilibrium states of thermodynamic systems. We observed that these states correspond to special values of the thermodynamic parameters, specifically, to a maximum in the entropy and a minimum in the (free) energy. The former favors disorder, while the latter encourages order.

We noted earlier that the free energy seeks a minimum at equilibrium. We observe from Eq. (2.47) that the Helmholtz free energy A is the difference between the internal energy $\langle E \rangle$ and the entropic potential TS. The change in free energy for a system in contact with a heat bath at the same temperature T is

$$\Delta A = \Delta \langle E \rangle - T \Delta S \tag{3.7}$$

At high temperatures the second term dominates the expression for the free energy change. The first term contains contributions from the kinetic and interaction energies. At high energies, and/or in dilute media, the kinetic energy is more important than the interaction potential. However, as the temperature is lowered, the short-range interactions become progressively more important, and eventually the system seeks a configuration in which the potential energy is a minimum. Both T and the entropy change decline in value, and highly ordered configurations, corresponding to mimima in the interaction energy, are formed.

3.2.2 Phase Transitions

Ordinarily fluctuations in the parameters of a system, such as temperature, volume, or density, bring about processes that restore the equilibrium in that system. If a system is brought into a condition where it loses its stability against fluctuations in its thermodynamic parameters, that entity will cease to be homogeneous and will split into several portions. This destabilization is known as a *phase transition*. Familiar examples of phase transitions are those of water as it changes from a vapor to a liquid, and from a liquid to a solid. In condensing from vapor to liquid, the lower temperatures enable the intermolecular forces to bring the molecules closer together. In the solid phase the

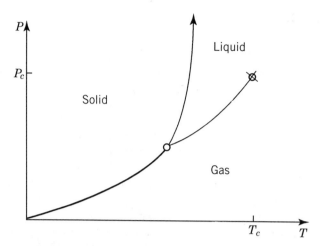

Figure 3.1. Typical phase diagram obtained by projecting the pressure-volume-temperature surface onto the pressure-temperature (PT) plane. The solid lines separating the solid-gas, solid-liquid, and liquid-gas phases are called the sublimation, fusion, and vapor pressure curves, respectively. These curves denote the locus of points where the two phases coexist. The coordinates of the critical point are labeled as P_c, and T_c.

molecules assume fixed locations, and the kinetic energy is limited to vibrations about the lattice positions. These transitions can be represented by a phase diagram of the form shown in Fig. 3.1. The solid lines in the pressure-temperature phase diagram delineate the different phases. The point where the three phases coincide is known as the *triple point*, and the place where the curve separating the liquid phase from the gas phase terminates is an example of a *critical point*.

Below the liquid-gas critical point there is a difference between the densities of gas and liquid. This difference, $\rho_l - \rho_g$, vanishes at the critical point. Above it, the difference in densities is zero, and there is no distinction between liquid and vapor. If we begin well above the critical point and cool the system, droplets will start to appear. These droplets will increase in size as the critical point is approached, approaching and exceeding the dimensions of the wavelength of visible light. In such a mixture, light will be strongly scattered, giving rise to the phenomenon of *critical opalescence*. In the vicinity of the critical point, bubbles of liquid and vapor will be intermixed at all length scales from atomic scales up to macroscopic dimensions. Fluctuations occurring over all scales of length are not unique to the liquid-gas system near the critical point. Turbulent fluid flow producing atmospheric fluctuations spanning scales of length up to continental dimensions is a well-known and dramatic example of this manifestation of cooperative behavior. The composite internal structure of elementary particles over length scales down to zero is another profound consequence of cooperative, multiscale fluctuations.

In an ordinary, or first-order, phase transition, there are discontinuities in the molar parameters of state such as the molar entropy, energy, and volume. The entropic change $T\Delta S$ associated with a first-order phase transition is known as a *latent heat*. This quantity represents the heat absorbed in transforming a given amount of material from one phase to another, and the entropy difference ΔS arises from the increase in configuration order that occurs in the more condensed phase. If there is a change in the spatial symmetry of the system, there will be an associated discontinuity in the heat capacity. At temperatures above the liquid-gas critical point, the symmetry properties no longer differ in the two phases, and the system can continuously pass from one phase to the other.

In a continuous, or second-order, phase transition, the molar parameters are continuous, and there is no latent heat. This type of phase transition is characterized by an order parameter and by singularities in the heat capacity. Continuous phase transitions are associated with strongly cooperative interactions. Representative second-order phase transitions are the paramagnetic-ferromagnetic transitions, the order-disorder transitions in binary alloys such as β-brass, the beginning of superfluidity in liquid helium, and the appearance of superconductivity. The heat capacity for iron is displayed in Fig. 3.2 as a function of the temperature. The divergence in the heat capacity as the critical temperature for the paramagnetic-ferromagnetic phase transition is approached can be observed in this graph.

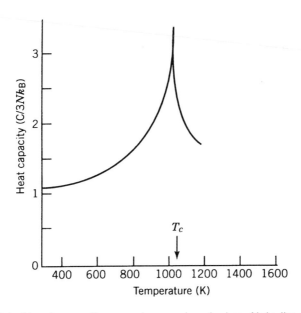

Figure 3.2. Plot of heat capacity versus temperature for iron. Note the sharp spike in the heat capacity at the Curie temperature T_c.

3.2.3 Order Parameters

The critical temperature for a phase transition marks the onset of the appearance of long-range order. This cooperative property can be described by an order parameter. This parameter is a fluctuating thermodynamic variable whose average provides a measure of the amount of order present in the system. The spontaneous magnetization is the order parameter for the ferromagnet, the pairing gap parameter is the order parameter for a superconductor, and the density minus the critical density is the order parameter for the liquid-gas phase transition near the critical point. The spontaneous magnetization, displayed in Fig. 3.3, has a typical thermal dependence. The threshold for the onset of long-range order occurs at the critical temperature. The order parameter vanished above the critical temperature, and it reaches its maximum value at zero temperature. The behavior of the order parameter near the critical temperature can be used to characterize the order of the phase transition. Specifically, if the order parameter approaches zero continuously from below, as it does for the spontaneous magnetization, the transition is not first order.

The order parameter can assume a number of distinct values below the critical temperature. For example, the spontaneous magnetization of a ferromagnet may be either positive or negative, and the density minus the critical density of a fluid may reflect either the liquid or gas phase. These different

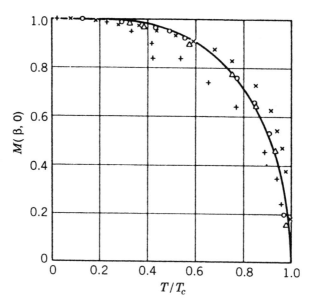

Figure 3.3. Experimental results on the thermal dependence of the spontaneous magnetization. The observed behavior is typical of an order parameter. Shown are data for iron ($\times$), nickiel ($\circ$), cobalt ($\triangle$) and magnetite ($+$). (From Wannier[9]. © 1996 John Wiley & Sons, Inc.)

possibilities exist under identical physical conditions as represented by the hamiltonian. For example, in the absence of a external magnetic field, the hamiltonian for a ferromagnet is symmetic with respect to orientation. The selection of a particular choice for the direction is triggered either by a vanishingly small external field or equivalently by a random fluctuation. This natural selection process may be viewed as the *spontaneous breaking of spatial symmetry*, and it occurs as the system attempts to restore stability near the phase transition point.

3.2.4 Critical Exponents and Scaling

In the vicinity of a critical point, the behavior of the thermodynamic parameters can be described by their critical exponents. The basic idea is that near the critical temperature, the free energy (the logarithm of the partition function), from which all thermodynamic parameters are computed, may be expanded in a power series in the order parameter and in the quantity

$$\varepsilon = \frac{T - T_c}{T_c} \tag{3.8}$$

As the critical temperature is approached from above or below the critical temperature, the leading term will increasingly dominate the power series expansion. The exponent of this leading term is called the *critical exponent*. For a ferromagnet, critical exponents may be defined for the heat capacity, the magnetization, and the magnetic susceptibility (the differential increase in magnetization with increasing field strength). By convention, their critical exponents are designated as α, β, γ:

$$C(T) \approx \varepsilon^{-\alpha}, \quad M(T) \approx \varepsilon^{-\beta}, \quad \chi(T) \approx \varepsilon^{-\gamma} \tag{3.9}$$

In many instances the behavior of the thermodynamic parameters observed as the critical temperature is approached from below differs from that found as the critical temperature is approached from above. In these cases two sets of critical exponents are defined, with $-\varepsilon$ replacing ε, and a "prime" (e.g., β') appended, to describe the approach from below toward the critical temperature.

Since the thermodynamic parameters are related to one another through the free energy, the critical exponents for a particular system might be expected to obey some set of relationships. In addition to these intrasystem relationships, there are striking similarities in the critical behavior observed in different systems. The intra- and extrasystem findings support the notion that the essential properties of many phase transition are independent of the detailed interactions, except for the dependence on the dimensionality of the system. This independence is known as *universality*.

3.2.5 Correlation Lengths

Let us now consider the mechanics underlying the initiation of a phase transition. The way a system probes its environment to find a deeper minima in the free energy at a different location in configuration space is through fluctuations. Typically the fluctuations in the thermodynamic parameters are small, and considerable time may elapse before a large enough fluctuation occurs at some small place in the system to induce a transition at that location to a new and deeper mimimum. That portion of the system undergoing a phase transformation becomes a seed for further changes. Cooperative interactions give rise to correlations between the fluctuations a different points in space. Let **r** and **r'** denote the positions of two lattice sites. Then we may define a correlation function $g(\mathbf{r}, \mathbf{r}')$, that provides us with an estimate of the spatial extent of the correlated fluctuations. When the distances between the sites are large, this correlation function is of the general form[10]

$$g(\mathbf{r}, \mathbf{r}') \approx \frac{e^{-|\mathbf{r}-\mathbf{r}'|/\xi}}{|\mathbf{r} - \mathbf{r}'|} \qquad (3.10)$$

with

$$\begin{aligned} \xi &\approx (T - T_c)^{-1/2}, & T &> T_c \\ \xi &\approx (T_c - T)^{-1/2}, & T &< T_c \end{aligned} \qquad (3.11)$$

The variable ξ is called the *correlation length*. It provides a measure of the size of the region, or the range, over which correlated fluctuations in the thermodynamic variables occur. We now see that range of the correlation function increases rapidly in the vicinity of a critical point and becomes infinite at the critical temperature. In other words, in the vicinity of a critical point, the cooperative interactions promote a rapid onset of large-scale, correlated fluctuations in the thermodynamic variables representative of the order in the system. These fluctuations in turn produce divergences in the range of correlations at the critical temperature that lead to singularities in the thermodynamic derivatives.

3.3 THE ISING MODEL

3.3.1 Ferromagnetism

There are several forms of magnetism. *Paramagnetism* is the term used to denote the oriented response to an external, applied magnetic field of atoms or ions possessing a permanent magnetic moment. The applied field produces a torque that leads to an alignment of the moments in the direction of the field. *Ferromagnetism* is the name given to a cooperative form of magnetism in which

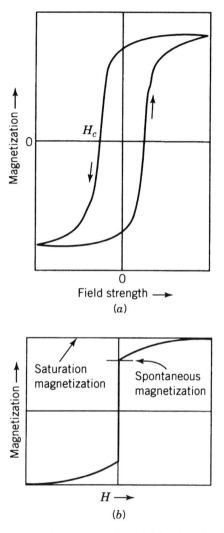

Figure 3.4. Plots of magnetization versus external field strength: (*a*) Hysteresis curve for ferromaqgnetic substance; (*b*) reversible magnetization in a single ferromagnetic domain. At zero field there are two possible values, +M and −M, for the spontaneous magnetization. (From Wannier[9]. © 1996 John Wiley & Sons, Inc.)

a alignment of the moments can be sustained in the absence of an external field. The amount of alignment present when the external field strength is zero is called the *spontaneous magnetization.* Typical ferromagnetic materials are Fe, Ni, Co, and their alloys and compounds. The ferromagnetic behavior of these substances possess a characteristic thermal dependence. Above a certain critical temperature, called the *Curie point*, the materials are no longer ferromagnetic but, instead, become paramagnetic.

In a ferromagnetic material the magnetization is not a unique function of the external, applied field, but rather depends on its previous history. This irreversibility, known as *hysteresis*, is illustrated schematically in Fig. 3.4a. In this plot of magnetization versus field strength, we see that the response is delayed, but once started, the change in magnetization is rapid until the saturation plateau is reached. The response properties change if we restrict ourselves to a single domain. We may recall that a ferromagnetic substance is composed of a number of interlocking magnetic domains. The response curve for a single ferromagnetic domain is shown in Fig. 3.4b. In this situation the changes in magnetization are reversible, and there is no hysteresis.

Ferromagnetism is quantum mechanical in origin. Pauli exchange coupling gives rise to an interaction between unpaired electronic spins of nearby particles i and j of the general form

$$\phi = J \mathbf{s}_i \cdot \mathbf{s}_j \tag{3.12}$$

The exchange coupling J is a function of distance between the spins. At large distances the coupling decreases exponentially, and we may approximate the coupling by one acting between nearest neighbors. If we restrict the exchange interactions to those operating in one spatial dimension, such as in the direction of an external field, then the exchange interaction takes the Ising form $J \cdot s_i \cdot s_j$.

3.3.2 The Classical Limit and XY-Model

In more detail, if we allow the couplings to vary with i and j, then the total energy E obtained by adding together the contributions from all pairwise interactions over the lattice is

$$E = -\sum_{i<j} J_{ij} \mathbf{s}_i \cdot \mathbf{s}_j = -\sum_{i<j} J_{ij}(s_{ix}s_{jx} + s_{iy}s_{jy} + s_{iz}s_{jz}) \tag{3.13}$$

To recover the Ising model, we take $J_{ij} = J > 0$, and only consider the projection of the spin along an arbitrarily defined z-axis. Then, as is customary, we write $s_{iz} \cdot s_{jz}$ as $s_i \cdot s_j$.

Another limit that is useful to consider is the classical, or continuous, limit in which the spin components are not quantized and can assume any one of a continuous range of values. If we rescale the interaction by replacing J_{ij} with J_{ij}/s^2, where s is the magnitude of the spin vector, and only retain the transverse portion of the interaction, then the hamiltonian assumes the form

$$E = -\sum_{i<j} J_{ij}(s_{ix}s_{jx} + s_{iy}s_{jy}) \tag{3.14}$$

This interacting spin model is known as the (Heisenberg) XY model. As a result of our rescaling, the spin vectors are normalized to unity, and we can write the coupling interaction as

$$\begin{aligned} E &= -\sum_{i<j} J_{ij}(\cos\phi_i \cos\phi_j + \sin\phi_i \sin\phi_j) \\ &= -\sum_{i<j} J_{ij} \cos(\phi_i - \phi_j) \end{aligned} \quad (3.15)$$

where ϕ_i is the angle between the direction of the ith spin vector in the lattice and the x-axis. We will encounter in later chapters several lattice models containing interactions analogous to that of Eqs. (3.14) and (3.15). In chapter 9 we will study the response properties of arrays of coupled rotators representing model oscillatory neurons. Expressions similar to the above appear in which the variables ϕ_i denote the phases of the oscillatory units.

3.3.3 Historical Development of the Ising Model

The Ising model, which we will discuss at some length in this chapter, is the prototypical nearest-neighbor model of cooperative phenomena. It was initially developed by Ernst Ising to study whether iron or nickel atoms affixed rigidly in position, but not in spatial orientation, in a metallic lattice could self-organize to produce an observable magnetic field. In his dissertation, summarized in his 1925 paper,[11] Ising showed that there is no phase transition to a ferromagnetic ordered state at any temperature for a one-dimensional lattice. Although Ising incorrectly concluded that the same negative result would hold in more than one dimension, an argument that the Ising model would in fact exhibit a phase transition in two or more dimensions was put forth by Peierls.[12] This argument, completed by Griffiths,[2] provides revealing insights into the difference between the one- and two-dimensional situations. The existence of a well-defined phase transition for a two-dimensional Ising system was formally demonstrated by Kramers and Wannier,[13,14] and by Montroll,[15] using a transfer matrix method which they had developed. Onsager's exact solution[16] for the heat capacity followed shortly thereafter, to be followed by further developments by Kaufman,[17] Kaufman and Osanger,[18] and by Yang.[19]

The importance of nearest-neighbor models of cooperative phenomena in describing order-disorder transitions in metallic alloys and phase separations in liquid-gas systems was recognized almost immediately. In binary alloys the two atomic species play the role of the $+1$ and -1 spins of the Ising ferromagnet. In a lattice gas, atoms are constrained to occupy and move between well-defined lattice sites. The short-range repulsive forces are taken into account by restricting the occuaption of a given lattice site to at most one atom, and the occupancy status of a given lattice site corresponds to the up-down orientation of the Ising spins. Ordered lattice (superlattice) models of binary alloys were developed by Bragg and Williams,[20,21] Bethe,[22] and Peierls,[23] and the equivalence between the lattice gas and Ising models was proved by Lee and Yang.[24]

3.3.4 The Ising Hamiltonian and Partition Function

Let us consider a system of N atoms arranged on a regular lattice. In the Ising model each atom on the lattice has an electronic spin coordinate s and an associated magnetic moment μ. The spin can assume one of two values, either $+1$ or -1, depending on whether it is aligned parallel or antiparallel to an external the magnetic field H. This situation for a one-dimensional array was illustrated in Fig. 2.4. As discussed in Section 2.10.1, a state, or configuration, of the spin system is specified by giving the value for the spin at each lattice site. For the array of N atoms or spin elements there are $2N$ configurations of the form $(s_1, s_2, \ldots, s_N)$. We designate these configurations by the notation $\{s_i\}$.

In one dimension the $(i-1)$th and $(i+1)$th lattice sites are nearest neighbors of the ith site. In two dimensions there are four nearest neighbors. The interaction energy E_{jk} between atoms located at the jth and kth lattice sites is defined as

$$E_{jk} = \begin{cases} -Js_js_k, & j,k = (j,k) \\ 0 & \text{otherwise} \end{cases} \quad (3.16)$$

where the notation (j,k) is used to indicate that j and k are nearest-neighbor lattice sites. The constant J is a measure of the strength of the exchange interaction. For a ferromagnetic interaction, J is positive, and for an antiferromagnetic interaction, it is negative. For J positive, the energy is lowered whenever neighboring atoms have the same spin and is raised whenever their spins are in opposition. There is also an energy term, E^{ext}, representing the interaction of the external magnetic field with the lattice atoms, of the form

$$E_j^{ext} = -\mu H s_j \quad (3.17)$$

The total energy $E(H)$ is the sum of the spin exchange and atom-field interaction terms,

$$E(H) = -J \sum_{(j,k)} s_j s_k - \mu H \sum_j s_j \quad (3.18)$$

The sum in the first term extends over all nearest-neighbor sites, while the sum in the second term is over the $j = 1, \ldots, N$ lattice sites. The quantity $E(H)$ represents the total energy for a particular configuration of the system. The total energy is sometimes written as $E(\{s_i\})$ to emphasize this aspect of its definition.

The canonical partition function is dependent upon the Lagrange multiplier $\beta = (k_B T)^{-1}$, and the external magnetic field H; that is, $Z = Z(\beta, H)$. To evaluate the partition function, we sum over all spin configurations of the system:

$$Z(\beta, H) = \sum_{\{s_i\}} \exp\left(\beta J \sum_{(j,k)} s_j s_k + \beta \mu H \sum_j s_j\right) \quad (3.19)$$

This summation includes contributions from the two spins states $+1$ and -1 of each element in the array. We can exhibit the summation explicitly by writing the partition function as

$$Z(\beta, H) = \sum_{s_1 = \pm 1} \sum_{s_2 = \pm 1} \cdots \sum_{s_N = \pm 1} \exp\left(\beta J \sum_{(j,k)} s_j s_k + \beta \mu H \sum_j s_j\right) \quad (3.20)$$

3.3.5 Thermodynamic Parameters

As we saw in Chapter 2, the quantities of thermodynamic interest can be obtained formally by differentiating the logarithm of the partition function or equivalently, recalling Eq. (2.46), by differentiating the Helmholtz free energy. In particular, we can determine the internal energy $\langle E \rangle$ by differentiating the logarithm of the partition function with respect to β:

$$\langle E \rangle = -\frac{\partial \ln Z}{\partial \beta} = -\frac{1}{Z}\frac{\partial Z}{\partial \beta} = k_B T^2 \frac{1}{Z}\frac{\partial Z}{\partial T} \quad (3.21)$$

and we can find the magnetization by differentiating the logarithm of the partition function with respect to external magnetic field H:

$$\langle M \rangle = \frac{\partial \ln Z}{\partial H} = \frac{1}{Z}\frac{\partial Z}{\partial H}$$

$$= \frac{1}{Z} \sum_{\{s_i\}} \left(\beta \mu \sum_j s_j\right) \exp\left(\beta J \sum_{(j,k)} s_j s_k + \beta \mu H \sum_j s_j\right) \quad (3.22)$$

$$= \beta \left\langle \mu \sum_j s_j \right\rangle$$

Two properties of considerable interest for determining the presence or absence of a phase transition are the heat capacity and the spontaneous magnetization. We can determine the heat capacity from the internal energy. The spontaneous magnetization is the value in the limit of zero external magnetic field of the magnetization $M(\beta, 0)$.

3.4 THE ISING MODEL IN ONE DIMENSION

In one dimension we can simplify the problem of evaluating the sums in the partition function. The interaction energy for a one-dimensional lattice assumes the form

$$E(H) = -J \sum_j s_j s_{j+1} - \frac{1}{2}\mu H \sum_j (s_j + s_{j+1}) \qquad (3.23)$$

In writing the second term in a form which is symmetric in the indexes j and $j+1$, we have imposed the periodic boundary condition that

$$s_{N+1} = s_1 \qquad (3.24)$$

That is, we have closed the one-dimensional lattice upon itself to form a ring in which the last element equals the first one. To evaluate the partition function for this interaction energy, we will employ an approach known as the *transfer matrix method*. In the transfer matrix procedure the partition function is related to the largest characteristic value (eigenvalue) of a matrix whose elements represent single nearest-neighbor interactions. To formulate this procedure, we first write out the partition function

$$Z(\beta, H) = \sum_{s_1 = \pm 1} \sum_{s_2 = \pm 1} \cdots \sum_{s_N = \pm 1} \exp\left(\beta J \sum_j s_j s_{j+1} + \frac{1}{2}\beta\mu H \sum_j (s_j + s_{j+1})\right) \qquad (3.25)$$

Let us now examine a single term in the summation over the two possible spin values $+1$ and -1. For a pair of nearest neighbors s and s', there are four possible values of the exponent. We will arrange these four values in a 2×2 matrix $P_{ss'}$:

$$P_{ss'} = e^{\beta J s s' + \beta\mu H(s+s')/2} \qquad (3.26)$$

The matrix elements are generated by letting s and s' independently assume the values $+1$ and -1. We can easily tabulate the results of making this substitution:

s	s'	$\ln P_{ss'}$
$+1$	$+1$	$\beta J + \beta\mu H$
$+1$	-1	$-\beta J$
-1	$+1$	$-\beta J$
-1	-1	$\beta J - \beta\mu H$

We may write this matrix as

$$P = \begin{pmatrix} e^{\beta J + \beta \mu H} & e^{-\beta J} \\ e^{-\beta J} & e^{\beta J - \beta \mu H} \end{pmatrix} \qquad (3.27)$$

The partition function can be rewritten in terms of transfer matrix P as

$$Z(\beta, H) = \sum_{S_1}\sum_{S_2}\cdots\sum_{S_N} P_{S_1 S_2} P_{S_2 S_3} \cdots P_{S_N S_1} \qquad (3.28)$$

We now observe that the partition function consists of a sequence of matrix products:

$$\sum_k P_{jk} P_{kl} = (P^2)_{jl}$$

$$\sum_k \sum_l P_{jk} P_{kl} P_{lm} = (P^3)_{jm}$$

and finally,

$$Z(\beta, H) = \sum_{S_1} (P^N)_{S_1 S_1} = \text{Tr}(P^N) \qquad (3.29)$$

In the above expression, Tr denotes the trace (sum of the diagonal elements) of a matrix. To evaluate the trace, we recall the following properties of matrices: The eigenvalues of an $(n \times n)$ matrix A are the values Λ that are solutions of the equation

$$\sum_j A_{ij} x_j = \Lambda x_i \qquad (3.30)$$

This set of equations will possess a nontrivial solution provided that

$$\det(A - \Lambda I) = 0 \qquad (3.31)$$

where I is the identity matrix, and det signifies the determinant of the bracketed matrix. The n (not-necessarily-distinct) roots Λ_n of this equation are the eigenvalues. These eigenvalues are independent of the choice of basis. That is, we may select another basis

$$x' = S^{-1} x \qquad (3.32)$$

$$A' = S^{-1} A S \qquad (3.33)$$

and find that

$$A' x' = \Lambda x \qquad (3.34)$$

Therefore the eigenvalues remain unchanged when there is a switch in representation. Both the determinant and the trace are invariant under changes in basis. Noting that $\mathrm{Tr}(AB) = \mathrm{Tr}(BA)$, we have

$$\mathrm{Tr}(A') = \mathrm{Tr}(S^{-1}AS) = \mathrm{Tr}(AS^{-1}S) = \mathrm{Tr}(A) \tag{3.35}$$

Similarly, observing that $\det(AB) = \det(A)\det(B)$, we find that

$$\det(A') = \det(S^{-1}AS) = \det(S^{-1})\det(A)\det(S) = \det(A) \tag{3.36}$$

These properties imply that we can select a basis in which the transfer matrix P is diagonal with matrix elements equal to the eigenvalues. That is, we can put

$$P = \begin{pmatrix} \Lambda_+ & 0 \\ 0 & \Lambda_- \end{pmatrix} \tag{3.37}$$

with the eigenvalues Λ_+ and Λ_- given as solutions to the secular equation

$$\begin{vmatrix} e^{\beta J + \beta\mu H} - \Lambda & e^{-\beta J} \\ e^{-\beta J} & e^{\beta J - \beta\mu H} - \Lambda \end{vmatrix} = 0 \tag{3.38}$$

In this representation we can easily evaluate the trace. The result is

$$\mathrm{Tr}\, P^N = \mathrm{Tr}(S^{-1}PS)^N = \mathrm{Tr}\begin{pmatrix} \Lambda_+^N & 0 \\ 0 & \Lambda_-^N \end{pmatrix} \tag{3.39}$$

and we have completed the evaluation of the partition function. Thus

$$Z(\beta, H) = \sum_{s_1} (P^N)_{s_1 s_1} = \mathrm{Tr}(P^N) = (\Lambda_+^N + \Lambda_-^N) \tag{3.40}$$

with eigenvalues obtained by solving the secular equation

$$(e^{\beta J + \beta\mu H} - \Lambda)(e^{\beta J - \beta\mu H} - \Lambda) - e^{-2\beta J} = 0 \tag{3.41}$$

namely

$$\Lambda_\pm = e^{\beta J}[\cosh\beta\mu H \pm \sqrt{\cosh^2\beta\mu H - 2e^{-2\beta J}\sinh\beta J}] \tag{3.42}$$

It is customary to examine the behavior in the thermodynamic limit in which N is assumed to be large. In this limit only the larger of the two

THE ISING MODEL IN ONE DIMENSION

eigenvalues contribute. For $\Lambda_+ > \Lambda_-$ we have

$$\frac{1}{N}\ln Z(\beta, H) = \ln \Lambda_+ + \frac{1}{N}\ln(1 + (\Lambda_-/\Lambda_+)^N) \to \ln \Lambda_+ \quad \text{as } N \to \infty \quad (3.43)$$

We can then calculate the magnetization per spin $M(\beta, H)$ in the thermodynamic limit

$$M(\beta, H) = \frac{1}{N}\frac{\partial \ln Z}{\partial H} = \frac{\partial \ln \Lambda_+}{\partial H} \quad (3.44)$$

and find that

$$M(\beta, H) = (\mu \sinh \beta\mu H)(\sinh^2 \beta\mu H + e^{-4\beta J})^{-1/2} \quad (3.45)$$

The magnetization is plotted as a function of magnetic field for several representative temperature values in Fig. 3.5. As can be seen in these plots, the magnetization vanishes as the magnetic field goes to zero. Thus there is no spontaneous magnetization for a one-dimensional Ising chain.

It is straighforward to calculate the internal energy and heat capacity of the one-dimensional Ising system. Displayed in Fig. 3.6 are plots of the heat capacity as a function of temperature for the one-dimension Ising model. We observe that the heat capacity is a smoothly varying function of temperature. In the absence of an external magnetic field, the internal energy per spin in the one-dimensional Ising model,

$$\langle E \rangle = -\frac{1}{N}\frac{\partial \ln Z}{\partial \beta} = -\frac{\partial \ln \Lambda_+}{\partial \beta} \quad (3.46)$$

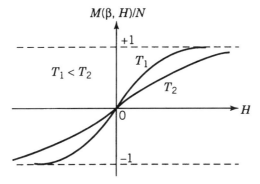

Figure 3.5. Magnetization versus external field for an Ising chain. The magnetization disappears when the external field goes to zero.

80 COOPERATIVITY IN LATTICE SYSTEMS

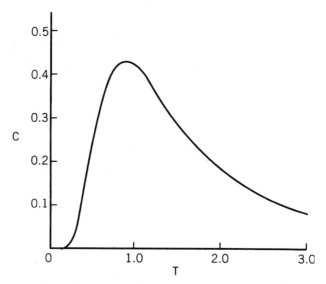

Figure 3.6. Heat capacity expressed as C_v/k_B versus temperature expressed as the ratio $k_B T/J$ for an Ising chain. This curve varies smoothly at all temperatures.

assumes the simple form

$$\langle E \rangle = -J \tanh \beta J \tag{3.47}$$

The heat capacity, obtained from the internal energy by differentiating with respect to the temperature, is

$$c(\beta, 0) = \frac{J^2}{k_B T^2} \operatorname{sech}^2 \beta J \tag{3.48}$$

3.5 THE ISING MODEL IN TWO DIMENSIONS

Let us now formulate the problem for a two-dimensional Ising lattice in a manner analogous to that for the one-dimensional case. We accomplish this by constructing a two-dimensional lattice from a series of m one-dimensional Ising chains. These chains form the rows, of length n, of our two-dimensional Ising lattice. We proceed by introducing the set v_j of the spin coordinates of the jth row. That is, the set v_j is defined by enumerating the internal coordinates of the n lattice points in the jth row:

$$v_j = (s_1, s_2, \ldots, s_n)_j \tag{3.49}$$

and a configuration of our Ising system is defined by enumerating the set $\{v_1, v_2, \ldots, v_m\}$. The assumption of nearest-neighbor interactions means that the jth row interacts only with the $(j-1)$th and $(j+1)$th rows. The total interaction energy of a particular configuration is

$$E = \sum_{j=1}^{m-1} V(v_j, v_{j+1}) + \sum_{j=1}^{m} V(v_j) \tag{3.50}$$

The first term describes the interactions between the layers of the lattice, and the second term represents the internal energy within the jth layer. This second term is the same as that for the one-dimensional chain, containing both nearest-neighbor and external field contributions. If we once again impose periodic boundary conditions for the rows, and for the layers to produce a toroidal topology, we can write the partiton function as

$$Z(\beta, H) = \sum_{v_1} \sum_{v_2} \cdots \sum_{v_m} \exp\left(-\beta \sum_j \left\{ V(v_j, v_{j+1}) + \left(\frac{1}{2}\right)[V(v_j) + V(v_{j+1})] \right\} \right) \tag{3.51}$$

The next step is to define matrix elements connecting the layers. As before, the partition function is given as a sum of the characteristic values of the transfer matrix. It is quite apparent at this point that the problem being solved as far more complex than that encountered in solving a one-dimensional chain. Instead of presenting the steps taken by Onsager[16] to solve this problem, let us summarize his results together with those of Yang[19] for the spontaneous magnetization.

We begin with the internal energy. The solution for the internal energy is usually written as

$$\langle E \rangle = -J \coth 2\beta J \left\{ 1 + \frac{2}{\pi} (2 \tanh^2 2\beta J - 1) K_1(\kappa) \right\} \tag{3.52}$$

It is useful to compare this expression to the internal energy in the zero field limit for the one-dimensional lattice. To facilitate this comparison, we recast Eq. (3.52) into the alternative form[9]

$$\langle E \rangle = -2J \tanh 2\beta J - J \frac{\sinh^2 2\beta J - 1}{\sinh 2\beta J \cosh 2\beta J} \left(\frac{2}{\pi} K_1(\kappa) - 1 \right) \tag{3.53}$$

In the above equations K_1 is the elliptic integral of the first kind,

$$K_1(\kappa) \equiv \int_0^{\pi/2} \frac{d\phi}{(1 - \kappa^2 \sin^2 \phi)^{1/2}} \tag{3.54}$$

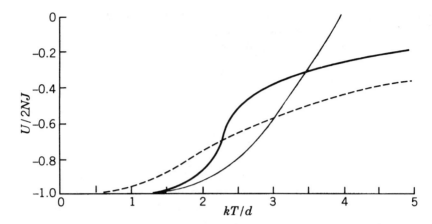

Figure 3.7. Internal energy versus temperature for two-dimensional Ising lattice. The heavy solid curve is the result from Eq. (3.53). At the Curie point the slope (heat capacity) becomes infinite. Shown for comparison are the contributions from the first term alone (dashed curve) and from the Weiss molecular field approximation (Section 3.7). (From Wannier[9]. © 1996 John Wiley & Sons, Inc.)

with

$$\kappa = \frac{2 \sinh 2\beta J}{\cosh^2 2\beta J} \tag{3.55}$$

The first term is identical to that for the one-dimensional situation, except for the factor of two arising from an additional exchange interaction for the second lattice dimension. The second term takes into account the cooperativity arising in the two-dimensional system. When plotted as a function of temperature (Fig. 3.7), the first term resembles the noncooperative kinetics shown in Fig. 1.1. The second term enhances the response in a cooperative manner, converting the overall dependence on temperature into a sigmoidal-like one with a discontinuity in slope at the critical temperature. Mathematically the elliptic integral has a logarithmic infinity at the critical point for which $\kappa = 1$. In the vicinity of the critical point,

$$\langle E \rangle \propto (T - T_c) \ln |T - T_c| \tag{3.56}$$

Therefore the heat capacity, the slope of internal energy versus temperature curve, will diverge at the critical point. The critical temperature corresponds to

$$\sinh 2\beta J = 1 \tag{3.57}$$

which yields the critical value

$$\frac{k_B T_c}{J} = 2.269 \tag{3.58}$$

The spontaneous magnetization for a two-dimensional Ising lattice is given by the expression

$$M(\beta, 0) = \mu \left[\frac{\cosh^2 2\beta J}{\sinh^4 2\beta J} (\sinh^2 2\beta J - 1) \right]^{1/8} \tag{3.59}$$

The nonzero value for this quantity, the long-range order parameter for a ferromagnet, is evidence that a continuous phase transition occurs in this system. As stated in our previous discussion, $M(\beta, 0)$ attains its maximum value at $T = 0$ and vanishes at the critical temperature.

3.6 STRONG COOPERATIVITY AND PEIERLS'S ARGUMENT

Let us now consider the meaning of the term "strong cooperativity" in relation to the differences in behavior found between the one- and two-dimensional Ising models. We recall that energy minimization favors ordered states in which all spins are aligned. Entropy maximization, on the other hand, leads to disordered states. The overall process is one in which the free energy, containing both energetic and entropic terms, is minimized. When the cooperative ordering is stronger than the entropic disordering, the result is spontaneous magnetism. We may regard an Ising system as being decomposed into a collection of distinct regions, each contains spins of the same sign, either all up or all down. These regions are separated from one another by borders. The number of ways this partitioning can be accomplished determines the entropy, while the length of the borders fixes the internal energy. A border separating an "up" region from a "down" region will be preserved if the entropy gained by its presence is greater than the energy recovered from its removal. In the case of a one-dimensional Ising chain, there is no value for the ratio $k_B T/J$ for which the energy gain exceeds the entropy loss associated with the removal of a border. For a two-dimensional Ising lattice, both entropic and energetic terms vary linearly with border length, and therefore a temperature can always be found for which the energy gain is greater than the entropic loss. At sufficiently low temperature, ordered states will be favored by free energy minimization. Thus the extent of the nearest-neighbor connectivity is an important aspect of the cooperativity. For the one-dimensional chain, there are insufficient numbers of nearest neighbors to generate the requisite cooperativity.

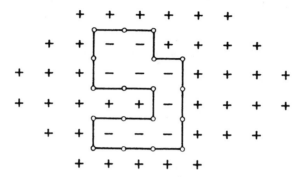

Figure 3.8. Schematic illustration of a boundary separating a positive region of Ising spins from a negative region. (From Wannier[9]. © 1966 John Wiley & Sons, Inc.)

Consider a two-dimensional Ising lattice in which all boundary spins are positive. As illustrated in Fig. 3.8, we may regard this lattice as being partitioned into regions each containing spins of the same sign. We observe that the length L of a boundary separating a region of negative spin from its surrounding positive spins determines the energy expenditure $2JL$ required to establish that region of negative spin. We now wish to compare this energy change to the number of ways a border of length L can be constructed enclosing a particular lattice site of negative spin. The logarithm of this multiplicity factor gives the entropic change associated with the construction of the region of negative spin. That is,

$$\Delta A = \Delta \langle E \rangle - k_B T \ln \Omega \tag{3.60}$$

To determine the multiplicity factor, we first note that once an initial border segment of unit length is drawn, there are three ways of adding each additional unit, and therefore $4(3)^{L-1}$ ways of constructing a border of length L. Since the site of interest can be anywhere within this enclosed region, we have to append a factor of $(L/4)^2$ representing the area of a square of perimeter L. This is an overestimate, since nonsquares of perimeter L will enclose smaller areas, and we have not considered the effect of other areas, nor have we built in a constraint representing closure of the area. We must divide the product of these two terms by $2L$. This factor takes into account the number of ways we can choose our starting point and direction and still end up with the same border. Combining these three factors yields the multiplicity

$$\Omega = \left(\frac{L}{24}\right) 3^L \tag{3.61}$$

the logarithm of which is proportional to L. This can be compared to the energetic change

$$\Delta\langle E\rangle = 2JL \tag{3.62}$$

We see that a temperature can always be found that will make the energy change greater than the entropic one. This situation is profoundly different from that encountered for a one-dimensional chain. There, if we repeat the same argument,[2,12] we observe that a boundary may be established isolating a region of negative spin from a surrounding region of positive spin. However, in the one-dimensional case the boundary can be moved to any other location in the chain without producing any change in the energy. Thus the configuration multiplicity associated with the energy change $2(2J)$, is N, the size of the chain. In the one-dimensional case there is no temperature below which free energy minimization will stabilize an ordered state against fluctuations.

3.7 MEAN-FIELD THEORY

We will now introduce the notion of a mean-field. This is an important tool in the study of cooperative phenomena. In a mean-field approach we replace the detailed effects of a set of neighboring elements upon a given element by a single, effective field that expresses the joint cooperative influences of those elements. A number of mean-field approximations have been developed for treating Ising lattice systems. Analogous approximations have been applied to spatial lattice models in image reconstruction (Chapter 6), to the construction of biologically motivated neural networks (Chapter 8), and to the collective dynamics of oscillator communities (Chapter 9).

The first and simplest mean-field approach is known as the Weiss molecular field approximation. We will examine this approach in some detail in this section, giving first a heuristic derivation and then rederiving the results by a random mixing argument due to Bragg and Williams.[20,21] We will then sketch the main idea behind the first-order approximation due to Bethe,[22] Peierls,[23] and Weiss.[25] A number of higher-order methods, such as that of Kramers-Wannier[13,14] and Kikuchi,[26] have been developed too. These will not be discussed here, but instead, the interested reader is referred to the excellent review by Domb.[27] We will follow these derivations by a brief overview of the lattice gas model used to describe critical phenomena in fluids and the van der Waals picture of liquid-solid phase transformations.

3.7.1 The Weiss Molecular Field Equation

Consider an individual spin element s_i at lattice site i. This element feels a force exerted on it by its neighboring spin elements, through the exchange interaction, and by the external field H. For an Ising system we may write the local field as

$$H_i = J \sum_{\langle i,j \rangle} s_j + \mu H \tag{3.63}$$

where $\langle i,j \rangle$ denotes a sum over of nearest-neighbor spin elements of s_i. The total field (interaction energy) is obtained by summing over the contributions from all sites. That is,

$$E(H) = -\sum_i H_i s_i \tag{3.64}$$

In general, the orientations in space of the neighboring spin elements will continually fluctuate. We will neglect these variations and replace the local field by its average value

$$H_0 = Jv\langle s \rangle + \mu H \tag{3.65}$$

where v is the number of nearest neighbors. In this expression we have dropped the index i, since we assume that the same average value is maintained at every point in space. Thus we neglect explicit couplings between the local field and the individual orientations of the spin elements at each distinct lattice site. It is then straightforward to compute the value of $\langle s \rangle$:

$$\langle s \rangle = \frac{\sum_{s=\pm 1} s_i e^{\beta H_0 s}}{\sum_{s=\pm 1} e^{\beta H_0 s}} = \frac{e^{\beta H_0} - e^{-\beta H_0}}{e^{\beta H_0} + e^{-\beta H_0}} \tag{3.66}$$

or

$$\langle s \rangle = \tanh \beta(Jv\langle s \rangle + \mu H) \tag{3.67}$$

Let us now explore the behavior of this self-consistent field equation in the limit of vanishing external field. At low temperatures, a solution to the above equation corresponds to $\langle s \rangle$ close to unity. Setting $\langle s \rangle$ equal to unity on the right-hand side yields the expression

$$\langle s \rangle = 1 - 2e^{-2T_c/T}, \quad T \ll T_c \tag{3.68}$$

where

$$T_c \equiv \frac{Jv}{k_B} \tag{3.69}$$

At zero temperature we recover $\langle s \rangle = 1$. Similarly there is a self-consistent solution at zero temperature for $\langle s \rangle = -1$.

In the vicinity of the critical temperature T_c, we may expand $\tanh(\langle s \rangle T_c/T)$ in a Taylor's series about $T = T_c$:

$$\langle s \rangle = \tanh\left(\langle s \rangle \frac{T_c}{T}\right)$$

$$= \tanh\langle s \rangle + \frac{\langle s \rangle}{\cosh^2\langle s \rangle}\left(1 - \frac{T}{T_c}\right) \quad (3.70)$$

We now expand out tanh and cosh, retaining terms up to cubic, to obtain the result

$$\langle s \rangle = \sqrt{3}\left(1 - \frac{T}{T_c}\right)^{1/2}, \quad T \cong T_c \quad (3.71)$$

The power 1/2, to which the factor $(1 - T/T_c)$ is raised in the above equation, is the critical exponent for spontaneous magnetization in the Weiss mean-field approximation to the Ising ferromagnet. The constant T_c is the Curie temperature. We can exhibit the significance of the critical temperature by graphically solving for the self-consistent value of the mean spin given by Eq. (3.67). When we do so, we find that when $T > T_c$ the only self-consistent solution is $\langle s \rangle = 0$. When $T < T_c$, we have a positive-negative pair of nontrivial solutions as well. Thus the constant T_c is the threshold for the onset of spontaneous magnetization.

3.7.2 The Bragg-Williams (Random Mixing) Method

Let us recall our initial sketch of the lattice system presented at the beginning of this chapter in Section 3.1. The basic idea of the mean-field approaches is to avoid the difficult task of evaluating the density of states, $\Omega(N_1, N_{12})$, by eliminating N_{12} from the problem. To specialize the general notation of Section 3.1 to an Ising ferromagnet, we observe that the pair energies ε_{11}, ε_{12}, and ε_{22} are equal to $-J$, $+J$, and $-J$, respectively. Upon appending a term describing the interaction of the lattice spins with an external field, we have

$$E = -\left(\frac{1}{2}\right)(vN_1 - N_{12})J + N_{12}J - \left(\frac{1}{2}\right)(vN_2 - N_{12})J + \mu(N_1 - N_2)H$$

$$= \left[-\left(\frac{1}{2}\right)vJ - \mu H\right]N + 2JN_{12} + 2\mu HN_1 \quad (3.72)$$

We now eliminate N_{12} from the problem by replacing it with the average value. That is, we assume that the spin orientations are completely random.

Under this assumption of random mixing, we have

$$N_{12} = \frac{\nu N_1 N_2}{N} \tag{3.73}$$

If we now introduce the variables $x_1 = N_1/N$ and $x_2 = N_2/N$ (note that $x_1 + x_2 = 1$), we find that the energy term can be cast into the simple form

$$E = -\left(\frac{1}{2}\right)\nu J(x_1 - x_2)^2 + \mu H(x_1 - x_2) \tag{3.74}$$

This expression can be further simplified, by replacing the difference $x_1 - x_2$ with $\langle s \rangle$ to yield

$$E = -\left(\frac{1}{2}\right)\nu J \langle s \rangle^2 + \mu H \langle s \rangle \tag{3.75}$$

With our simplification of random mixing, the partition function assumes the form

$$Z = \sum_{N_1, N_2} \Omega(N_1, N_2) e^{-\beta E(N_1, N_2)} \tag{3.76}$$

Since we assume that the spin orientations are distributed in an uncorrelated manner, the counting factor in the above expression is simply the number of ways N_1 spins pointing up and N_2 spins pointing down can be arranged on the lattice of size N, namely

$$\Omega(N_1, N_2) = \frac{N!}{N_1! N_2!} \tag{3.77}$$

To carry out the evaluation of the partition function, we replace the summation with the maximum value. This approximation will be valid for sufficiently large N. Applying Stirling's approximation to Ω yields the result

$$\frac{1}{N} \ln \Omega(N_1, N_2) = -x_1 \ln x_1 - x_2 \ln x_2 \tag{3.78}$$

We now observe that the configuration entropy is related to Ω by

$$S = k \ln \Omega(N_1, N_2) \tag{3.79}$$

and the Helmholtz free energy per spin is

$$\frac{1}{N} A = -\left(\frac{1}{2}\right) vJ(x_1 - x_2)^2 + \mu H(x_1 - x_2) + k_B T(x_1 \ln x_1 + x_2 \ln x_2) \quad (3.80)$$

Differentiating the free energy per spin with respect to the independent variable $\langle s \rangle$, and setting the result equal to zero, gives us the desired maximum value for the partition function. The result of carrying out this operation is

$$-vJ\langle s \rangle + \mu H + \left(\frac{1}{2}\right) k_B T \ln\left(\frac{1 + \langle s \rangle}{1 - \langle s \rangle}\right) = 0 \quad (3.81)$$

This formula for $\langle s \rangle$ reduces to the Weiss molecular field equation. Upon inserting this condition for maximality into our formula for the Helmholtz free energy, we find that

$$\frac{1}{N} A = \left(\frac{1}{2}\right) vJ\langle s \rangle^2 + \frac{k_B T}{2} \ln\left(\frac{1 - \langle s \rangle^2}{4}\right) \quad (3.82)$$

In the mean-field approximation the internal energy is related to the spontaneous magnetization. Above the Curie temperature, the spontaneous magnetization vanishes, and so does the internal energy. Below the critical temperature, the internal energy is

$$\langle E \rangle = -\left(\frac{1}{2}\right) vJ\langle s \rangle^2 \quad (3.83)$$

where we have set the external magnetic field equal to zero. Differentiating this quantity with respect to the temperature yields the specific heat as a function of temperature.

3.7.3 The Bethe Approximation and Synopsis of Mean-Field Results

To recapitulate, our main result in this section is the existence of a critical point below which there are two nontrivial solutions to the mean-field equation. The existence of a spontaneous magnetization, and the singular behavior of the specific heat, imply that there is a phase transition, marking the onset of long-range order in the lattice system. Thus the construction of a mean-field, expressing the cooperative influences of the lattice elements, captures the essential properties of the assembly. This result is an important one, despite the fact that many of the detailed numerical predictions of the approximation are in error. For example, the vanishing of the internal energy and specific heat at temperatures above the Curie point is not correct, the critical exponent is wrong, and the Curie point is too high. The mean-field approach is a zeroth-order approximation, in which short-range order or fluctuations are neglected. A more sophisticated, first-order method, known as the *Bethe-Peierls-Weiss approximation*, yields results in closer agreement with the exact solutions.

In this first-order approximation, we treat more accurately the interactions of a given spin with its nearest neighbors. To construct this approximation, we consider a small sublattice, or elementary cluster, consisting of a central lattice element s_0 and its nearest neighbors s_j. We treat exactly the interactions of s_0 with its nearest neighbors, and with an applied external field H_0. We then approximate the interactions of s_j with the external field, and with all spin elements other than s_0, by a mean field H_1. In this model the interaction energy for an elementary cluster is of the form

$$H_v = -J \sum_{j=1}^{v} s_0 s_j - \mu H_0 s_0 - \mu H_1 \sum_{j=1}^{v} s_j \qquad (3.84)$$

where, as before, v is the number of nearest neighbors. The mean field H_1 is determined by imposing a self-consistency condition that the average magnetization of the central atom in the external field is equal to that of its nearest neighbors. The critical point is the temperature below which the internal field does not disappear when the external field vanishes. In the Bethe-Peierls-Weiss approximation, the specific heat does not disappear at temperatures just above the critical value, and the Curie point is closer to the correct value. A

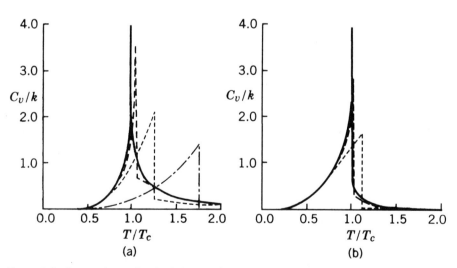

Figure 3.9. Comparison of calculations of the heat capacity for an Ising lattice in (a) two dimensions, and (b) three dimensions. Shown are exact results, and predictions of the mean field, Bethe and Kramers-Wannier and Kikuchi approximations (After Domb[27]).

comparison of the specific heat for the two-dimensional Ising lattice obtained by the mean-field, Bethe-Peierls-Weiss, and Kramers-Wannier, and Kikuchi approximation methods, and by Onsager, is shown in Fig. 3.9.

In comparing the mean-field approximations to the exact results, we find that the dimensionality of the lattice is an important consideration. For a one-dimensional lattice there is no phase transition. The mean-field prediction that there is a phase transition is entirely wrong. This erroneous result is a consequence of the neglect of energy fluctuations, which distroy the long-range order in the one-dimensional system. As we go to a two-dimensional system, and then to a three-dimensional one, the agreement of the various approximations with one another, and with the exact results, improves. In particular, the critical points are closer together for a three-dimensional lattice than for a two-dimensional system.

3.8 THE LATTICE GAS MODEL OF FLUIDS

Figure 3.10 illustrates the radial dependence of a typical intermolecular potential, such as a Lennard-Jones 6–12 potential, used to describe interactions between neighboring molecules of a fluid. There are three characteristics of this potential. First, the potential is strongly repulsive at small distances, reflecting a hard-sphere-like, minimal interpenetrability of the molecules. Second, the potential is attractive at separations of one or two molecular diameters or so. Third, the potential vanishes for large radial separations of the centers of the two molecules. Our goal in this section is to derive an equation of state for a fluid system possessing an interaction potential of the this type. We may recall that this is the same problem as was addressed by Metropolis et al. (Section 2.10) through the Monte Carlo method. In this section we treat the problem using a lattice gas model. We will see that the model is equivalent to the Ising model for a ferromagnet, and we will then establish a correspondence between the van der Waals equation of state and Weiss molecular field.

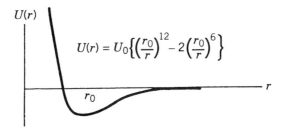

Figure 3.10. Characteristic radial dependence of an intermolecular potential for a fluid. (From Stanley[28]. Reprinted with permission of Oxford University Press.)

3.8.1 The Lattice Gas

In the lattice gas model, we partition a fluid, confined to a volume V, into cells that are arranged in a regular lattice. Each cell of the lattice gas is approximately equal in size to that of one of the molecules. A cell is regarded as being occupied if the center of a molecule lies within its boundaries. Two molecules cannot occupy the same cell. This restriction is our way of incorporating a hard-sphere repulsion into the model. A cell of the lattice is therefore either occupied by one molecule or empty. The two states of any cell i are described by the occupation number n_i. This number assumes the value 1 or 0, depending on whether the cell is occupied or empty. We are then able approximate the second and third characteristics of the intermolecular potential by a nearest-neighbor interaction of the form

$$E_{ij} = \begin{cases} -\varepsilon n_i n_j, & i,j = \langle i,j \rangle \\ 0 & \text{otherwise} \end{cases} \qquad (3.85)$$

where, as before, $\langle i,j \rangle$ denotes that i and j are nearest neighbors and ε is the average strength of the intermolecular potential.

We will shortly establish a correspondence between the lattice gas and Ising models. However, before doing so, we observe one difference between the two systems. For an Ising ferromagnet, since spins may flip from one orientation to the other, the number of up spins of one orientation is not fixed. In contrast, for a lattice gas, while molecules may move from one lattice site to another, they may not be created or distroyed. Therefore, in a lattice gas, we must introduce constraints on the mean values of both the energy and particle number. The appropriate partition function for the lattice gas is the grand canonical ensemble, defined in Section 2.8.5:

$$Z_G = \sum_i e^{-\beta E_i + \beta \mu_{ch} N_i} \qquad (3.86)$$

The total number of molecules N can be expressed in terms of the occupation numbers by summing the latter over the number of cells N_{ce}:

$$N = \sum_{j=0}^{N_{ce}} n_j \qquad (3.87)$$

A given configuration of the lattice gas can be described by an array of occupation numbers, each of which can take on one of two values, 0 or 1, and

$$Z_G = \sum_{\{n_i = 0,1\}} e^{-\beta \varepsilon \sum_{\langle i,j \rangle} n_i n_j + \beta \mu_{ch} \sum_j n_j} \qquad (3.88)$$

THE LATTICE GAS MODEL OF FLUIDS 93

To establish our correspondence between Ising ferromagnet and lattice gas, we observe that the occupation numbers and spin variables can be transformed into one another by the relationship

$$n_i = \left(\frac{1}{2}\right)(1 + s_i) \tag{3.89}$$

Using this relationship, we can rewrite the Ising hamiltonian in terms of occupation numbers:

$$E_{ij} = -J \sum_{\langle i,j \rangle} s_i s_j - \mu H \sum_j s_j$$
$$= -J \sum_{\langle i,j \rangle} [4n_i n_j - 2(n_i + n_j) + 1] - \mu H \sum_j (2n_j - 1)$$

and therefore

$$E_{ij} = 4J \sum_{\langle i,j \rangle} n_i n_j - 2N(2Jv - \mu H) \tag{3.90}$$

where we have dropped an unimportant additive constant from the last expression. If we now compare this energy term appearing in the argument of the exponential in the partition function for the Ising ferromagnet to the argument of the exponential in the grand partition function for the lattice gas we are able to establish a correspondence between the two, namely

$$4J \leftrightarrow \varepsilon$$
$$2(2Jv - \mu H) \leftrightarrow \mu_{ch} \tag{3.91}$$

We further observe that the density ρ of a lattice gas is

$$\rho = \frac{N}{N_{ce}} \tag{3.92}$$

This quantity is analogous to the density of up (+1) spins of an Ising ferromagnet, and since

$$\langle s \rangle = \frac{N_+ - N_-}{N_{ce}} = 2\frac{N_+}{N_{ce}} - 1 \tag{3.93}$$

we deduce another equivalence,

$$\langle s \rangle \leftrightarrow 2\rho - 1 \tag{3.94}$$

3.8.2 The van der Waals Equation

We can now transpose the mean-field formula for an Ising spin system into an equation describing a lattice gas. The resulting expression is

$$2\rho - 1 = \tanh \beta(\varepsilon_0 \rho + \mu_{ch}) \tag{3.95}$$

where ε_0 replaces the product $4Jv$. This relationship can be rewritten as

$$-\beta\mu_{ch} = \beta\varepsilon_0\rho + \ln\frac{1-\rho}{\rho} \tag{3.96}$$

The pressure can be found by differentiating the chemical potential with respect to the density, keeping the temperature fixed. That is,

$$\left(\frac{\partial \mu_{ch}}{\partial \rho}\right)_T = \left(\frac{1}{\rho}\right)\left(\frac{\partial p}{\partial \rho}\right)_T \tag{3.97}$$

To derive Eq. (3.97), we need to use some results from classical thermodynamics, namely the Gibbs-Duhem equation for a one-component system,

$$vdp - sdT = d\mu_{ch} \tag{3.98}$$

where v is the molar volume and s is the molar entropy. We treat the chemical potential as a function of the pressure and temperature, $\mu_{ch} = \mu_{ch}(p, T)$, and therefore

$$d\mu_{ch} = \left(\frac{\partial \mu_{ch}}{\partial p}\right)_T dp + \left(\frac{\partial \mu_{ch}}{\partial T}\right)_p dT \tag{3.99}$$

Identifying the molar volume with the indicated partial derivative, and then applying the chain rule, yields

$$\left(\frac{\partial \mu_{ch}}{\partial p}\right)_T = v\left(\frac{\partial p}{\partial \rho}\right)_T \tag{3.100}$$

Upon noting that $v = 1/\rho$, we obtain the desired expression, Eq. (3.97). Finally, differentiating both sides of Eq. (3.96) with respect to density and using Eq. (3.97) yields the equation

$$\left(\frac{\partial p}{\partial \rho}\right)_T = -\rho\varepsilon_0 + \frac{k_B T}{1-\rho}$$

or

$$p = -\left(\frac{\varepsilon_0}{2}\right)\rho^2 - k_B T \ln(1 - \rho) \qquad (3.101)$$

This result may be compared to van der Waals equation of state

$$p + \frac{a}{v^2} = \frac{k_B T}{v - b} \qquad (3.102)$$

The comparison is facilitated by setting $b = 1$, and $a = \varepsilon_0/2$. We then have for the van der Waals equation

$$p = -\left(\frac{\varepsilon_0}{2}\right)\rho^2 + \frac{k_B T \rho}{1 - \rho} \qquad (3.103)$$

These two expressions are quite similar with the only difference being the manner of treatment of the excluded volume term. We see from this close similarity that the van der Waals equation is analogous to the Weiss mean-field equation for an Ising ferromagnet.

3.8.3 The Triple and Critical Points

The van der Waals equation of state was originally introduced as an empirical expression designed to describe the behavior of imperfect fluids near the critical point. In comparing the van der Waals equation of state to the ideal gas law $PV = Nk_B T$, which describes a gas of noninteracting molecules, we note that there are two modifications. The first is the addition of an internal pressure term, a/v^2 that takes into account the attractive intermolecular forces. The second is the incorporation of an excluded volume term, of the form $v - b$, that reflects the short-range, repulsive interactions, or incompressibility, arising from Pauli exclusion. The attractive component stabilizes the system at a particular density and presssure. It is treated through the mean-field. The repulsive component is responsible for the detailed structure and packing properties of the system and is the crucial feature of the van der Waals picture. This picture is particularly applicable at the triple point, where there is little difference in densities between solid and liquid phases and where fluctuations, neglected in the mean-field approach, are damped by the high densities.[29,30,31]

The behavior of the system at the critical point provides an interesting contrast to that at the triple point. In the vicinity of the critical point, the fluid density is far lower than at the triple point and, as already discussed, long-range correlations (fluctuations) are important. These correlations are due to the attractive component, whereas near the triple point the important correlations are associated with the short-range, repulsive component. If the attractive component is missing near the critical point, the system will be disordered, except for the exclusion of multiple occupancy. Here the repulsive component primarily establishes a scale of distance for the cell size. Critical point phenomena can be treated using the full lattice gas model. In the two-dimensional case, the analytic results of Onsager and Yang for an Ising ferromagnet can be transcribed into lattice gas predictions. In the three-dimensional case, numerical calculations provide results in excellent agreement with the observations for real fluids.[29]

3.9 THE RENORMALIZATION GROUP

3.9.1 Coupled Degrees of Freedom

In Section 3.8 we observed that mean-field theory, in general, and van der Waals theory, in particular, provides a useful description of phase transitions near the triple point. Phenomena encountered in the vicinity of critical points are far more difficult to treat. In the vicinity of a critical point, the behavior of a system is qualitatively different from that encountered in other regimes. Far from critical points, the hamiltonian and the coupling strengths determine the response properties of the system. Near critical points the behavior of a system depends mostly on the existence of cooperativity itself, and on the nature of the degrees of freedom, rather than on the detailed character of the hamiltonian.

The correlation length represents the size of the smallest subunit of a system, which can be considered when modeling the behavior of a system, without altering those very properties under consideration. In our brief remarks on critical opalescence, we noted that fluctuations on all scales of length were important near a critical point. This means that the correlation length becomes very large. Systems possessing many coupled degrees of freedom within a correlation length are extraordinarily difficult to treat. The renormalization group method of Wilson[7,8] provides a way of treating cooperative systems with many coupled degrees of freedon distributed over many scales of length. In this section we will discuss this recent approach.

The basic idea[1,32] underlying the renormalization group approach is that the many length scales are locally coupled. For example, in a magnet, fluctuations with wavelengths in a given range of values are primarily affected by fluctuations with nearby wavelengths. Fluctuations with wavelengths much less or much greater than that given range are less important. The result of this assumption is that there is a cascade effect in the whole system. Atomic fluctuations influence fluctuations at somewhat larger wavelengths, which influence fluctuations at still larger wavelengths, and so on. There are two principle consequences of the cascade picture. The first is *scaling*. The sense here of this term is that fluctuations at intermediate wavelength regimes tend to be identical except for a change of energy scale. This idea appears to be valid except at the limits, for example, it fails at wavelengths near a length parameter. The second feature emerging from the couplings is amplification and deamplification. This means that small temperature changes are amplified by the development of a cascade, so a small temperature change produces macroscopic changes at large wavelengths. Deamplification means that while two different materials may have different atomic characteristics, these differences decrease in importance as the cascade develops. At large scales the two systems behave in a similar manner. This leads to the notion of universality—all critical phenomena possess a common set of characteristics.

3.9.2 The Kadanoff Block Spin Construction

Let us consider again the Ising spin system in two dimensions. Suppose[6] that we divide the Ising lattice into blocks of spins, each block having linear dimensions much less than the correlation length as shown in Fig. 3.11. The total spin of the Jth block is given by the sum of the individual contributions,

$$\tilde{S}_J = \sum_{j \in J} s_j \qquad (3.104)$$

This block of spins can be treated as a single-spin element, which can take on the usual values of $+1$ or -1 by introducing the appropriate multiplicative factor f, which takes into account the fact that not all spins within a block will point in the same direction.

$$\tilde{S}_J = f \cdot S_J \qquad (3.105)$$

The hamiltonian can be written in terms of the block spin variable as

$$E = -J_{bl} \sum_{\langle I,J \rangle} S_I S_J - H_{bl} \sum_J S_J \qquad (3.106)$$

where J_{bl} is a new coupling constant and $H_{bl} = fH$.

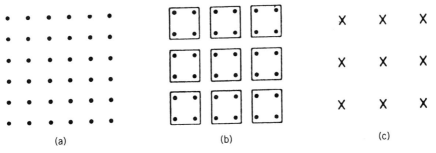

Figure 3.11. The Kadanoff block spin construction. The original lattice (a) is divided into blocks of size 2 × 2 (b), which is replaced by a second lattice of effective spin elements (c). (From Wilson and Kogut[1]. Reprinted with permission of Elsevier Science, The Netherlands.)

3.9.3 The Renormalization Group and Fixed Points

In the renormalization group approach, we generate new free energies (partition functions) from the old ones by progressivly integrating out degrees of freedom at each length scale. At the end of each step, we are left with a hamiltonian representing the interactions over the length scales not yet treated. The key step for us is to find a transformation, relating the old and new free energies, that leaves the form of the partition function alone, absorbing the effect of the summation into a new effective coupling strength. In terms of the block spin picture, this means we have found the relationship between J and J_{bl}. Once we establish the transformation, we have a simple recursion (renormalization) relation decribing the successive integration steps. Critical phenomena manifests itself through the appearance of fixed points. These are values for the coupling strength where recursion no longer produces any change in the coupling strength, so the further removal of degrss of freedom has no effect.

The renormalization group method can be illustrated[33] fairly easily for a one-dimensional Ising spin system. Let us absorb the Boltzmann constant and temperature into the definition of a coupling constant K:

$$K \equiv J\beta = \frac{J}{k_B T} \tag{3.107}$$

and let us assume that there is no external applied magnetic field. The partition function involves the evaluation over all spin configurations of the sum

$$Z = \sum_{\{s_i\}} e^{K(s_1 s_2 + s_2 s_3 + s_3 s_4 + \cdots)} = \sum_{\{s_i\}} e^{K(s_1 s_2 + s_2 s_3)} e^{K(s_3 s_4 + s_4 s_5)} \cdots$$

$$= \sum_{s_1 = \pm 1} \sum_{s_3 = \pm 1} \cdots [e^{K(s_1 + s_3)} + e^{-K(s_1 + s_3)}][e^{K(s_3 + s_5)} + e^{-K(s_3 + s_5)}] \cdots \tag{3.108}$$

In the first step, we have explicitly written out the summation over the spin elements contributing to the interaction energy, grouping together those terms sharing a common even index, such as s_2 and s_4. In the second step, we performed the summation over the even indexes, leaving the remaining odd indexes as yet unevaluated. Next we note that if we can find a spin-independent transformation $f(K)$, that satisfies the relationship

$$e^{K(s_1 + s_3)} + e^{-K(s_1 + s_3)} = f(k) e^{K' s_1 s_3} \tag{3.109}$$

for all possible values of the spin variables, then we have a way of relating the partition function for N spins to that for $N/2$ spins, which preserves the form of the partition function. To derive the form of the transformation, we first set $s_1 = s_3 = 1$ to give

$$e^{2K} + e^{-2K} = f e^{K'} \tag{3.110}$$

Now, if we set both spin variables equal to -1, we obtain the same expression. However, setting one of the spin variables equal to $+1$ and the other equal to -1 yields a different expression, namely

$$2 = fe^{-K'} \tag{3.111}$$

Upon solving Eqs. (3.110) and (3.111) for the two unknowns f and K', we obtain the results

$$f(K) = 2\cosh^{1/2}(2K) \tag{3.112}$$

and

$$K' = \left(\frac{1}{2}\right)\ln\cosh(2K) \tag{3.113}$$

With these last equations we have the desired recursion relation between partition functions:

$$Z(N, K) = \sum_{\{s_1, s_3, s_5, \ldots\}} f(K)e^{K'(s_1 s_3)} f(K)e^{K'(s_3 s_5)} \ldots$$

$$= [f(K)]^{N/2} \sum_{\{s_1, s_3, s_5, \ldots\}} e^{K'(s_1 s_3 + s_3 s_5 + \ldots)}$$

or

$$Z(N, K) = [f(K)]^{N/2} Z\left(\frac{N}{2}, K'\right) \tag{3.114}$$

The recursion formula above and others like it are called *Kadanoff transformations*.

Finally, let us consider the free energy. This quantity is proportional to the size of the system. We may introduce a free energy per spin, $\phi(K)$, which is size independent, through the formula

$$\ln Z(N, K) = N\phi(K) \tag{3.115}$$

Substituting this last expression into the recursion formula yields the result that

$$\phi(K) = \left(\frac{1}{2}\right)\ln f(K) + \left(\frac{1}{2}\right)\phi(K') \tag{3.116}$$

and replacing $f(K)$ with its value as given in Eq. (3.112),

$$\phi(K') = 2\phi(K) - \ln[2\cosh^{1/2}(2K)] \tag{3.117}$$

100 COOPERATIVITY IN LATTICE SYSTEMS

Equations (3.112), (3.113), and (3.117) are our main results and are known as the *renormalization group equations*. In this form the new coupling constant K' is always less than the original quantity K. We can derive an alternative set of RG equations that produces a recursion in the opposite direction. The resulting equations are

$$K = \left(\frac{1}{2}\right) \cosh^{-1}(e^{2K'}) \qquad (3.118)$$

and

$$2\phi(K) = \phi(K') + K' + \ln 2 \qquad (3.119)$$

To see how they may be used to evaluate the partition function $\phi(K)$, let us choose a value for K' that is small such as $K' = 0.01$. In this situation the couplings between the spins are so small that we can put $Z = 2^N$, and $\phi(0.01) = \ln 2$. We can calculate K from Eqs. (3.118) and (3.119), finding that its value is 0.100334. This in turn leads to a value for ϕ equal to 0.698142, and so on. Table 3.1 contains a list of the successive values for K, ϕ, and the exact result for ϕ calculated by means of the transfer matrix method. We observe that the renormalization group values rapidly converge to the exact results for the partition function. We can plot the values for K obtained in succeeding iterations. The result is the flow diagram presented in Fig. 3.12 (upper diagram). The only fixed points, where recursion does not produce any change in K, is at zero and infinity. These are termed *trivial fixed points*, and the absence of nontrivial fixed points means that there is no phase transition, as we found before.

TABLE 3.1

K	$\phi(K)$ (RG)	$\phi(K)$ (Exact)
0.01	ln 2	0.693197
0.100334	0.698147	0.698172
0.327447	0.745814	0.745827
0.636247	0.883204	0.883210
0.972710	1.106299	1.106302
1.316710	1.386078	1.386080
1.662637	1.697968	1.697968
2.009049	2.026876	2.026877
2.355582	2.364536	2.364537
2.702147	2.706633	2.706634

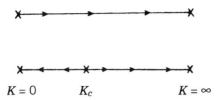

Figure 3.12. RG flow diagrams for the one- and two-dimensional Ising systems. *Upper:* The one-dimensional Ising chain with trivial fixed points at zero and infinity. *Lower:* The two-dimensional Ising latice with one fixed point at the critical value, K_c.

In a two-dimensional Ising spin system, integrating out degrees of freedom leads to progressively more complicated interactions. This increasing complexity is a consequence of the strong cooperativity, or topology, of the two-dimensional lattice system, which is absent in the one-dimensional case. To obtain a useful recursion relation, the complicated interactions must be approximated. When this is done using some inspired guesswork (the reader is referred to the paper by Maris and Kadanoff[33] for details), we find a flow diagram that is dramatically different from that of a one-dimensional system. Shown in Fig. 3.12 (lower diagram) is the flow for a two-dimensional spin system. We observe that the flow diagram is partitioned into two distinct regions by the presence of a nontrivial fixed point. If we start with a value for K below the fixed point, the system will evolve toward $K = 0$. If we start above the fixed point, the system will evolve toward ever increasing values for K. The fixed point value $K_c = 0.506$ found using the RG method compares favorably with the exact fixed point value $K_c = 0.441$.

3.9.4 Widom-Kadanoff Scaling

Let us recall that we noted in Section 3.3 that the behavior of thermodynamic variables in the vicinity of critical points are described in terms of critical exponents. We stated that the temperature dependence can be expanded in powers of $\varepsilon = (T - T_c)/T_c$ in the neighborhood of a critical point, and we remarked that the exponent of the leading term in these expansions is known as a critical exponent. As we have seen, all three thermodynamic quantities are derived from the partition function and are therefore not independent of one another. If the critical exponents are found to obey relationships such as

$$2\beta = 2 - \alpha - \gamma \qquad (3.120)$$

then the system is said to exhibit scaling. Evidence has been acquired for many systems demonstrating just this sort of behavior.[34] Let us now see how this type of relationship among the critical exponents can be obtained under the Widom-Kadanoff homogeneity hypothesis.

Let us recall that a function $h(x)$ is said to be homogeneous if it obeys a relationship of the form

$$h(Lx) = L^p h(x) \tag{3.121}$$

for all values of the parameter L. The quantity p is called the *degree of homogeneity*. This type of function has the interesting property that knowledge of $h(x)$ at one point, and of p, provides a complete description of the function everywhere. A generalized homogeneous function is a function that satisfies the scaling relationship

$$h(L^a x, L^b y) = L h(x, y) \tag{3.122}$$

where a and b are arbitrary constants. The Widom-Kadanoff scaling hypothesis[46] asserts that the Gibbs (or Helmholtz) free energy per spin g is a generalized homogeneous function. If we consider the free energy as a function of ε and H, then the hypothesis means that

$$g(L^a \varepsilon, L^b H) = L^d g(\varepsilon, H) \tag{3.123}$$

where, for convenience, we have introduced a constant d. To understand the consequences of the homogeneity assumption, let us differentiate both sides of the equation with respect to the magnetic field. This gives

$$L^b M(L^a \varepsilon, L^b H) = L^d M(\varepsilon, H) \tag{3.124}$$

Let us now make the selection

$$L = \left(\frac{-1}{\varepsilon}\right)^{1/a} \tag{3.125}$$

In the limit as H goes to zero, we have

$$M(\varepsilon, 0) = (-\varepsilon)^{(d-b)/a} M(-1, 0) \tag{3.126}$$

But as ε approaches zero from below,

$$M(\varepsilon, 0) \approx (-\varepsilon)^\beta \tag{3.127}$$

Replacing the right-hand side of the previous expression with this limiting value yields the relation

$$\frac{d-b}{a} = \beta \tag{3.128}$$

We may continue this type of reasoning, calculating the derivatives leading to the specific heat and susceptibility. The result is that we find that the critical exponents obey the scaling relationship given in Eq. (3.120). In the renormalization group method these relations emerge from an analysis of the RG transformation near the fixed point. If we linearize the transformation about the fixed point, the eigenvalues of the resulting transformation matrix are directly related to the critical exponents.

3.10 SUMMARY

Fluctuations in thermodynamic parameters such as temperature, volume, or density ordinarily bring about processes that restore the equilibrium in that system. When a system is brought into a condition where it loses its stability against fluctuations in its parameters, phase transitions occur. Singularities or discontinuities in the specific heat and nonzero order parameters signify the presence of these consequences of cooperativity. The specific heat provides a measure of the fluctuations, while an order parameter marks the onset of ordering in a cooperative system at the critical temperature. This ordering increases as the temperature is lowered and becomes maximal at the lowest possible temperatures. In the vicinity of a critical point, cooperative interactions promote a rapid onset of large-scale, correlated fluctuations in the thermodynamic variables representative of the order in the system. These fluctuations in turn produce divergences in the range of correlations at the critical temperature, which lead to singularities in the thermodynamic derivatives.

We observed in our examination of the Peierls-Griffiths argument that for a two-dimensional lattice a temperature can always be found that will make the energy change associated with producing order greater than the entropic ones identified with generating disorder. Thus the free energy may be stabilized against fluctuations. This situation is profoundly different from that encountered for a one-dimensional chain. In the one-dimensional case, a boundary isolating a region of negative spin from a surrounding region of positive spin can be moved to any other location in the chain without producing any change in the energy. In the one-dimensional case, there is no temperature below which free energy minimization will stabilize an ordered state against fluctuations. These results are consistent with our analytic deductions using the transfer matrix method that while there is no phase transition for a one-dimensional chain, a two-dimensional Ising lattice will undergo a phase transformation.

In a mean-field theory detailed interactions among the elements of the assembly are assumed to generate an effective internal field. Local fluctuations are neglected, and the order is long range, of the form N_1/N. The overall effect of including the internal field is to add to the internal and free energies a term quadratic in the order parameter. In more sophisticaled mean-field theories, such as the Bethe and Kikuchi approximations, some attempt is made to include local order of the form N_{12}/N by replacing isolated spins by small clusters of elements. In one dimension, the neglect of fluctuations in mean-field approaches produces erroneous results. In two dimensions, the results are in qualitative agreement with exact predictions, and in three dimensions, there is an impressive convergence of the mean-field predictions toward quantitatively agreement.

In the lattice gas model and in the van der Waals picture, we introduced for the first time both short-ranged, harshly repulsive interactions and longer-ranged attractive forces. In the van der Waals picture we observed the order-producing influences of repulsive interactions. The relative importance of the attractive and repulsive interactions changes as we alter our focus from the triple point, where fluid densities are high, to the critical point where densities are lower and long-range fluctuations in the attractive component are an essential ingredient. The van der Waals picture of fluids is a mean-field theory. The van der Waals theory treats the attractive component as a mean-field, thereby neglecting fluctuations. This approach is most appropriate near the triple point, while the full lattice gas model may be used to treat the fluid system in the vicinity of the critical point.

The organizational principles elucidated by the renormalization group approach differs from those of the mean-field and van der Waals theories. Designed to treat critical phenomena, we find in a renormalization group treatment that at each summation stage, cooperativity generates a progressively more complex set of interactions. This complexity arises in systems possessing many coupled degrees of freedom operating locally over many scales of length. The detailed character of the cooperative interactions is not important in these situations; instead, we encounter scaling and universality.

In the remainder of this book, we will explore the consequences of cooperative behavior in a variety of physical, biological, and engineering systems and situations. We began this chapter by noting the opposing tendencies of energetic and entropic influences, observing that at sufficiently low temperatures, strongly cooperative interactions promote the formation of highly ordered stable states corresponding to a minimum in the free energy. Our main objective in the next chapter will be to apply the universal principles and mechanisms surrounding energy minimization, phase transitions, the onset of order, and critical phenomena to the study and solution of problems in other settings of interest to us such as neurophysiological and engineering ones. Our primary tool in achieving this goal will be the Monte Carlo method, the fourth approach toward treating complex interacting systems. This class of techniques will provide a way of searching for highly ordered states and studying the overall evolution of the system toward these optimal configurations. Our initial formulation, the Metropolis algorithm, will be broadened into the powerful technique known as *simulated annealing*. We will start with an introductory look at binary alloys, then introduce the method as applied to engineering design problems, and finally, model self-organization during development of the central nervous system.

3.11 ADDITIONAL READING

Insightful discussions of the Ising and lattice gas models can be found in the texts by Ma,[35] Chandler,[36] and Wannier.[9] Additional information, particularly on the transfer matrix and combinatorial methods, can be found in the review papers by Montroll and Newell[37] and by Domb.[27] The articles by Widom[28] and by Chandler[31] et al. provide a good overview of the modern theory of liquids and the van der Waals picture. The monographs by Brout[38] and Stanley[28] contain especially clear discussions of phase transitions, mean-field theory and critical phenomena in a variety of cooperative physical systems. Many of the topics omitted in the present chapter, such as the Landau and Ornstein-Zernike theories and superconductivity, are treated by these authors. In addition there are a number of excellent and thorough review articles on critical phenomena and the renormalization group. Anong these are papers by Kadanoff et al.,[10] Fisher,[39] Wilson and Kogut,[1] and Wilson.[40] The presentation in this chapter of fixed points in the renormalization group follows closely that of Maris and Kadanoff,[33] and their paper is highly recommended.

3.12 REFERENCES

1. Wilson, K. G., and Kogut, J. (1974). The renormalization group and the ε-expansion. Phys. Rep., **C12**, 75–200.
2. Griffiths, R. B. (1964). Peierls proof of spontaneous magnetization in a two-dimensional Ising ferromagnet. Phys. Rev., **A136**, 437–439.
3. Changeux, J.-P., Thiery, J., Tung, J., and Kittel, C. (1967). On the cooperativity of biological membranes. Proc. Nat. Acad. Sci. USA, **57**, 335–341.
4. Widom, B. (1965a). Surface tension and molecular correlations near the critical point. J. Chem. Phys., **43**, 3892–3897.
5. Widom, B. (1965b). Equation of state in the neighborhood of the critical point. J. Chem. Phys., **43**, 3898–3905.
6. Kadanoff, L. (1966). Scaling laws for Ising models near T_c. Physics, **2**, 263–272.
7. Wilson, K. G. (1971a). Renormalization group and critical phenomena, I. Renormalization group and Kadanoff scaling picture. Phys. Rev., **B4**, 3174–3183.
8. Wilson, K. G. (1971b). Renormalization group and critical phenomena, II. Phase-space cell analysis of critical behavior. Phys. Rev., **B4**, 3184–3205.
9. Wannier, G. H. (1966). *Statistical Physics*. New York: Wiley.
10. Kadanoff, L. P., Götze, W., Hamblen, D., Hecht, R., Lewis, E. A. S., Palciauskas, V. V., Rayl, M., and Swift, J. (1967). Static phenomena near critical points: Theory and experiment. Rev. Mod. Phys., **39**, 395–431.
11. Ising. E. (1925). Beitrag zur Theorie des Ferromagnetismus. Z. Phys., **31**, 253–258.
12. Peierls, R. (1936). On Ising's model of ferromagnetism. Proc. Camb. Phil. Soc., **32**, 477–481.
13. Kramers, H. A., and Wannier, G. H. (1941a). Statistics of the two-dimensional ferromagnet. Part I. Phys. Rev., **60**, 252–262.

14. Kramers, H. A., and Wannier, G. H. (1941b). Statistics of the two-dimensional ferromagnet. Part II. Phys. Rev., **60**, 263–276.
15. Montroll, E. W. (1941). Statistical mechanics of nearest neighbor systems. J. Chem. Phys., **9**, 706–721.
16. Onsager, L. (1944). Crystal statistics I: A two-dimensional model with an order-disorder transition. Phys. Rev., **65**, 117–149.
17. Kaufman, B. (1949). Crystal statistics, II: Partition function evaluated by spinor analysis. Phys. Rev., **76**, 1232–1243.
18. Kaufman, B., and Onsager, L. (1949). Crystal statistics, III: Short-range order in a binary Ising lattice. Phys. Rev., **76**, 1244–1252.
19. Yang, C. N. (1952). The spontaneous magnetization of a two-dimensional Ising model. Phys. Rev., **85**, 808–816.
20. Bragg, W. L., and Williams, E. J. (1934). The effect of thermal agitation on atomic arrangement in alloys. Proc. Roy. Soc. Lond., **A145**, 699–730.
21. Bragg, W. L., and Williams, E. J. (1935). The effect of thermal agitation on atomic arrangement in alloys — II. Proc. Roy. Soc. Lond., **A151**, 540–566.
22. Bethe, H. A. (1935). Statistical theory of superlattices. Proc. Roy. Soc. Lond., **A150**, 552–575.
23. Peierls, R. (1936). Statistical theory of superlattices with unequal concentrations of the components. Proc. Roy. Soc. Lond., **A154**, 207–222.
24. Lee, T. D., and Yang, C. N. (1952). Statistical theory of equations of state and phase transitions. II. Lattice gas and Ising model. Phys. Rev., **87**, 410–419.
25. Weiss, P. R. (1948). The application of the Bethe-Peierls Method to ferromagnetism. Phys. Rev., **74**, 1493–1504.
26. Kikuchi, R. (1951). A theory of cooperative phenomena. Phys. Rev., **81**, 988–1003.
27. Domb, C. (1960). On the theory of cooperative phenomena in crystals. Adv. Phys., **9**, 149–361.
28. Stanley, H. E. (1971). *Introduction to Phase Transitions and Critical Phenomena*. New York: Oxford University Press.
29. Widom, B. (1967). Intermolecular forces and the nature of the liquid state. Science, **157**, 375–382.
30. Longuet-Higgins, H. C., and Widom, B. (1964). A rigid sphere model for the melting of argon. Mol. Phys., **8**, 549–556.
31. Chandler, D., Weeks, J. D., and Andersen, H. C. (1982). Van der Waals picture of liquids, solids, and phase transformations. Science, **220**, 787–794.
32. Wilson, K. G. (1975). The renormalization group: Critical phenomena and the Kondo problem. Rev. Mod. Phys., **47**, 773–840.
33. Maris, H. J., and Kadanoff, L. P. (1978). Teaching the renormalization group. Am. J. Phys., **46**, 652–657.
34. Ho, J. T., and Litster, J. D. (1969). Magnetic equation of state of $CrBr_3$ near the critical point. Phys. Rev. Lett., **22**, 603–606.
35. Ma, S.K. (1985). *Statistical Mechanics*. Singapore: World Scientific.
36. Chandler, D. (1987). *Introduction to Modern Statistical Mechanics*. New York: Oxford University Press.

37. Newell, G. F., and Montroll, E. W. (1953). On the theory of the Ising model of ferromagnetism. Rev. Mod. Phys., **25**, 353–389.
38. Brout, R. (1965). *Phase Transitions*. New York: W. A. Benjamin.
39. Fisher, M. E. (1974). The renormalization group in the theory of critical behavior. Rev. Mod. Phys., **46**, 597–616.
40. Wilson, K. G. (1983). The renormalization group and critical phenomena. Rev. Mod. Phys., **55**, 583–600.

4

SIMULATED ANNEALING

4.1 INTRODUCTION

4.1.1 Energy Landscapes

We observed in the last chapter that the spontaneous magnetization of a two-dimensional Ising ferromagnet can be either positive or negative. At temperatures below the Curie point, there are two minima in the free energy, corresponding to positive or negative ordering. These minima, as shown in Fig. 4.1a, are separated from one another by a barrier that prevents passage from one mimimum to the other. In the absence of an applied external field, the free energy versus magnetization curve is symmetric. The addition of an external field, however small, introduces an asymmetry and serves as an initial condition that selects one of the two minima as the stable state of the system. This situation is illustrated in Fig. 4.1b.

The energy landscape for a ferromagnet is a fairly simple one. This situation changes dramatically if we allow the sign of the coupling constants J_{ij} to vary in a random way from site to site. Let us examine what happens in an Ising lattice with fixed, random couplings. We assume that the interaction energy is of the standard form

$$E = - \sum_{\langle i,j \rangle} J_{ij} s_i s_j \qquad (4.1)$$

where the spin variables can assume the values $+1$ or -1 and $\langle i,j \rangle$ denotes that we are summing over nearest-neighbor pairs. In contrast to a ferromagnet, where all coupling constants were positive, or an antiferromagnet, where they are all negative, we now choose $J_{ij} = +J$ or $-J$ randomly over the lattice. A

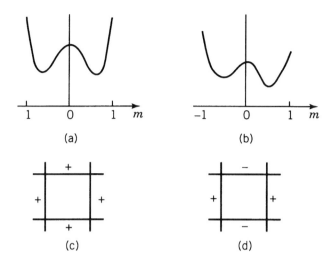

Figure 4.1. Unfrustrated and frustrated ferromagnets: (a) Spontaneous magnetization in a ferromagnet below the Curie temperature; (b) symmetry breaking in the presence of an external magnetic field; (c) ferromagnetic square; (d) frustrated square.

magnet with this property is said to be *disordered*. Let us now consider a small portion of such a lattice as illustrated in Fig. 4.1d. In this schematic picture the J_{ij}'s are represented as bonds linking the lattice sites at the corner of the square. For a positive bond the lowest energy state is one in which both spins have aligned, while for a negative bond the minimum energy corresponds to opposing spin orientations. We see that it is not possible to select a set of spins that simultaneously minimizes all contributions to the total energy for the square depicted in Fig. 4.1d. This inability to completely satisfy the energy minimization constraint is known as *frustration*.

An important consequence of frustration in the creation of a multiplicity of low-energy states. This aspect can be illustrated by comparing the lowest-energy states of the frustrated square to those of the ferromagnetic square shown in Fig. 4.1c. For the ferromagnet there are two lowest-energy states corresponding to the two configurations where all spins point in the same direction, up or down. The energy in either case is $-4J$. For the frustrated square there are eight, shallower energy minima, each with energy $-2J$.

The concurrent presence of positive and negative couplings in a lattice is not the criteria for frustration. Frustration in our disordered magnet is present if the frustration measure[1]

$$\Phi = \prod_{C(i,j)} J_{ij} \qquad (4.2)$$

is negative, where $C(i, j)$ denotes a closed contour along the sequence of bonds under consideration. Thus, if $\Phi = +1$ the loop is not frustrated and, conversely, if $\Phi = -1$ the sequence is frustrated. For example, a square with two positive couplings and two negative ones is not frustrated, having two minima, each with energy $-4J$, as was the case for the ferromagnet.

4.1.2 Multiple Constraints and Selectivity

The frustrated square serves as a particularly simple example of the more general and widespread phenomenon of multiple reinforcing, and sometimes conflicting, constraints (interactions) in complex nonlinear systems. These constraints take many forms and serve a number of useful purposes in physical, artificial, and biological systems. In the traveling salesman, graph partitioning and other NP-hard problems, competing constraints, frustration, and a multiplicity of local energy minima are essential features. The structure of these problems bears a great similarity to the statistical mechanics description of a dilute disordered magnet known as a *spin glass*. One of our goals in this chapter will be to explore these connections and accompanying solutions.

A second broad class of problems in which multiple constraints appear are those encountered in image processing. Digital images can be regarded as highly organized spatial lattice systems composed of grey-level pixel values. We can endow these structures with an artificial dynamics. When we do this, we must employ two types of constraints to guide the evolution of the image systems toward desired, highly ordered states. One class of constraint preserves spatial structure, while a second group restricts the hypothesis space to permit convergence to physical meaningful stable states. We will examine two-dimensional images modeled as Markov random fields in Chapter 6.

The third class of processes of interest to us are those involved in the self-organization of the central nervous system. As noted in the Introduction, it is abundantly clear that multiple mechanical, chemical, and electrical interactions operate simultaneously and sequentially to achieve many developmental goals. The agents, mechanisms, strategies, and principles associated with these interactions are common to both central nervous system development and adult physiological function. We find that neurophysiological systems are endowed with many stable states that are exploited as the need arises in response to changes in the external and internal environment. A particularly dramatic instance of responsiveness, and multiplicity, is provided by the rich spectrum of neural firing patterns which can be experimentally evoked. In the next chapter we will explore an equally dramatic set of findings regarding developmental and functional plasticity in the retinotectal projection of lower vertebrates such as *Xenopus*, and in Chapter 9 we will look at the dynamic modulation of neural firing states.

4.1.3 Searching for Optimal Solutions

The search for optimal solutions — that is, for highly ordered, low-energy, stable states — is a particularly difficult one in systems possessing many local minima, since it is relatively easy to get trapped in nonoptimal configurations. Three observations from Chapters 2 and 3 provide the necessary clues for developing a strategy for finding optimal solutions in such systems. The first notion is that low-energy, stable states stand out at low temperatures, whereas they are buried among the multitude of disordered states at high temperatures.

The second observation from the last chapter is that fluctuations are a useful mechanism, in that they allow a system to probe its surrounding solution space for better energy configurations. A particular fluctuation may temporarily produce either an increase or a decrease in the configuration energy. A transient increase in the total energy will enable a system to hop out of a shallow, nonuseful, energy basin or minimum on the way toward finding a better one.

The third idea is that difficulties in obtaining exact, analytic solutions to nearest-neighbor models of lattice cooperativity can be alleviated using the Monte Carlo importance sampling method such as the Metropolis[2] algorithm. In our initial discussion in Chapter 2, we noted that the Metropolis algorithm generates a sequence of configurations that converges uniquely to the stable states characteristic of the temperature under consideration. Our primary goal in this chapter will be to examine the generalization of this algorithmic process to include a decreasing temperature. This generalization, simulated annealing, will enable us to search for optimal, low-energy configurations of a system. Although not discussed in Chapter 2, we will make use of an important built-in property of the Metropolis and related algorithms, namely that they permit increases in energy as well as decreases. Thus they can exploit fluctuations and noise to avoid trapping in local, nonoptimal minima.

4.2 OBJECTIVES

Metropolis and Metropolis-like algorithms have been used to explore the kinetics and equilibrium properties of a wide variety of systems. In these studies a number of alternatives to the sampling procedure introduced by Metropolis et al. were devised. It is instructive to examine the different ways of performing random sampling that satisfy the symmetry condition and generate Markov chains of states obeying the detailed balance condition. We will begin the remainder of this chapter by examining two algorithms of this type. The first sampling method we will look at is the Glauber[3] heat bath algorithm used in conjunction with the kinetic Ising model. We will sketch this approach in Section 4.3. The second set of alternative procedures is related to the treatment of binary alloys. Monte Carlo importance sampling methods have been applied to the study of order-disorder phenomena in binary alloys first by Fosdick,[4] then by Flynn and McManus,[5] and by others.[5] The binary alloy studies contributed to the evolution of the Markov random field methods to be presented in Chapter 6 and to the models of central nervous system development to be examined in Chapter 5. We will outline the algorithms for treating binary alloys in Section 4.4.

Spin glasses and the general problems encountered in treating systems with frozen-in, or *quenched*, disorder helped inspire the development of the simulated annealing method. We will begin Section 4.5 with a discussion of the distinction between quenched and annealed random variables. We will follow these introductory remarks with a sketch of the replica method of Edwards and Anderson[6] and the application of this method by Sherrington and Kirkpatrick[7] to a model of an infinite-ranged spin glass. We will complete our survey of spin glasses with a discussion of ultrametricity and the topological organization of the spin glass states.

The problem of finding the ground-state a three-dimensional spin glass is intractable, as are many of the problems in combinatorial optimization. These problems are termed NP-hard, and we precede our discussion of simulated annealing with a brief overview of the world of NP-completeness (Section 4.6). We will examine the simulated annealing algorithm of Kirkpatrick, Gelatt, and Vecchi,[8] and Cerny[9] in detail in Sections 4.7 and 4.8. We have encountered all the essential elements of the method in earlier sections and chapters. We will connect these elements together and study the application of the approach to the solution of NP-hard problems. In the last two sections of the chapter, we will study two other annealing methods. The first of these is microcanonical annealing that is based upon the microcanonical ensemble of statistical mechanics. The second is continuous simulated annealing that is founded upon the Langevin equation first introduced as a model of Brownian motion. We will describe the microcanonical annealing approach in Section 4.9 and outline the continuous simulated annealing method in Section 4.10.

4.3 KINETIC ISING MODEL

In lattice systems atoms require a source of energy to change spins, exchange positions with other atoms, or move from one lattice site to another. This energy is provided by the thermal vibrations, or phonons, that are in equilibrium at temperature T with the lattice system and serve as a heat bath. We will now describe two classes of kinetic Ising models. In the first, due to Glauber,[4] spins may randomly flip from one state to the other. In the spin model neither the total energy nor the magnetization is conserved. In the second class of models, molecules may exchange positions with one another or migrate to empty lattice sites. Particle (molecule) number, the analog of magnetization, is conserved in these kinetic processes. The particle models have been used to study cooperative phenomena underlying order-disorder transformations in binary alloys and liquid-gas and solid-liquid phase transformations.

4.3.1 The Master Equation

Let us consider a lattice system in which a lattice variable x may assume one of a number of values at each of the $i = 1, \ldots, N$ lattice sites. Let $w(\{x\}, t)$ denote the probability that a system has assumed a particular configuration $\{x\} = (x_1, \ldots, x_N)$ at time t. The time evolution of the system can be described by a master equation of the form

$$\frac{dw(\{x\}, t)}{dt} = \sum_{\{x'\}} [p(\{x'\}, \{x\})w(\{x'\}, t) - p(\{x\}, \{x'\})w(\{x'\}, t)] \quad (4.3)$$

where $p(a, b)$ is the transition probability from state a to state b. The first term on the right-hand side gives the sum of all transitions into the state under consideration, and the second term is the sum of all transitions out of the specified state. At equilibrium the time derivative is zero:

$$\frac{dw(\{x\}, t)}{dt} = 0 \quad (4.4)$$

and the transition probabilities are equal to the equilibrium values:

$$w(\{x\}, t) = w_{eq}(\{x\})$$

The master equation then reduces to the detailed balance condition discussed in Chapter 2. In the above notation the detailed balance condition is

$$p(\{x'\}, \{x\})w_{eq}(\{x'\}) = p(\{x\}, \{x'\})w_{eq}(\{x\}) \quad (4.5)$$

We now wish to choose the transition probabilities in a manner that will guarantee that this kinetic model will approach the same equilibrium distribution as that of the (static) Ising model. This will happen if, as was done for the Metropolis algorithm, we use the detailed balance condition to define the transition probability ratios.

4.3.2 Spin Kinetics

Let us consider a simple case where the variable for a single lattice site j may change its state. We will further assume that the lattice variables may assume one of two values, $+1$ and -1. We will then denote the configuration of the system by the value x_j and the alternative, or flipped, value $-x_j$. The detailed balance condition can be written as

$$p(-x_j \rightarrow x_j)\pi(-x_j) = p(x_j \rightarrow -x_j)\pi(x_j) \quad (4.6)$$

where, as in Chapter 2, we write the equilibrium distribution $w_{eq}(x)$ as $\pi(x)$. If we regard the lattice variables as spin elements whose interactions are given by an Ising hamiltonian, we may put

$$\pi(x_j) = \frac{1}{Z} e^{-\beta E(H)} \tag{4.7}$$

with the Hamiltonian

$$E(H) = -J \sum_{\langle i,j \rangle} x_i x_l - \mu H \sum_i x_i \tag{4.8}$$

Defining the local energy at the lattice site of interest as

$$E_j = J \sum_{\langle j,l \rangle} x_l + \mu H \tag{4.9}$$

allows us to express the stationary distribution ratio as

$$\frac{\pi(x_j)}{\pi(-x_j)} = e^{-2\beta E_j x_j} \tag{4.10}$$

Since these are the only two states of the system, the normalization condition for the transition probabilities takes the form

$$\frac{p(x_j \to -x_j)}{p(-x_j \to x_j)} = \frac{p(x_j \to -x_j)}{1 - p(x_j \to -x_j)} = e^{-2\beta E_j x_j} \tag{4.11}$$

where we have used the detailed balance condition in the last step. This result provides us with an alternative to the Metropolis algorithm. In the notation of Chapter 2, we then have the sampling algorithm

$$p_{ij} = \tau_{ij} \frac{\pi_i}{\pi_i + \pi_j} \tag{4.12}$$

This result is sometimes called the *heat bath algorithm*, and for the one-dimensional case, it is known as *Glauber dynamics*. This algorithm is similar to the Gibbs sampler that we will introduce when we discuss image restoration and Markov random fields. The heat bath may be compared to the Metropolis algorithm, which we recall (Eq. 2.68) is

$$p_{ij} = \begin{cases} \frac{\tau_{ij} \pi_j}{\pi_i}, & \frac{\pi_j}{\pi_i} < 1 \\ \tau_{ij}, & \frac{\pi_j}{\pi_i} \geq 1 \end{cases} \tag{4.13}$$

An algorithm simulating an Ising spin system is:

1. Initialize by arranging spins into the lattice in an ordered arrangement, in a completely disordered arrangement, or randomly.
2. Select a lattice site.
3. Calculate the transition probability corresponding to flipping the spin.
4. Choose a random number. If it is less than the transition probability, flip the spin; otherwise, do not flip the spin.

Several steps are involved when using the Glauber heat bath and Metropolis algorithms to simulate an interacting system. We must first define the configurations $\{x\}$ of the system. We must next define an energy, or objective function, that returns a scalar value upon quantifying the results of the various interactions and constraints for any configuration. We have to define a procedure for generating trial configurations in a manner that is unbiased and satisfies the condition that $\tau_{ij} = \tau_{ji}$, and then we must use the Glauber, Metropolis, or some other algorithm for deciding whether to accept or reject the trail move. Last, we must have a criteria for deciding when to stop. This is typically done when trial moves begin to be rejected with a high probability, indicating that $dw/dt = 0$ and that equilibrium has been reached.

4.4 ORDER-DISORDER TRANSITIONS IN BINARY ALLOYS

Binary alloys are an important class of lattice systems. In a binary alloy the relative concentration of the alloying material is an important factor in determining the characteristics and presence of a specific phase transition. Metallurgical techniques used in preparing binary alloys, such as annealing and quenching, have an important influence on the properties of the product and serve as models for the simulated annealing processes we will discuss in the next few chapters.

4.4.1 Order-Disorder Transitions

Beta brass is an alloy composed of copper and zinc in equal amounts and arranged into a body-centered cubic lattice. In the completely ordered state, all copper atoms are nearest neighbors of zinc atoms, and vice versa. As the temperature increases, the probability for finding the correct neighbor decreases toward 1/2, and at temperatures above the critical temperature, copper and zinc placements on the lattice are completely random. The Ising model was first applied to order-disorder transitions in face-centered cubic binary alloys by Fosdick,[4] and then to transitions in body-centered cubic lattices representing β-brass and β-AgZn by Flynn and McManus.[5]

There are two types of ordered states in these binary alloys. The lattice ordering described above resembles an antiferromagnetic domain, and it arises when the interaction between the two kinds of atoms is attractive. If the interaction is repulsive, or alternatively if one introduces an attraction between like atoms, a different kind of ordering, that of a two-phase system, is produced. In a phase-separated system, there are regions where atoms of one type only are found. Thus phase separation is analogous to ferromagnetism.[10]

4.4.2 Particle Kinetics and Binary Alloys

Let us consider a system composed of two different kinds of atoms, called A and B, arranged in a regular lattice. A configuration of the system of N_A atoms of type A and N_B atoms of type B, with $N_A + N_B = N$, would correspond to an array $(\mu_1, \mu_2, \ldots, \mu_N)$ where $\mu_i = +1$ if an atom of type A is present at lattice site i, and $\mu_i = -1$ if an atom of type B is located at that site. For nearest neighbors we introduce interaction energies $V_{AA}^{(1)}$, $V_{BB}^{(1)}$, and $V_{AB}^{(1)}$, and similarly for second nearest neighbors we have $V_{AA}^{(2)}$, $V_{BB}^{(2)}$, and $V_{AB}^{(2)}$. Let us form the quantities

$$u_1 = \frac{V_{AA}^{(1)} + V_{BB}^{(1)}}{2} - V_{AB}^{(1)} \qquad (4.14)$$

and

$$u_2 = \frac{V_{AA}^{(2)} + V_{BB}^{(2)}}{2} - V_{AB}^{(2)} \qquad (4.15)$$

We now generate configurations by interchanging pairs of nearest-neighbor atoms, A on lattice site S and B on site S'. The energy difference is then given by

$$\Delta E = 2(N_B^{(1)} - N_B^{(1)'} - 1)u_1 + 2(N_B^{(2)} - N_B^{(2)'})u_2 \qquad (4.16)$$

where $N_B^{(1)}$ and $N_B^{(1)'}$ are the numbers of first nearest-neighbor B atoms of S and S', respectively, before the interchange, and similarly for $N_B^{(2)}$ and $N_B^{(2)'}$. In the above we observe that if u_1 is positive, then the ordered state has a lower energy than the disordered one, whereas if u_1 is negative, then the energetics would favor phase separation of the A and B atoms. Second nearest neighbors are likely to be alike in the ordered state. Therefore negative values of u_2 enhance the ordering tendencies of positive u_i's.

A Metropolis algorithm simulating the kinetics of a binary alloy is:

1. Initialize by arranging A and B atoms into the lattice in an ordered arrangement, in a completely disordered arrangement, or randomly.
2. Select a central lattice site at random.

3. Select one of the 12 nearest-neighbor sites at random.
4. If the nearest-neighbor atom selected differs from the central atom,
   calculate the energy change resulting from an exchange of atoms;
   if negative, exchange particles;
   if positive, calculate the probability;
   select a random number — if less than the probability, switch particles;
   otherwise, do not exchange particles, but count as a new configuration.
5. If the atoms are identical, do not exchange, but count as a new configuration.

We observe that in this and the previous sampling algorithm, only one or two elements of the configuration are changed in each iteration. Methods based on making only small local changes in a configuration at each step are known as *iterative improvement approaches*. A shortcoming of such methods is that if they are performed in a deterministic or "greedy" manner, the system will become trapped in a local minimum that may be suboptimal. Examples of "greedy" algorithms are (1) accept only changes that lower the energy, or (2) check several nearest neighbors, and only switch with the one that produces the largest decrease in energy. The sampling algorithms that we are using have the important property that they accept moves that increase the energy. This stochastic property permits escape from local minima that otherwise would trap the system in undesirable states. For this reason sampling methods of the Metropolis type are sometimes called *probabilistic hill-climbing algorithms*. This ability plays an important role in systems that possess many such local minima, such as spin glasses and systems undergoing combinatorial optimization.

4.5 SPIN GLASSES

4.5.1 Introduction

Spin glasses are materials in which interactions between elements are random and conflicting. Spin glass phenomena were first identified in certain dilute metallic alloys with substitutional magnetic impurities (e.g., CuMn, AuFe, and AgMn). They have been found since then in a wide variety of substitutionally and topologically disordered systems. In dilute magnetic alloys, the term "spin glass" refers to a low-temperature phase in which local magnetic moments are frozen into spatially random equilibrium orientations. In a metallic spin glass, conduction electrons mediate an indirect exchange interaction between the local moments. This exchange interaction, known as the Ruderman-Kittel-Kasuya-Yosida (RKKY) interaction, has an oscillatory spatial dependence. As a result of this interaction the moments experience a distribution of competing ferromagnetic and antiferromagnetic forces. Below the critical temperature the orientations appear random, and there is no long-range order, but each spin is in fact aligned in a preferred direction consistent with that of the energy minimum.

Spin glasses been studied intensively since the late 1960s. A crucial finding was the observation in 1972 by Cannella and Mydosh[11] of a sharp cusp in the magnetic susceptibility for AuFe, signaling the onset of magnetic ordering. These and other related experimental observations were followed by a series of studies that established a theoretical framework for the phenomena. In the first of these, by Edwards and Anderson,[6] the existence of a spin glass phase was demonstrated using a replica procedure. That work was closely followed by the theoretical investigations of Grinstein and Luther,[12] and by Emery,[13] using replica plus renormalization group methods, and by the investigation, by Sherrington and Kirkpatrick,[7] in which a mean-field theory was defined as an infinite-ranged version of the Edwards-Anderson model. These papers were followed by the study of Thouless, Anderson, and Palmer,[14] in which low-temperature problems in the Sherrington-Kirkpatrick model were removed, a longer paper by Kirkpatrick and Sherrington,[15] and the replica solutions of Parisi.[16,17]

In the Edwards-Anderson and Sherrington-Kirkpatrick models, the spatial oscillations are neglected, but the important aspects, the randomness and the competing ferromagnetic and antiferromagnetic interactions, are retained. Let us suppose that we have a set of classical dipoles arranged on a regular lattice. These dipoles, or spins, are characterized by a spin vector s_i, where the index i denotes the lattice site. In the Edwards-Anderson (EA) model, the hamiltonian is of the Hiesenberg form with nearest-neighbor interactions

$$H_{\{J\}}(\{s\}) = - \sum_{\langle i,j \rangle} J_{ij} \mathbf{s}_i \cdot \mathbf{s}_j \qquad (4.17)$$

here, as before, the symbol $\langle i,j \rangle$ indicates that only nearest-neighbor interactions are considered. In the Sherrington-Kirkpatrick (SK) model, the starting point is the Ising hamiltonian

$$H_{\{J\}}(\{s\}) = -\sum_{i,j} J_{ij} s_i s_j - H_{ext} \sum_i s_i \qquad (4.18)$$

and the summation extends over all pairs of spins and is therefore said to be *infinite-ranged*. In the SK model the spin variables can take on the usual $+1$ or -1 values, and H_{ext} is an external magnetic field. In both formulations the exchange integrals, J_{ij}, are assumed to be independently and identically distributed according to the Gaussian probability distribution

$$p(J_{ij}) = [(2\pi)^{1/2} \sigma_J]^{-1} \exp\left(-\frac{J_{ij}^2}{2\sigma_J^2}\right) \qquad (4.19)$$

4.5.2 Annealed and Quenched Random Variables

In the spin glass models the exchange integrals J_{ij} are treated as quenched (immobile) random variables. These are fixed at some set of selected values for all calculations, and the goal is to determine the thermodynamic quantities of interest, such as magnetization, susceptibility, internal energy, and specific heat, in a manner that is independent of the particular set of exchange integrals. To understand how this is done, let us first discuss the distinction in statistical mechanics between annealed and quenched random variables.

Suppose that we have two sets of random variables $\{S\}$ and $\{J\}$, where $\{S\}$ is the spin degree-of-freedom and $\{J\}$ is the impurity degree-of-freedom. In an annealed system, $\{J\}$ and $\{S\}$ are in thermal equilibrium with one another, and both are free to participate in the kinetic process. The partition function Z can be written as

$$Z = \text{Tr}_{\{s\},\{J\}} \exp(-\beta E(\{s\}, \{J\})) \tag{4.20}$$

In this expression we are using the trace (Tr) as a shorthand notation for the sum over all values of the random variables $\{s\}$ and $\{J\}$. Suppose now that $\{J\}$ are quenched random variables. They are not in thermal equilibrium with the $\{S\}$ variables and may no longer participate in the kinetic process. The partition function does not contain summations over configurations of the quenched variables, and it takes the form

$$Z(\{J\}) = \text{Tr}_{\{s\}} \exp(-\beta E(\{s\}, \{J\})) \tag{4.21}$$

The free energy A or logarithm of the partition function is

$$A(\{J\}) = -\beta^{-1} \ln Z(\{J\}) \tag{4.22}$$

To obtain results that are independent of the particular set of $\{J\}$ values, we make use of the macroscopic extensive character of the free energy and consider the sample to be composed of a large set of subsamples. Neglecting surface interactions, the total partition function factors into a product of partition functions for the subunits. Consequently we have

$$\langle A\{J\}\rangle = -\beta^{-1}\langle \ln Z(\{J\})\rangle \tag{4.23}$$

and upon dividing by the number of subsystems, we obtain a $\{J\}$-independent free energy that has been averaged over the quenched random variable.[18] This procedure is analogous to the construction of the canonical ensemble from the microcanonical ensemble.

4.5.3 Replicas

Introducing a probability distribution $p(J)$ for the J parameters which are distributed among the lattice sites in an unknown manner, we may rewrite the

J-averaged free energy as

$$A = \langle A(J) \rangle = \sum_J p(J) A(J) \tag{4.24}$$

The form of this expression for the free energy averaged over the quenched variable is difficult to evaluate. Edwards and Anderson avoided this difficulty by using the relation

$$\ln Z = \lim_{n \to 0} \frac{Z^n - 1}{n} \tag{4.25}$$

thereby converting the problem of evaluating the logarithm of the partition function into one requiring the evaluation of the nth power of the partition function. Formally, we introduce the quantity

$$A_n = -(\beta n)^{-1} \sum_J p(J) \{Z(J)\}^n \tag{4.26}$$

and then using the identity, Eq. (4.25), it follows that

$$\lim_{n \to 0} A_n = A \tag{4.27}$$

and we have

$$A = \langle A(J) \rangle = -\lim_{n \to 0} (\beta n)^{-1} \sum_J p(J) \{Z(J)\}^n \tag{4.28}$$

For integral values of n, we can write the nth power of the partition function as

$$\{Z(J)\}^n = \sum_{\{s^1\}} \sum_{\{s^2\}} \cdots \sum_{\{s^n\}} \exp(-\beta E[\{s^\alpha\}, \{J\}])$$

$$= \prod_{\alpha=1}^{n} \mathrm{Tr}_{\{s^\alpha\}} \exp(-\beta E[\{s^\alpha\}, \{J\}]) \tag{4.29}$$

$$= \prod_{\alpha=1}^{n} Z^\alpha$$

In the expression, the $Z^1, Z^2, \ldots, Z^n$ are a sequence of partition functions describing n identical noninteracting systems, each with the same set of J couplings. These are the replicas of the original system. Thus in the replica method there are two sets of indexes labeling the elements of the spin glass. Superscripts denote the n replicas, and subscripts indicate the N lattice sites.

For the SK model, we have the hamiltonian

$$E = -\sum_{(i,j)} J_{ij} s_i^\alpha s_j^\alpha - H_{ext} \sum_i s_i^\alpha \qquad (4.30)$$

where the summation is over all pairs of indexes and the J_{ij} are distributed according to Eq. (4.19). There are two limits that need to be taken. One is the thermodynamic limit as N goes to infinity, where we note that the quantity to be evaluated is the free energy per spin A/N. The other, usually taken at the end of the calculation, is the replica limit as n goes to zero.

4.5.4 The TAP Equations

An alternative way of solving the SK model is through the mean-field technique of Thouless, Anderson, and Palmer.[14] We recall from Chapter 3 that in the Weiss molecular field approximation the mean spin is given by a self-consistent field equation of the form (Eq. 3.64)

$$k_B T \tanh^{-1}\langle s \rangle = Jv\langle s \rangle + \mu H \qquad (4.31)$$

In the first-order, Bethe approximation, elementary sublattices are considered in which there is a central site and its nearest neighbors. The interactions of the central site element with its nearest neighbors and external field H are treated exactly, while the interactions of the neighbors with the remainder of the lattice and external field are approximated by a mean-field.

In a spin glass the J_{ij} are not the same at each site, and the medium is no longer homogeneous. As a result the mean spin at a particular site is site dependent. Thouless et al.[14] derived a mean-field approximation for the no-external-field, infiniteranged, SK spin glass, expressed in terms of the mean spin,

$$m_i \equiv \langle s \rangle_i \qquad (4.32)$$

at site i. The TAP mean-field was initially derived using a Bethe approximation[14] but was later derived in a more direct manner.[19] The mean-field must be adjusted by a factor $(1 - m_i^2)$ to remove the response at the other sites j to the spin at the central site. The resulting TAP mean-field equations are[14,19]

$$k_B T \tanh^{-1} m_i = \sum_{j=1}^{N} J_{ij} m_j - m_i \beta \sum_{j=1}^{N} J_{ij}^2 (1 - m_j^2) \qquad (4.33)$$

where the right-hand side is the mean-field. In the remainder of this section we present a synopsis of the main findings on SK spin glasses obtained using these analytic approaches, supplemented by results on critical phenomena derived through Metropolis Monte Carlo plus renormalization group methods.

4.5.5 The Parisi Order Parameter and Ultrametricity

In a spin glass, long-range order of the type found in an Ising ferromagnet does not occur. If the system is in a state a and we add together the local magnetizations from each site, we obtain a global magnetization per spin,

$$M^{(a)} = \left(\frac{1}{N}\right) \sum_{i=1}^{N} \langle s_i^{(a)} \rangle \tag{4.34}$$

that vanishes in the thermodynamic limit. However, as noted by Edwards and Anderson, the sum of the squares of the mean spin from each lattice site

$$q_{EA}^{(aa)} = \left(\frac{1}{N}\right) \sum_{i=1}^{N} \langle s_i^{(a)} \rangle^2 \tag{4.35}$$

does not vanish, and this is known as the Edwards-Anderson order parameter. A more general family of order parameters can be defined by introducing the overlap $q^{(ab)}$, between two states a and b:

$$q^{(ab)} = \left(\frac{1}{N}\right) \sum_{i=1}^{N} \langle s_i^{(a)} \rangle \langle s_i^{(b)} \rangle \tag{4.36}$$

This quantity can be interpreted geometrically as a distance between two (pure) states in phase space. If we include the weights p_a and p_b of these two states, then we can associate a probability distribution $p_J(q)$ with the overlap by forming the sum

$$p_J(q) = \sum_{a,b} p_a p_b \delta(q - q^{(ab)}) \tag{4.37}$$

This distribution of distances has been shown by Parisi[17] to serve as an order parameter for a spin glass. The Parisi order parameter and the space of states were studied further by Mézard et al.[20,21] and by Mézard and Virasoro.[22] For an Ising ferromagnet in zero external field, this order parameter has a single peak at temperatures above the Curie temperature (i.e., one state with no local magnetization) and two peaks below the critical temperature, at $+M^2$ and $-M^2$:

$$p(q) = \delta(q + M^2) + \delta(q - M^2) \tag{4.38}$$

For a spin glass the order parameter in the SK model in zero magnetic field, averaged over the samples, has two peaks, at $+q_M$ and $-q_M$, plus a continuum of intermediate values corresponding to an infinite number of equilibrium states.

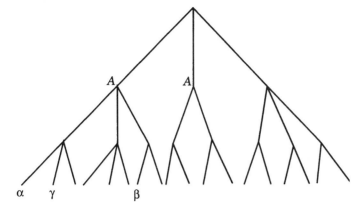

Figure 4.2. The ultrametric space of equilibrium states of a spin glass. The states $\alpha, \beta, \gamma, \ldots$ are the extremities of the branches of the tree. The distances between two states is a monotonic function of the number of steps one has to climb along the tree to find a common ancestor (From Mézard and Virasoro[22]. Reprinted with permission of *Les Editions de Physique*, France.)

The equilibrium states of the spin glass, whose overlaps are described by the Parisi order parameter, are organized in a hierarchical manner. This organization, termed *ultrametric*, is illustrated in Fig. 4.2. The general features of this structure is that if three states α, β, γ, are picked at random, the probability for their mutual overlap will be nonvanishing only if either (1) $q^{(\alpha\beta)} = q^{(\beta\gamma)} = q^{(\alpha\gamma)}$ or (2) $q^{(\alpha\beta)} = q^{(\alpha\gamma)} = q$ and $q^{(\beta\gamma)} > q$. In the first case we have an equilateral triangle in the space of states, and in the second case we have an isoceles triangle in which the unequal side is smaller than the other two. In Fig. 4.2 the extremity of each branch represents a state. The overlap (distance) between two states depends only on the number of steps we need to ascend to find their oldest common ancestor; in other words, the higher we go in the tree, the greater the distance and the smaller the overlap. As the temperature is lowered below the critical value, the various equilibrium states are generated by a process of bifurcation as depicted in Fig. 4.2.

4.5.6 Critical Behavior

In spin glasses, as in other magnetic systems, dimensional considerations are important. The results and models we have been discussing pertain to three-dimensional glasses. In contrast to the three-dimensional system, two-dimensional Ising spin glasses do not exhibit a transition to an ordered phase at a finite temperature.[23,24] Metropolis algorithms with Glauber dynamics, together with renormalization group methods, have been used in several studies to explore critical behavior of random spin glasses. The heat capacity, critical temperature, and critical exponents for two- and three-dimensional Ising

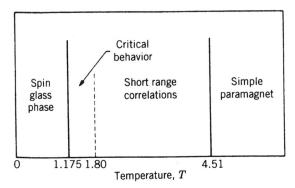

Figure 4.3. Thermal regimes of a three-dimensional spin glass. (From Ogielski[26]. Reprinted with permission of the American Physical Society.)

systems have been explored using these methods by McMillan,[25] Bray and Moore,[24] Huse and Morgenstern,[23] and Ogielski.[26]

In a detailed study of the nearest-neighbor Ising spin model with $J_{ij} = \pm J$ distributed over a simple cubic lattice, Ogielski[26] found that there are three distinct thermal regimes, each possessing a distinctive static and dynamic behavior. The three regimes are depicted in Fig. 4.3. We observe in this plot that there is a paramagnetic regime above a critical temperature, $T_c = 4.51$, a middle-, short-range correlation domain at lower temperatures merging into a critical regime at temperatures below $T_p = 1.80$, and a spin glass phase below $T_g = 1.175$. As shown in Fig. 4.4, the heat capacity for a three-dimensional spin glass has a rounded peak in the heat capacity. The middle thermal regime corresponds to the broad peak in the heat capacity. At temperatures at and

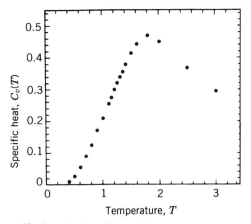

Figure 4.4. The specific heat of a three-dimensional spin glass as a function of temperature. (From Ogielski[26]. Reprinted with permission of the American Physical Society.)

below the maximum in the heat capacity, there is a rapid onset of changes in the static and dynamic properties of the three-dimensional system.

4.5.7 The Energy Landscape and Ergodicity Breaking

In an Ising ferromagnet the spontaneous magnetization is either positive or negative depending on which low-energy valley and ground-state is selected. Once the temperature is lowered so that the barrier separating the two ferromagnetic valleys can no longer be surmounted, any sequence of state transitions generated by sampling algorithms will remain within a given valley, and the larger space of states is no longer accessible. This restriction of the hypothesis or state space to a particular region is termed *ergodicity breaking*. In an SK spin glass the situation is even more complex. There are an infinite number of equilibrium states not related to one another by any obvious symmetry relation. The hypothesis space is partitioned into a large number of low-energy valleys below the phase transition temperature T_g, and these valleys are separated from one another by infinitely high barriers. In finite systems we may expect the energy landscape to be less starkly constructed, but the essential features of many valleys separated by difficult to impossible to surmount barriers would persist.

In the language of optimization theory, the task of finding the energy minima of three-dimensional Ising spin glasses is an NP-hard problem.[27] The energy minima of two-dimensional Ising spin glassses can be found in polynomial time, but if external magnetic fields are present, this problem is NP-hard too. There are close connections between combinatorial optimization problems, such as finding the optimal configurations of the traveling salesman tour and graph partitioning, and statistical physics problems, such as the identification of the energy minima of spin glasses, and we will explore these connections further in the remainder of this chapter.

The general situation emerging from studies of spin glasses may be summarized as follows: Our goal was to minimize a function, the energy, in the presence of competing interactions and randomness. In the case of a spin glass, the competing interactions were generated by the simultaneous presence of positive and negative J_{ij}'s. Disorder was present, since the couplings reflected either a random choice or one that has large complexity and so appeared to be random. These couplings did not participate in the dynamics and were said to be quenched random variables. The results of analyses of the TAP mean-field equations, as well as the infinite-ranged SK spin glass, show that there is a critical temperature below which there is a well-defined spin glass phase. In the spin glass phase there is a multiplicity of low-energy minima rather than one or two deep minima. These minima may be thought of as arising from the presence of interactions or couplings that drive the energy in different directions, thereby necessitating compromises. As a consequence (1) there are many very different configurations of the spin glass that have energies that are near the optimal, minimal value and (2) ergodicity is broken; that is, the system will become trapped in one region of configuration space, unable to escape to other valleys of comparable or lower energy. In this type of situation, an approximate method such as simulated annealing that finds near-optimal minima in polynomial time is desirable.

4.6 COMBINATORIAL OPTIMIZATION AND NP-COMPLETENESS

4.6.1 Combinatorial Optimization

In solving a combinatorial optimization problem, we must find the solution that is in some sense best from among a finite or countably infinite set of alternatives. These problems are either minimization or maximization problems that are specified by a set of problem instances. A problem instance is obtained by choosing a particular set of problem parameters. Let us consider the traveling salesman problem as a representative example. In the traveling salesman problem (TSP), the parameters are the list of N cities to be visited, and the distances d_{ij}, between any pair of cities i and j. Once these are fixed, we have a problem instance. We recall that in a tour we sequentially visit each city, each exactly once, and return to the starting city directly from the last city visited. The optimal solution is the tour, or permutation P, of the cities that gives the smallest total length L of travel through all the cities. These tours are the configurations of the system. Each configuration is an element of the set of all cyclic permutations $P = \{P(1), P(2), \ldots, P(N)\}$ where $P(i)$ for $i = 1, \ldots, N$ denotes the successor city in tour P, and $P(N + 1) = P(1)$. The optimal solution is the configuration that minimizes

$$L = \sum_{i=1}^{N} d_{i,P(i)} \tag{4.39}$$

We observe in our traveling salesman example that an optimization problem (1) contains a set of instances, (2) has a finite set of candidate solutions for each instance, and (3) has a function that assigns to each problem instance and each candidate solution a positive rational number called the solution value, cost, or energy E. An optimal solution E_0, satisfies the relation $E_0 \leq E$ for all solutions, for the chosen problem instance. For the traveling salesman problem the candidate solutions are the configurations, and the solution values are the lengths.

4.6.2 The World of NP-Completeness

There exists a hierarchical classification scheme[28] that describes the relative hardness or complexity of the various algorithms that have been found for solving combinatorial problems such as the TSP. The hierarchy includes the classes P (polynomial), NP (nondeterministic polynomial), NP-complete, and NP-hard. The optimization form of the traveling salesman problem described above as well as others such as the graph-partitioning problem (GPP), which we will discuss shortly, are NP-hard problems. These problems are considered to be intractable, meaning that no algorithm for solving any of these problems in polynomial time has been found to date. In this classification scheme the measure of hardness is taken as the time (number of elementary operations) needed by an algorithm to solve a particular instance of the problem, and the variation of that time with the problem size. The hierarchy and accompanying terminology are as follows:

A time complexity function $T(n)$ for an algorithm is the largest amount of time needed by the algorithm to solve a problem instance of a particular size or input length. Let us recall that a function $f(n)$ is said to be of $O(g(n))$ (order $g(n)$) whenever there exists positive constants c and n_0 such that $|f(n)| \leq c|g(n)|$ for all c and $n > n_0$. A polynomial-time algorithm (class P) is one whose time complexity function is of $O(n^k)$, where n denotes the input length and k is an integer constant. Any algorithm whose time complexity function cannot be so bounded is loosely called an exponential algorithm. For a $T(n)$ of exponential order there is of course a rapid increase in complexity with increasing problem size. For this reason problems requiring exponential algorithms for their solution are considered intractable.

The class of NP problems can be illustrated in terms of the following type of traveling salesman problem. Given a set of cities, the distances between them, and a bound B, does there exist a tour having length B, or less? This problem is not solvable in polynomial time. But suppose that we are given a tour, and we are asked to determine in polynomial time whether it satisfies the bound on its length. This can be done in polynomial time. We observe that verification of a proposed solution is easier than finding a valid solution. In general, the class NP may be defined as decision problems that contain two stages, a guessing stage and a polynomial-time verification stage. Again, referring to our traveling salesman problem, a guessing stage would involve picking a tour, and a verification stage would involve our taking the problem instance (input), our guess, and our executing in polynomial time the aforementioned test on the length of the tour. This test would return an affirmative answer for some guess if and only if there exists a tour with the desired length property in the problem instance. A problem is said to be of class NP if there exists some guess that leads to the deterministic checking stage to respond "yes" in polynomial time for every problem instance containing at least one solution with the desired characteristics.

All decision problems solvable by a polynomial-time deterministic algorithm (P) can also be solved by a polynomial-time nondeterministic algorithm (NP). To see that this is the case, we have only to note that if we have a polynomial-time deterministic algorithm, we may use it for the verification stage and discard the guessing portion. Thus class P is a subset of class NP, or $P \subseteq NP$. It is widely believed, but not proved, that the the two classes of problems are not the same, namely $P \neq NP$. The problems in NP but not in P are the intractable ones.

To understand the character of the class NP-complete, we invoke the notion of polynomial transformability. A polynomial transformation is a 1–1 mapping, of all instances of one NP problem into the corresponding instances of another, that (1) can be computed in polynomial time and (2) satisfies the condition that for all problem instances, there is a "yes" decision response for the problem instance prior to the mapping if and only if there is a "yes" response for the problem instance subsequent to the mapping. NP-complete problems are NP problems with the property that all other NP problems can be mapped into them through suitably chosen polynomial transformations. Thus, by solving the post-transformation NP-complete problem, we solve the NP problem that was mapped into it. In other words, the class of NP-complete problems contain those NP problems to which all others may be reduced by means of polynomial transformations, and they are therefore at least as hard as any NP problem.

In more detail we introduce the notation $\Pi_1 \to \Pi_2$ to denote that a decision problem Π_1 can be reduced to a decision problem Π_2 by polynomial transformation. We can show that if $\Pi_1 \to \Pi_2$, then $\Pi_2 \in P$ implies that $\Pi_1 \in P$ (and $\Pi_1 \notin P$ implies that $\Pi_2 \notin P$). Furthermore we can prove that if $\Pi_1 \to \Pi_2$, and $\Pi_2 \to \Pi_3$, then $\Pi_1 \to \Pi_3$. Two decision problems are considered to be polymonially equivalent if $\Pi_1 \to \Pi_2$ and $\Pi_2 \to \Pi_1$. The relation "$\to$" is formally an equivalence relation. The class P forms an equivalence class of the least equivalent (easiest) problems in NP. The class NP-complete is an equivalence class that contains the most equivalent (hardest) problems in NP. A problem is defined to be NP-complete if $\Pi \in NP$ and for all other $\Pi' \in NP$ we have $\Pi' \to \Pi$. The significance of the equivalence relation for demonstrating NP-completeness is that we do not actually have to show that all NP problems can be reduced to a candidate NP-complete problem. Instead, if Π is our candidate NP-complete problem, we simply have to demonstrate that $\Pi \in NP$ and then take a known NP-complete problem Π_c and show that $\Pi_c \to \Pi$.

To complete our discussion of the hierarchy of problem complexity, we return to our original optimization problem. We have defined the class NP in terms of decision problems. A problem to which an NP-complete problem can be reduced cannot be solved in polynomial time unless $P = NP$. This statement is valid irregardless of whether or not the problem under discussion lies in class NP. A problem is considered to be NP-hard if the existence of a polynomial algorithm for its solution implies the existence of such an algorithm for all NP-complete problems. Two representative examples of NP-hard problems are the optimization forms of the TSP and GPP.

4.7 OPTIMIZATION BY SIMULATED ANNEALING

4.7.1 Introduction

When we first described the Gibbs distribution in Section 2.8, we introduced the exponential quantity $E_i/k_B T$, where E_i denoted the energy of the ith configuration of the system. We then noted that wep could write the partition function in two different ways. In the first expression, we summed over all configurations of the system. In the second, we sorted out the various states according to their energy and then wrote the partition function as a sum over energy of terms of the form $\Omega(E) \exp(-E/k_B T)$, where E is the configuration energy and $\Omega(E)$ is the statistical weight or number of states of that energy. We then noted that the statistical weights $\Omega(E)$ increase rapidly with increasing configuration energy. These weights are related to the entropy, in particular, $S(E) = \ln \Omega(E)$, and since there are many more disordered states than ordered low-energy ones, at typical elevated temperatures the factor $\Omega(E)$ favors disordered states. At low temperatures the situation changes. The statistical weights decrease in importance, and the low-energy-ordered states are more strongly favored by the Boltzmann factor. As T is lowered, a Monte Carlo simulation based on these distributions will sample configurations with energies progressively nearer to the ground-state energy provided that the energy changes are sufficient compared to $k_B T$ and to the statistical weights.

These observations form the basis for simulated annealing in which we endow the system under study with an artificial stochastic dynamics and gradually lower the temperature to locate optimal low-energy configurations. Simulated annealing was formally introduced by Kirkpatrick, Gelatt, and Vecchi,[8] and independently by Cerny,[9] as a Monte Carlo–Metropolis method for finding the global minima of objective, or cost, functions of large and complex systems. The approach is based on the above-mentioned observation that the formalism of statistical mechanics used to study thermodynamic properties of physical systems can be applied to solve hard, NP-complete combinatorial optimization problems in other types of systems.

The Metropolis criterion plays an important role in the statistical physics model of optimization. We recall from our earlier discussion at the end of Section 4.4 that configurations that result in an increase, ΔE in the energy of the system are accepted with probability $\exp(-\Delta E/k_B T)$. Moves of this type that raise the energy are valuable because they allow the system to escape from shallow, local minima, whereas, if the only moves allowed were those that lower the energy, trapping might occur. These occasional escapes are most likely to occur at elevated temperatures, and in this regime many configurations and valleys may be sampled. This ability to locate deep minima while avoiding metastable states is particularly important in systems possessing complicated energy landscapes, such as the spin glass systems just discussed. In the next section we will explore two NP-hard problems of this type — the traveling salesman problem and the graph-partitioning problem. In each of these problems, we will exploit the connection between combinatorial optimization and statistical mechanics. Before we begin these examinations, we will formalize our annealing algorithm.

4.7.2 The Simulated Annealing Algorithm

The first element of our simulated annealing algorithm is the familiar Gibbs probability distribution and accompanying partition function. In adapting these forms to optimization, the Boltzmann constant will be neglected. We then have the probability distribution

$$p(E = E_i) = \frac{1}{Z} \exp\left(-\frac{1}{T} E_i\right) \qquad (4.40)$$

and partition function

$$Z = \sum_i \exp\left(-\frac{1}{T} E_i\right) \qquad (4.41)$$

In simulated annealing, we interpret the energies E_i as a numerical cost that assigns to each configuration a scalar value representing the desirability of that configuration. The temperature T becomes a control parameter. The next ingredient is the identification of configurations, and the generation of new ones from previous ones in a local manner. This is our iterative improvement or small-step kinetics discussed earlier. As done for the binary allow and spin systems, we must sample configurations in an unbiased manner so that the a priori transition probabilities satisfy the condition $\tau_{ij} = \tau_{ji}$. We again use the Metropolis, heat bath, or equivalent sampling algorithm for selecting or rejecting the moves to new configurations. If, for clarity, we omit to write the a priori transition probabilities, our algorithm assumes the familiar form

$$p_{ij} = \begin{cases} 1, & \Delta E_{ij} \leq 0 \\ \exp\left(-\frac{\Delta E_{ij}}{T}\right) & \Delta E_{ij} > 0 \end{cases} \qquad (4.42)$$

where $\Delta E_{ij} = E_j - E_i$ is the change in cost involved in the transition from state i to state j. Thus we do not have to evaluate the partition function in simulated annealing. We may summarize the correspondence between the terminology of statistical physics and combinatorial optimization by the entries in Table 4.1.

TABLE 4.1 Correspondence between statistical physics and combinatorial optimization

Physics	Optimization
Sample	Problem instance
Configuration (state)	Configuration
Hamiltonian (energy)	Cost function
Temperature	Control parameter
Ground-state energy	Minimum cost
Ground-state configuration	Optimal configuration

In metallurgy, annealing is a process of heating and subsequent cooling of a metal or metal alloy to modify its microstructure and endow the finished product with a set of desired mechanical and physical characteristics. There are several important parameters that govern metallurgical annealing. These include the temperature to which the metal or alloy is heated (annealing temperature), the duration of time that the elevated temperature is maintained (annealing time), and the rate at which the temperature is lowered (cooling schedule). These parameters will vary with the type of material to be annealed and with the sets of desired properties. These considerations have their counterparts in simulated annealing. We must select an appropriate initial temperature, final temperature, and annealing schedule for the problem under consideration.

4.7.3 The Cooling Schedule and Convergence of the Algorithm

We recall that at a given temperature the Metropolis algorithm generates a sequence of transition probabilities which forms a homogeneous Markov chain. These sequences of transition probabilities, constructed by means of the Metropolis, heat bath, or equivalent method, converge to unique Gibbsian stationary distributions of the form given by Eq. (4.40). There are two ways in which we can devise a Metropolis-like algorithm for simulating an annealing process in which we lower the temperature. One method is to construct a sequence of homogeneous Markov chains, one for each temperature. In every sequence configurations of the system would be sampled until equilibrium is reached. The second approach is to lower the temperature at regular intervals, thereby generating a single large nonhomogeneous Markov chain. In either method, departures from equilibrium can be made minimal by taking small temperature steps.

The nonhomogeneous and homogeneous Markov chains will converge to a set of configurations of minimal energy provided that the temperature is lowered no faster than logarithmically. That is, if the temperature T_k of the kth temperature step in the iteration satisfies the condition

$$T_k \geq \frac{\Gamma}{\log(1 + k)} \quad (4.43)$$

where Γ is a positive constant, then the transition probabilities for a given problem instance will converge to a stationary distribution corresponding to a global minimum. This result was first derived by Geman and Geman.[29] The convergence properties of the simulated annealing algorithm was investigated further by Gidas,[30] Romeo and Sangiovanni-Vincentelli,[31] Hajek,[32] and others. A summary and discussion of some of the annealing convergence theorems has been presented by Laarhoven and Aarts.[33] It is already clear from our discussion of spin glasses that the situation for NP-complete systems is complex. Factors identified in these studies that influence the annealing schedule and the nature of the low-energy states found include ergodicity breaking, the heights of the local minima, and the presence or absence of phase transitions.

The simplest cases are those where the Markov chains are ergodic so that all states are accessible from any starting configuration and where there is a single global minimum. In these situations the logarithmic schedule will ensure convergence to a set of configurations of minimal energy. If the systems are not ergodic, as in the case of spin glasses (Section 4.5.7), the logarithmic annealing schedule may still guarantee convergence to a global minimum. However, the resulting stationary distributions may depend on initial configuration or problem instance; in other words, the Markov chains will retain some memory of their starting points. Various cases and the convergence properties of the corresponding Markov chains are discussed by Gidas.[30]

Cooling schedules have been studied in spin glasses and other disordered systems, and in the traveling salesman problem, using Monte Carlo simulations techniques. In the studies by Grest, Soukoulis, and Levin,[34] Huse and Fisher,[35] and Randelman and Grest,[36] the annealing schedule was explored by examining the residual energy as a function of the cooling rate r defined as

$$r = \frac{\Delta T}{m} \quad (4.44)$$

where ΔT is the change in temperature made after running the algorithm for a fixed number m of Monte Carlo steps (random rearrangements) per element. The residual energy $\varepsilon(\tau)$ is defined as

$$\varepsilon(\tau) = \langle E \rangle_\tau - E_0 \quad (4.45)$$

where $\langle E \rangle_\tau$ is the expected energy per spin upon reaching $T = 0$ in time τ. The residual energy is the amount by which the energy exceeds the true ground-state energy per spin E_0. For the NP-complete problems considered a logarithmic dependence of the residual energy upon the cooling schedule was found by Grest, Soukoulis, and Levin and by Randelman and Grest. A more general result was given by Huse and Fisher who find that for frustrated systems, the residual energy has a logarithmic dependence

$$\varepsilon(\tau) \approx (\ln \tau)^{-\kappa} \quad (4.46)$$

where the exponential factor κ depends upon the particular system under study. The essential result from the studies by Grest, Soukoulis, and Levin and by Huse and Fisher is that we find that NP-complete (hard) problems must have a slower than power-law relaxation with any physical dynamics. Finally, we observe in a study of a one-dimensional Ising spin glass by Ettelaie and Moore[37] that the residual entropy approaches zero logarithmically with the inverse cooling rate.

The second factor influencing the annealing schedule, the relative heights of the local minima, does so in several ways. It is clear that the temperature must be sufficiently elevated to allow passage in a reasonable number of trials over the barriers associated with the local minima. This provides a condition on the selection of the constant Γ appearing in Eq. (4.43). These considerations have been incorporated into several of the convergence theorems, most notably by Hajec.[32] Another link between the local minima and the annealing schedule was provided by Huse and Fisher[35] who give a heuristic argument for the logarithmic schedule in terms of the decay rate from metastable states at late times. (One further criterion that is useful in setting the initial temperature is to accept a substantial fraction of the proposed transitions at the beginning of the annealing process.)

The third property of the system under study that must be taken into account when devising an annealing schedule is the presence or absence of phase transitions. We recall from Chapters 2 and 3 that the specific heat C_E provides information about the fluctuations in energy that signal the onset of phase transitions. The relevance of this information for simulated annealing is that we must anneal slowly in the vicinity of phase transitions. From Chapter 2 we have

$$C_E = \frac{d\langle E \rangle}{dT} = \frac{\sigma_E^2}{T^2} \tag{4.47}$$

where σ_E represents the energy fluctuations. Several different kinds of phase behavior may take place, and in some instances the existence and nature of the transition may be difficult to establish. In some systems large-scale rearrangements in the vicinity of the critical temperature produce rapid fluctuations in cost (energy), resulting in a discontinuous mean cost. In other cases less drastic changes may take place in which the energy changes smoothly, and there are discontinuous changes in specific heat. We will discuss phase transitions again at the end of Section 4.9.

Still another consideration is the determination of a stop time. One quantity that may help in evaluating the approach of the simulated annealing process to an optimal configuration is the entropy. We recall that we may determine the entropy from the Helmholtz free energy or alternatively from the specific heat. Since

$$\frac{dS}{dT} = \frac{C_E(T)}{T} \tag{4.48}$$

we may determine the entropy at some temperature of interest by integrating the expression

$$S(T_1) - S(T_2) = \int_{T_1}^{T_2} \frac{C_E(T)}{T} dT \tag{4.49}$$

134 SIMULATED ANNEALING

after fixing the second temperature at a high level where the entropy is known. The entropy and specific heat are expected to become vanishingly small as the optimal configuration is approached.

4.8 THE TRAVELING SALESMAN AND GRAPH-PARTITIONING PROBLEMS

4.8.1 Traveling Salesman Problems

We will now use the traveling salesman and graph-partitioning problems to frame two representative simulated annealing algorithms. We recall from Chapter 1 that in the traveling salesman problem we have a set of N cities. We are given the distance between each pair of cities, and a tour, the quantity to be optimized, is a closed path visiting each city once and only once. In this problem the TSP tours as the configurations, the summed length L is our cost function or energy, and we assume that we have a problem instance. We must now devise an iterative method of generating valid configurations. One way of selecting configurations is to choose a pair of cities in the tour and then reverse the order that the cities in-between the selected pair are visited. This procedure, known as *2-opt*, generates new tours in a local manner, and satisfies our requirement on the a priori transition probabilities. The 2-opt procedure is illustrated in Fig. 4.5. This method supplemented by the ancillary requirement that the cities selected should either lie in the same subregion, or adjacent subregions,[38] completes our selection procedure. To summarize, in the traveling salesman problem we are given N points (cities), $i = 1, \ldots, N$, and a set of point-to-point distances d_{ij}. The objective is to find a tour or permutation of the order in which the cities are visited, that is, of minimal length. The total length of the tour L_P is

$$L_P = \sum_{i=1}^{N} d_{P(i)P(i+1)} \tag{4.50}$$

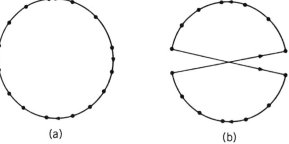

(a) (b)

Figure 4.5. The Opt-2 procedure illustrated for a sequence of cities arranged on circles: (a) Original tour in which each city is visited in turn in a clockwise manner; (b) a new tour generated by reversing the order of visitation of the cities in-between a pair of selected cities.

where P denotes a permutation and $P(N + 1) = P(1)$, and the partition function is

$$Z = \sum_P e^{-L_P/T} \qquad (4.51)$$

A simulated annealing algorithm for the TSP is as follows:

1. Establish an annealing schedule.
2. Initialize by randomly selecting a tour.
3. Use 2-opt supplemented by the pair subregion criterion to randomly choose city pairs.
4. Calculate the energy difference produced by reversing paths between the cities.
5. If ΔE is zero or negative, accept the new tour.
6. If ΔE is positive, accept the change with probability given by the Metropolis expression.
7. Select another city pair, and repeat steps 3 to 6 until the requisite number of iterations are completed.
8. Lower the temperature, and repeat steps 3 to 7.

The properties of the traveling salesman problem has been studied in some detail. In the studies by Kirkpatrick, Gelatt, and Vecchi and by Bonomi and Lutton,[38] the the city-to-city distances d_{ij} are euclidean distances between points uniformly distributed in a square. In another set of investigations, by Vannimenus and Mézard,[39] Kirkpatrick and Toulouse,[40] and Mézard and Parisi,[41] the triangular correlation is relaxed, and the d_{ij}'s are independent random variables. In both forms of TSP, we observe that the d_{ij} are quenched random variables.

In the work by Vannimenus and Mézard,[39] the replica method was used to show that there are two different temperature regimes in a solvable spin glass model. These differ from one another in the way the thermodynamic parameters scale with the number of points N. In the high-temperature regime, the free energy is given by the annealed approximation, there are no phase transitions, and the average tour length scales with N. In the low-temperature regime, a spin-glass-like phase transition is found, and the average tour length varies as $N^{(1-1/D)}$, where D is the dimension of the unit cube. In the study by Kirkpatrick and Toulouse,[40] evidence was presented for freezing due to frustration and for a hierarchical, ultrametric structure in configuration space. The existence of a phase transition at a critical temperature $T_c = 1$ was established by Mézard and Parisi[41] using the replica method. Below this critical temperature ergodicity is broken. From these studies it begins to appear that ergodicity breaking is a characteristic feature of NP-complete problems.

4.8.2 Graph Partitioning

In our graph-partitioning problem we want to distribute N circuits between a pair of computer chips. We introduce indexes i and j to denote any pair of circuits, and denote by α_{ij} the number of signals between these circuit elements. We then specify the chip upon which circuit i is located by the variable μ_i which carries one of two chip labels, $+1$ or -1. A configuration of our system of circuits is specified by listing the N circuit labels $\{\mu_i\}$. We are interested in finding an optimal circuit configuration, that is, a partitioning of the circuits among the two chips that satisfies the following two constraints. First, we wish to minimize the number of signals N_{cp} that must cross chip boundaries. This number is given by the expression

$$N_{cp} = \sum_{i<j} \frac{a_{ij}}{4} (\mu_i - \mu_j)^2 \tag{4.52}$$

Second, we want to maintain a balance between the number of circuits on the two chips. In the absence of the first constraint, the goal would be to have the same number of circuits on the two chips. The difference in the number of circuits on the chips D_{cp} is obtaining by counting the μ_i's:

$$D_{cp} = \sum_i \mu_i \tag{4.53}$$

If we square this second term and add a Lagrange multiplier λ to control the balance between the two constraints, we obtain an energy (cost function) E that embodies our two constraints

$$E = N_{cp} + \lambda D_{cp}^2 = \sum_{i<j} \left(\lambda - \frac{a_{ij}}{2}\right) \mu_i \mu_j \tag{4.54}$$

where we have dropped all constant terms, which emerge from taking sums, of the form

$$\sum_i \mu_i^2 \tag{4.55}$$

The energy function given in Eq. (4.54) is similar to that of a hamiltonian for an Ising ferromagnet, in which the α_{ij} play the role of the coupling strengths and in which a long-range, repulsive, antiferromagnetic interaction has been added.

In the above formulation of the graph-partitioning problem, we have incorporated all constraints into the hamiltonian. The second constraint, on the balance of chips between the two circuits, is not enforced absolutely but instead is imposed with strength λ relative to the first term. This form of optimization, called *constrained optimization*, will be encountered in the next two chapters where we discuss the patterning of neural connections, which arises during development of the central nervous system, and image-processing problems. This soft-constraint formulation may be contrasted with our statement of the traveling salesman problem where we demanded absolutely that each configuration be a tour. At sufficiently high temperatures the constraint requiring the configuration to be a tour may be neglected. This approach,[8,42] when applied to disordered systems, corresponds to taking the annealed approximation in place of the quenched approximation. If we distinguish between interaction energies and constraint terms, calling the former E and the latter U, then the hamiltonian for constrained optimization assumes the form $E + \lambda U$. If we wish to enforce the constraints strongly, then in the simulated annealing process λ should become large (infinite) as the temperature is lowered. In the problems to be discussed in the next two chapters, the interaction energies express the constraints of the problem, and the Lagrange multipliers serve as control parameters that govern the relative strengths of the interactions.

The graph-partitioning problem has been analyzed using replica methods by Fu and Anderson.[42] We have just noted the two competing constraints in graph partitioning. These give rise to frustration and the accompanying spin glass phenomena. Fu and Anderson find that there is a spin-glass-like phase transition below which there is an ultrametric structure in configuration space and ergodicity is broken. These results are similar to those observed in the traveling salesman problem, and they serve to reinforce our inferences on the nature of the dynamics in NP-complete systems.

4.9 MICROCANONICAL ANNEALING

The annealing methods we have been studying are based on the Gibbs sampler, the Metropolis algorithm, or the heat bath. These are all canonical representations. An alternative method, based on a microcanonical ensemble, has been developed by Creutz.[43] In his approach a "demon" travels about the system, executing a random walk on the surface of a constant energy sphere in configuration space. In the process of wandering about, the demon picks up and deposits energy at the various sites.

We recall that the microcanonical counterpart of the canonical partition function Z is the number or density of states $\Omega(E)$ for a predetermined total energy of the system E. The sum over all states i of the various Boltzmann factors, $\exp(-\beta E_i)$, becomes the sum over all states of a delta function with argument $E - E_i$ that counts those states with the correct total energy:

$$\Omega(E) = \sum_i \delta(E_i - E) \qquad (4.56)$$

138 SIMULATED ANNEALING

We may modify this expression by introducing a demon that tranfers energy as it changes the dynamic variables. If we denote the demon energy as E_D, then the above sum is replaced by

$$\Omega(E) = \sum_i \delta(E_i + E_D - E) \qquad (4.57)$$

If the demon's energy remains much less than the total energy available to the system, then at equilibrium, the demon's energy will become exponentially distributed,

$$p(E_D) \propto \exp(-\beta E_D) \qquad (4.58)$$

The exponential dependence upon the demon energy is a general result that holds whenever a subsystem, of energy E_D is brought into thermal contact with a much larger system with energy E_i such that $E_D \ll E_i$. The energy in the subsystem may fluctuate as a result of the thermal contact, but the sum $E = E_D + E_i$ remains constant. If the subsystem is prepared in a state of energy E_D then the number of states accessible to the heat bath is

$$\Omega(E_i) = \Omega(E - E_D) \qquad (4.59)$$

The probability of finding the demon in a state with energy E_D is proportional to this number of states:

$$p(E_D) \propto \Omega(E_i) = \exp\{\ln \Omega(E - E_D)\} \qquad (4.60)$$

If we now expand the $\ln \Omega(E)$ in a Taylor's series retaining only the first two terms, we get

$$\ln \Omega(E - E_D) = \ln \Omega(E) - \frac{\partial \ln \Omega}{\partial E} E_D = \ln \Omega(E) - \beta E_D \qquad (4.61)$$

and since $\Omega(E)$ is a constant, Eq. (4.58) follows. In forming this Taylor's series, we chose to expand about the logarithm of $\Omega(E)$ rather than $\Omega(E)$. This choice was motivated by noting that the logarithm of $\Omega(E)$ is likely to be a more slowly varying function of E than is $\Omega(E)$.

The microcanonical annealing algorithm is as follows:

1. Initialize the system in a state i, and initialize the demon's energy at zero or some positive number.
2. Select a site at random, in the manner of a raster scan or some other way.
3. Make a local change in the state at that site, and calculate the corresponding energy difference ΔE.

4. If ΔE is negative, accept the change and add the energy ΔE to the demon's energy E_D.
5. If ΔE is positive, accept the change provided that $\Delta E \leq E_D$, and decrement the demon's energy by ΔE. (If $\Delta E > E_D$, reject the change.)
6. Select another site, and repeat steps 3 to 5 until an equilibration criterion is met.
7. Remove a small amount of energy from the demon, and repeat steps 2 to 6.

In microcanonical annealing, the total energy is allowed to decrease (step 7) in contrast to canonical annealing where the temperature is lowered. The microcanonical algorithm can be transformed into a canonical Metropolis procedure by replacing the demon's energy at each step by a new value randomly selected with a Boltzmann weight. Another way of generating a Metropolis algorithm is through the introduction of a large number of demons.

Microcanonical annealing is one of a number of alternatives to the Metropolis type of Monte Carlo algorithm. One of the problems encountered in using a Metropolis type of algorithm is the phenomenon of *critical slowing down* in the vicinity of a critical point. In Chapter 3 we discussed the appearance of long-range correlations (fluctuations) near critical points for second-order phase transitions. Two consequences of their appearance are that relaxation time τ for these correlations increases rapidly with the linear dimension of the system and the system equilibrates slowly. If L is the linear dimension of the system, then τ increases as

$$\tau \sim L^z \tag{4.62}$$

where z is a critical exponent whose value is generally near or greater than 2.0. In studying critical phenomena, the slow probing of local changes by Metropolis-type algorithms has led to the development of cluster methods by Swensen and Wang[44] and by Wolff[45] in which groups of spins are changed in a single move. In a further development by Creutz,[46] cluster methods were combined with the microcanonical transfer of energy to produce a fast hybrid algorithm. Other alternative methods that have been developed include multi-grid Monte Carlo methods[47] and approaches[48] based on the Langevin equation, which we will discuss in the next section and in Chapter 7.

Microcanonical annealing has been applied to a study of an Ising spin system by Bhanot et al.,[49] and has been used in a stochastic optimization approach to stereo matching by Barnard.[50] In the former study, difficulties associated with convergence and ergodicity of the algorithm were addressed. In the latter investigation, an interaction energy suggested by the regularization theory was used together with a multiscale technique to encode an disparity map that specifies the correspondence between stereo image pairs.

4.10 CONTINUOUS SIMULATED ANNEALING

We now examine a class of global optimization algorithms called *Langevin diffusions*, or *continuous simulated annealing*. These algorithms, like simulated annealing, generate sequences of transition probabilities that converge to equilibrium Gibbs distributions and that, given an appropriate cooling schedule, permit us to find those states corresponding to a near-optimal global minimum of a potential. This class of algorithms is based on finding a solution to the stochastic differential equation

$$dy(t) = -\nabla U(y)dt + \sqrt{2T(t)}\,dW \qquad (4.63)$$

In this equation the variable y, which can assume any real value in a specified range, characterizes the states of the system. The parameter t represents time, U is the potential to be minimized, and T is a time-dependent temperature that specifies the cooling schedule and controls the magnitude of a stochastic noise process W.

This stochastic differential equation is a generalization of the Langevin equation introduced as a mathematical model of brownian motion. We will discuss the Langevin equation and the associated Fokker-Planck equation describing the time evolution of the transition probabilities in detail in Chapter 7. The salient feature of this equation is the inclusion of a stochastic noise term along with the gradient of the potential U. In the absence of the random noise term, Eq. (4.63) would generate a sequence of states through a deterministic gradient descent resulting in one of a set of values for the variable y corresponding to a local minimum of U. The addition of the noise term converts the algorithm into one that permits uphill transitions, thereby making an escape from local minima possible.

In more detail, let us suppose that U is a twice differentiable function of y. Let us further suppose that the minimim value of $U(y)$ is zero and that $U(y)$ grows properly as $|y|$ is increased. That is, it satisfies the conditions

$$U(y) \to \infty, \quad |\nabla U(y)| \to \infty, \quad |y| \to \infty \qquad (4.64)$$

$$\lim_{|y| \to \infty} |\nabla U(y)|^2 - \Delta U(y) > -\infty \qquad (4.65)$$

where Δ is the Laplacian. We next define the Gibbs distribution $\pi_T(y)$ as

$$\pi_T(y) = \frac{1}{Z_T(y)} \exp\left(-\frac{U(y)}{T}\right) \qquad (4.66)$$

where $Z_T(y)$ is the partition function for the (multivariate) continuous variable

y, namely

$$Z_T(y) = \int_{\Re^n} \exp\left(-\frac{U(y)}{T}\right) dy < \infty \qquad (4.67)$$

and $0 < 2T(t) < 1$. It can be shown that as $T(t)$ approaches zero, $\pi_T(y)$ converges weakly to a distribution $\pi(y)$ which, as in the discrete case, approaches a delta function and concentrates the probabilities for different configurations on those corresponding to a minimum of the potential U. The main result that follows from the above is that the transition probabilities converge to $\pi(y)$ as t approaches infinity provided that $T(t)$ follows a logarithmic cooling schedule: $T(t) = c/\ln t$ for t large, and $c > c_0$, where c_0 is a positive constant. This result (theorem) tells us that the solution to the stochastic differential equation, namely the associated transition probabilities for the variable y, converges to an equilibrium Gibbs distribution as the temperature is lowered. The large random fluctuations present at early times enable the system to escape from shallow local minima, and the system eventually settles into a prominent minimum of the potential U.

Several theorems establishing these and related results were proved by Aluffi-Penzini et al.,[51] Gidas,[52] Geman and Hwang,[53] and Chiang et al.[54] Langevin diffusions were first used for global optimization by Aluffi-Pentini et al. A variety of potentials were considered. Some of these potentials had multiple local and global minima, while others possessed multiple local minima and a single global minimum. Geman and Hwang considered the problem of finding the global minima of $U:[0,1]^d \to \Re$. In their study they were able to guarantee convergence to a unique solution by introducing the notion of a reflecting boundary. Chiang et al. investigated the global minima of $U:\Re^d \to \Re$. In this work the authors provided insight into the determination of the constant c_0 appearing in the annealing schedule. The Langevin equation can be generalized to include the effects of correlated sampling noise. This type of dynamics was studied by Kushner[55] who used the theory of large deviations to examine mean escape times for neighborhoods of one metastable state to another. This dynamics was also studied by Gelfand and Mitter[56,57] who incorporated the effects of additive sampling noise upon ∇U.

Both the Markov chains generated by the Metropolis-like simulated annealing algorithm and the continuous-state chains produced by Langevin diffusions converge to Gibbs distributions at a fixed temperature. If we convert the Metropolis and heat bath algorithms into continuous-state methods, then the question arises as to the relationship between the dynamics generated by simulated annealing and Langevin diffusions. One answer to this question, found by Gelfand and Mitter,[58] is that there is a class of continuous-state versions of the simulated annealing algorithm that converge in distribution to Langevin diffusions. This result shows that the artificial stochastic dynamics introduced in simulated annealing is related to the dynamics of a physical system not far from equilibrium, such as a system of brownian particles in contact with a heat reservoir.

4.11 FURTHER READING

Monte Carlo sampling methods have been used extensively to study the thermodynamic properties of Ising and other cooperative physical systems. The free energy, heat capacity, critical temperature, and static and dynamic critical points have been mapped and, whenever possible, compared to exact numerical results. A good review of earlier Monte Carlo work is presented in the monograph edited by Binder.[59] The book by Mézard, Parisi, and Virasoro[60] contains a valuable review of spin glass theory. The authors discuss replica symmetry breaking and the cavity method, in addition to replicas and TAP theory. This volume also contains reprints of some of the main theoretical papers on the subject.

4.12 REFERENCES

1. Toulouse, G. (1977). Theory of the frustration effect in spin glasses: I. Commun. Phys. **2**, 115–119.
2. Metropolis. N., Rosenbluth, A. W., Rosenbluth, M. N., Teller, A. H., and Teller, E. (1953). Equation of state calculations by fast computing machines. J. Chem. Phys., **21**, 1087–1092.
3. Glauber, R. J. (1963). Time-dependent statistics of the Ising model. J. Math. Phys. **4**, 294–307.
4. Fosdick, L. D. (1959). Calculation of order parameters in a binary alloy by the Monte Carlo method. Phys. Rev., **116**, 565–573.
5. Flynn, P. A., and McManus, G. M. (1961). Monte Carlo calculation of the order-disorder transformation in the body-centered cubic lattice. Phys. Rev., **124**, 54–59.
6. Edwards, S. F., and Anderson, P. W. (1975). Theory of spin glasses. J. Phys. F. Metal Phys., **5**, 965–974.
7. Sherrington, D., and Kirkpatrick, S. (1975). Solvable model of a spin glass. Phys. Rev. Lett., **35**, 1792–1796.
8. Kirkpatrick, S., Gelatt, C. D., and Vecchi, M. P. (1983). Optimization by simulated annealing. Science, **220**, 671–680.
9. Cerny, V. (1985). Thermodynamical approach to the traveling salesman problem: An efficient simulation algorithm. J. Opt. Th. Appl., **45**, 41–51.
10. Flynn, P. A. (1974). Monte Carlo calculation of phase separation in a two-dimensional Ising system. J. Stat. Phys., **10**, 89–97.
11. Cannella, V., and Mydosh, J. A. (1972). Magnetic ordering in gold-iron alloys. Phys. Rev., **B11**, 4220–4235.

12. Grinstein, G., and Luther, A. H. (1976). Application of the renormalization group to phase transitions in disordered systems. Phys. Rev., **B13**, 1329–1343.
13. Emery, V. J. (1975). Critical properties of many-component systems. Phys. Rev., **B11**, 239–247.
14. Thouless, D. J., Anderson, P. W., and Palmer, R. G. (1977). Solution of "solvable model of a spin glass." Philos. Mag., **35**, 593–601.
15. Kirkpatrick, S., and Sherrington, D. (1978). Infinite-ranged models of spin-glasses. Phys. Rev. **B17**, 4384–4403.
16. Parisi, G. (1979). Infinite number of order parameters for spin glasses. Phys. Rev. Lett., **43**, 1754–1756.
17. Parisi, G. (1983). Order parameter for spin-glasses. Phys. Rev. Lett., **50**, 1946–1948.
18. Brout, R. (1959). Statistical mechanical theory of a random ferromagnetic system. Phys. Rev., **115**, 824–835.
19. Anderson, P. W. (1979). Lectures on amorphous systems. In R. Ballian, R. Maynard and G. Toulouse (eds.), *Ill-Condensed Matter, Les Houches XXXI*. Amsterdam: North-Holland, pp. 159–261.
20. Mézard, M., Parisi, G., Sourlas, N., Toulouse, G., and Virasoro, M. (1984). Nature of the spin-glass phase. Phys. Rev. Lett., **52**, 1156–1159.
21. Mézard, M., Parisi, G., Sourlas, N., Toulouse, G., and Virasoro, M. (1984). Replica symmetry breaking and the nature of the spin glass phase. J. Physique (Paris), **45**, 843–854.
22. Mézard. M., and Virasoro, M. A. (1985). The microstructure of ultrametricity. J. Physique (Paris), **46**, 1293–1307.
23. Huse, D. A., and Morgenstern, I. (1985). Finite-size scaling study of the two-dimensional Ising spin glass. Phys. Rev., **B32**, 3032–3034.
24. Bray, A. J., and Moore, M. A. (1985). Critical behavior of the three dimensional Ising spin glass. Phys. Rev. **B31**, 631–633.
25. McMillan, W. L. (1985). Domain-wall renormalization-group study of the three-dimensional random Ising model at finite temperature. Phys. Rev. **B31**, 340–341.
26. Ogielski, A. T. (1985). Dynamics of three-dimensional Ising spin glass in thermal equilibrium. Phys. Rev. **B32**, 7384–7398.
27. Barahona, F. (1982). On the computational complexity of Ising spin glass models. J. Phys. **A15**, 3241–3253.
28. Garey, M. R., and Johnson, D. S. (1979). *Computers and Intractability: A Guide to the Theory of NP-Completeness*. New York: W. A. Freeman.
29. Geman, S., and Geman, D. (1984). Stochastic relaxation, Gibbs distributions and the Bayesian restoration of images. IEEE Trans. Pattern Anal. Machine Intell, **PAMI-6**, 721–741.
30. Gidas, B. (1985). Nonstationary Markov chains and convergence of the annealing algorithm. J. Stat. Phys., **39**, 73–131.
31. Hajec, B. (1988). Cooling schedules for optimal annealing. Math. Operations Res., **13**, 311–329.
32. Romeo, F., and Sangiovanni-Vincentelli, A. (1985). Probabilistic hill climbing algorithms: Properties and applications. In *Proceedings 1985 Chapel Hill Conference on VLSI*, pp. 393–417.

33. Laarhoven, P. J. M. van, and Aarts, E. H. L. (1987). *Simulated Annealing: Theory and Applications*. Dordrecht: Kluwer.
34. Grest, G. S., Soukoulis, C. M., and Levin, K. (1986). Cooling-rate dependence for the spin-glass ground-state energy: Implications for optimization by simulated annealing. Phys. Rev. Lett., **56**, 1148–1151.
35. Huse, D. A., and Fisher, D. S. (1986). Residual energies after slow cooling of disordered systems. Phys. Rev. Lett., **57**, 2203–2206.
36. Randelman, R. E., and Grest, G. S. (1986). N-city traveling salesman problem: Optimization by simulated annealing. J. Stat. Phys., **45**, 885–890.
37. Ettelaie, R., and Moore, M. A. (1985). Residual entropy and simulated annealing. J. Physique Lett., **46**, L-893–L-900.
38. Bonomi, E., and Lutton, J.-L. (1984). The N-city travelling salesman problem: Statistical mechanics and the Metropolis algorithm. SIAM Rev., **26**, 551–568.
39. Vannimenus, J., and Mézard, M. (1984). On the statistical mechanics of optimization problems of the traveling salesman type. J. Physique Lett., **45**, L-1145–L-1153.
40. Kirkpatrick, S., and Toulouse, G. (1985). Configuration space analysis of traveling salesman problems. J. Physique, **46**, 1277–1292.
41. Mézard, M., and Parisi, G. (1986). A replica analysis of the traveling salesman problem. J. Physique (Paris), **47**, 1285–1296.
42. Fu, Y., and Anderson, P. W. (1986). Application of statistical mechanics to NP-complete problems in combinatorial optimization. J. Phys. A: Math. Gen., **19**, 1605–1620.
43. Creutz, M. (1983). Microcanonical Monte Carlo simulation. Phys. Rev. Lett., **50**, 1411–1414.
44. Swendsen, R. H., and Wang, J.-S. (1987). Nonuniversal critical dynamics in Monte Carlo simulations. Phys. Rev. Lett., **58**, 86–88.
45. Wolff, U. (1989). Collective Monte Carlo updating for spin systems. *Phys. Rev. Lett.*, **62**, 361–364.
46. Creutz, M. (1992). Microcanonical cluster Monte Carlo simulation. Phys. Rev. Lett., **69**, 1002–1005.
47. Goodman, J., and Sokal, A. D. (1986). Multigrid Monte Carlo method for lattice field theories. Phys. Rev. Lett., **56**, 1015–1018.
48. Batrouni, G. G., Katz, G. R., Kronfeld, A. S., Lepage, G. P., Svetitsky, B., and Wilson, K. G. (1985). Langevin simulations of lattice field theories. Phys. Rev., **D32**, 2736–2747.
49. Bhanot, G., Creutz, M., and Neuberger, H. (1984). Microcanonical simulation of Ising systems. Nucl. Phys. **B235**, 417–434.
50. Bernard, S. T. (1989). Stochastic stereo matching over scale. Int. J. Comput. Vis., **3**, 17–32.
51. Aluffi-Pentini, F., Parisi, V., and Zirilli, F. (1985). Global optimization and stochastic differential equations. J. Optim. Theory Appl., **47**, 1–16.
52. Gidas, B. (1986). The Langevin equation as a global minimization algorithm. In E. Bienenstock et al. (eds.), *Disordered Systems and Biological Organization*. Berlin: Springer-Verlag, pp. 321–326.

53. Geman, S., and Hwang, C.-Y. (1986). Diffusions for global optimization, SIAM J. Control Optim., **24**, 1031–1043.
54. Chiang, T.-S., Hwang, C.-Y., and Sheu, S.J. (1987). Diffusions for global optimization in $\Re^n$. SIAM J. Control. Optim., **25**, 737–753.
55. Kushner, H. J. (1987). Asymptotic global behavior for stochastic approximation and diffusions with slowly decreasing noise effects: Global minimization via Monte Carlo. SIAM J. Appl. Math., **47**, 169–185.
56. Gelfand, S. B., and Mitter, S. K. (1991). Recursive stochastic algorithms for global optimization in $\Re^d$. SIAM J. Control Optim., **29**, 999–1018.
57. Gelfand, S. B., and Mitter, S. K. (1993). Metropolis-type annealing algorithms for global optimization in $\Re^d$. SIAM J. Control and Optim., **31**, 111–131.
58. Gelfand, S. B., and Mitter, S. K. (1991). Weak convergence of Markov chain sampling methods and annealing algorithms to diffusions. J. Optim. Theory Appl., **68**, 483–498.
59. Binder, K., ed. (1979; 2nd ed., 1986). *Monte Carlo Methods in Statistical Physics*. Berlin: Springer-Verlag.
60. Mézard, M., Parisi, G., and Virasoro, M. A. (1987). *Spin Glass Theory and Beyond*. Singapore: World Scientific.

5

THE PATTERNING OF NEURAL CONNECTIONS

Our focus in this chapter is on the developing vertebrate visual system. This has been a favored area of study for some time because of its relative ease of experimental accessibility. As a result there exists a considerable body of data, anatomical, physiological, and biochemical, that guides and constrains our insights and ensuing mathematical models. The retinotectal projection in lower vertebrates such as goldfish and frogs is capable of regeneration following surgical manipulations. If, for instance, the optic nerve is severed and potions of the retina are ablated, the remaining retinal tissue will project a topographically ordered set of functional connections to the optic tectum. Experiments have been done in which several rudiments of the retina from donor and host were grafted together; other experiments have been done in which potions of the neural epithelium through which the growth cones must navigate were rotated and reattached, and still other data have been acquired where the tectal targets were either translated or removed. The exceptional plasticity exhibited by the retinotectal system permits us to probe the underlying neural mechanisms.

In our introductory remarks in Chapter 1, we noted that multiple cooperative interactions guide the development of the central nervous system. As was the case for the physical systems we have been examining, global order emerges from multiple interactions each stochastic in character taking place in the local microenvironment. These multiple interactions give rise in many instances to useful metastable states and a variety of less useful shallower local minima. The highly favored stable states may be exploited by a biological system as the need arises in response to changes in the internal and external environments. Another general feature of a self-organizing system is the role

THE PATTERNING OF NEURAL CONNECTIONS IN THE CENTRAL NERVOUS SYSTEM 147

that edge effects and local singularities play in the promoting and stabilizing global structure. We will explore the biological uses of multiple stable states in considerable detail in the retinotectal projection, and we will investigate the influences of singularities in the retinogeniculate pathway. In both instances we will use simulated annealing to evolve the system into the desired states while avoiding trapping in shallow, undesirable minima.

5.1 THE PATTERNING OF NEURAL CONNECTIONS IN THE CENTRAL NERVOUS SYSTEM

In this chapter we will study two developmental tasks, the creation of topographically ordered mappings in the retinotectal projection and the segregation of eye-specific afferents in the lateral geniculate nucleus (LGN). A

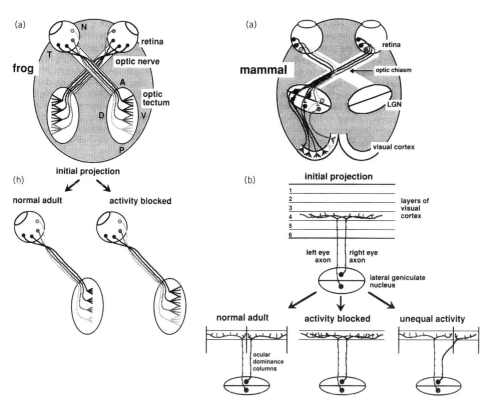

Figure 5.1. Primary visual pathways in vertebrates: (a) Connections between retina and optic tectum in the frog showing fine-grained topography in the normal adult and less precise patterning when electrical activity is blocked; (b) connections from the retina to the lateral geniculte nucleus and from the LGN to layer 4 of striate cortex under various conditions of electrical activity. (From Goodman and Shatz[1]. Reprinted with permission of Cell Press.)

state or configuration in these systems denotes a particular pattern of connectivity. Our goal is to understand how highly ordered spatial patterns evolve from initial loose associations of source and target cells. We will begin in this section by describing the initial and final patterns of connectivity, and the classes of interactions leading from one to the other. As noted in Chapter 1, the themes and organizational principles underlying these self-organizing processes are encountered elsewhere in the developing central nervous system and are present throughout adult physiological function.

Shown in Fig. 5.1 are pictoral representations of the primary visual pathways under consideration. The first panel depicts the projection from the retina of the frog to the optic tectum. The axons from the left eye project to the right optic tectum, and those from the right eye lead to the left optic tectum. The lateral geniculate nucleus is the thalamic nucleus that relays signals from the retinas of the two eyes to visual areas 17 and 18. The second panel illustrates that afferents from both eyes innervate each LGN and from the LGN axons establish connections with cells in layer 4 of the primary visual cortex.

5.1.1 Laminar Structure of the Mammalian Lateral Geniculate Nucleus

There is considerable variability in the structure of the LGN from species to species, but all mammalian systems share a number of essential features. First, the lateral geniculate nuclei are organized retinotopically. The retinotopic mapping is organized so that far more cells in the LGN are devoted to the central area of the retina than to the periphery. Second, the retinal inputs to the nuclei are segregated into eye-specific layers. Thus the retinotopic maps are stacked on top of one another in vertical register. Third, LGN cells possess concentric receptive fields resembling those of retinal ganglion cells. Fourth, the laminar structure develop from initially diffuse sets of connections, in which afferents from the two eyes are intermixed to a greater or lesser degree. In cat,[2,3] axons from the contralateral eye enter the LGN by E32, and those from the ipsilateral eye penetrate several days later. During the next 30 days or so, these inputs first intermix and then begin to segregate so that by birth (E65) eyes-pecific separation is nearly complete.

The primate LGN is organized into six layers. Three of the layers (1, 4, and 6) receive their input from the contralateral eye. The other three layers (2, 3, and 5) obtain their input from the ipsilateral eye. The first two layers (1 and 2) in the LGN contain large cells that are termed *magnocellular*, and the remaining four layers contain smaller cells that are called *parvocellular*. An additonal functional distinction is that the parvocellular layer 5 and 6 cells have ON polarity, while layer 3 and 4 cells have OFF polarity. These differing cell types project to different sublayers in the visual cortex. Therefore we will categorize the afferents by eye specificity — ipsilateral or contralateral, functional class — parvocellular or magnocellular, and polarity of the receptive field center. Although the detailed time course of segregation in primates differs

THE PATTERNING OF NEURAL CONNECTIONS IN THE CENTRAL NERVOUS SYSTEM 149

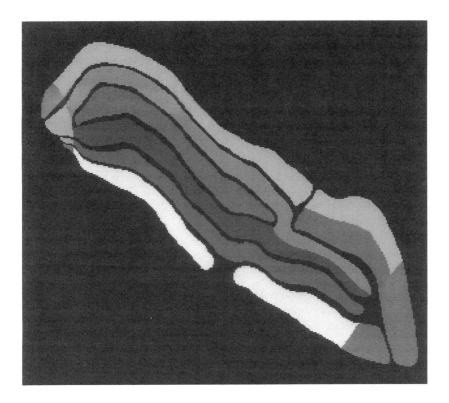

Figure 5.2. Laminar structure of the LGN in the rhesus monkey. From ventral to dorsal: Contralateral magnocellular (1), ipsilateral magnocellular (2), ipsilateral parvocellular OFF (3), contralateral parvocellular OFF (4), ipsilateral parvocellular ON (5), and contralateral parvocellular ON (6). (From Lee and Malpeli[8]. Reprinted with permission of Prof. Joseph G. Malpeli.)

from that in the cat, the overall sequence of developmental events is the same. In rhesus monkey,[5,6] the initial projections from the two eyes are nearly completely overlapping during early development (E64). Somewhat later, by E110, adultlike lamination has started to emerge, and by E124 characteristics of the adult pattern are quite strongly developed (birth is at E165). The stereotypic laminar structure of the primate lateral geniculate nucleus is depicted in Fig. 5.2.

The eye-specific laminar structure develops prenatally from diffuse sets of connections through a process in which terminal arbors from each eye expand into appropriate territory, while branches located in inappropriate regions are eliminated.[3] The segregation process requires electrical activity from both eyes. In primate systems, if one eye is removed, two layers form in place of the normal six, and abnormal retinal projections are formed.[6] Similar results are observed in the cat. Electrical activity consisting of action potentials and synaptic transmissions strengthen and preserve newly formed neural connections during fetal development, and drive the segregation process. The importance of electrical activity in the developing visual pathways of the cat can be demonstrated in blocking experiments. In one such study, by Shatz and Stryker,[9] tetrodotoxin (TTX), a voltage-sensitive sodium channel blocker, was used to prevent action potentials during the period of time when segregation into distinct eye-specific layers in the LGN would normally take place. Erroneous retinal projections resulting in a failure to form eye-specific layers in the LGN subsequent to the use of TTX were observed. As shown in Fig. 5.1, electrical activity also promotes the refinement of the sets of connections between geniculate and cortical cells of layer 4. We will discuss these cortical processes that take place later, during the critical period of postnatal development, in Chapter 8.

5.1.2 The Retinotectal Projection in Lower Vertebrates

The retinotectal projection is the primary visual pathway in lower vertebrates. As is the case elsewhere in the nervous system, the patterning of neural connections is precise. Neighboring retinal ganglion cells project to adjacent target neurons in the contralateral optic tectum, thereby preserving nearest-neighbor relationships, and the patterning of connections is orderly across the tectum, thereby maintaining topography.[4] As illustrated in Fig. 5.1, cells in the dorsal part of the retina send their axons to the lateral (ventral) region of the tectum, those from the ventral retina project to the medial (dorsal) tectum, those from the nasal (anterior) retina connect to the caudal tectum, and those from the temporal (posterior) retina attach to the rostral tectum.

In contrast to the situation encountered in the mammalian visual system, the initial connections between retina and optic tectum in lower vertebrates are not diffuse but instead are topographically ordered from the outset. Position information in the eyebud, tectum, extracellular matrix, and neural epithelium provide navigational cues for the axons. Terminal arbors from each axon span an appreciable area of the tectum, and arbors from neighboring ganglion cells largely overlap one another. These coarse-grained, overlapping connections are subsequently refined through electrical-activity dependent interactions.[10] In goldfish and frogs both retina and tectum continue to grow through larval development. In fish this growth is maintained well into the adult form. Synaptic connections are transient and are continually being formed and reformed while preserving the topographic mappings. In this continued growth, retinal cells are added circumferentially (annular rings) while tectal cells are generated caudomedially (crescentlike).

5.2 MULTIPLE INTERACTIONS

Three classes of interactions — mechanical, chemical, and electrical — drive the developmental processes. The chemical interactions fall into two groups: those mediated by diffusible molecules and those involving cell adhesion molecules such as NCAMs, L1, cadherins, and integrins. As was the case for the binary alloys and related order-disorder processes discussed in the last chapter, both attractive and repulsive interactions are required. These interactions encode position information in a genetically efficient manner and provide nearest-neighbor information needed for the proper development of retinotopy.

As mentioned in Chapter 1, the question of how cells migrate, aggregate, and establish highly specific, topographically ordered sets of connections was first addressed by Ramon y Cajal and Harrison. Studying growth cone behavior *in vitro*, Harrison made the important observation that nerve fibers must grow in contact with surfaces and are not able to move freely into fluid spaces. In the chemoaffinity hypothesis of Sperry,[11] it is proposed that axons are guided to their target structures by unique sets of biochemical labels, or markers, that are read by their growth cones. These labels are distributed in a graded fashion, both horizontally and vertically, across the source and target neural structures permitting cell-cell recognition. These markers are acquired during differentiation and uniquely code positional information through their concentrations. These markers, which take the form of diffusible molecules detected by the growth cones, are supplemented by local positional cues provided by cell surface molecules placed at decision points and elsewhere along the axonal pathways.

New techniques[12,13] have been developed that permit study of retinal arbor growth and remodeling in live subjects. In the study by O'Rourke, Cline, and Fraser,[14] retinal axons were labeled with the fluorescent dye DiI, and their growth imaged using a high-resolution laser-scanning confocal microscope. Three-dimensional reconstructions were then produced by superimposing sets of optical sections. Rapid remodeling of terminal arbors in the tecta of *Xenopus* was observed (Fig. 1.3). Extension and retraction of short spikes and longer branches was rapid—individual nerve arbors made about 200 extension-retraction cycles per day. Short spikes were particularly active and were able to rapidly test the target tissue for optimal locations for placing a stable axon branch. This activity is similar to the probing actions by growth cone filopodia.

The remodeling of terminal arbors and neurite outgrowth take place in response to signals from the external environment in accordance with intracellular genetic programs. The genetic programs determine which proteins are synthesized, and the proteins in turn modulate the axon's motility. In response to the appropriate signals growth cones are transformed into synaptic terminals, and when required, synaptic terminals are changed back to growth cones. Thus cells surface receptors just mentioned are dynamically regulated on growth cones. We will discuss some of the supporting evidence for their up- and down-regulation later in this chapter.

The cortical and geniculate blocking experiments mentioned earlier provide evidence that spontaneous electrical activity contributes to eye-specific afferent segregation in the visual pathways. This contention is supported by direct evidence by Galli and Maffei[15] and Maffei and Galli-Resta[16] that spontaneous activity is present in prenatal life. In *in vivo* studies of electrical discharges from retinal ganglion cells of prenatal rat, these authors observed electrical activity during this period when axons reach and form functional synapse with cells of the superior colliculus and lateral geniculate nucleus. An important property of vertebrate retinal ganglion cell electrical activity is that the discharges from neighboring cells are strongly correlated during prenatal life.[15,16] This property is encoded in the electrical-activity-dependent free energy in the multiple constraint model, and in an analogous interaction energy in the model of LGN lamination to be discussed in this chapter.

One of the most dramatic aspects of the correlated activity is its rhythmic characteristics. In the study by Maffei and Galli-Resta,[16] it was found that most retinal ganglion cells show rhythmic patterns of discharge, with each cell exhibiting its own rhythm. These rhythms are shared whenever two neighboring cells have correlated discharges. In studies by Meister et al.[17] and Wong et al.,[18] multielectrode arrays were used to record correlated bursting activity in neonatal ferrets. The authors observed patterns of periodically generated traveling waves of electrical activity across the retina. Ganglion cells fire spikes in nearly synchronous bursts lasting a few seconds separated by one or two minute in duration silent periods. The periods of synchronous bursting begin at one side of the retina and sweep across it in the form of several hundred micronwide waves of activity, propagating at rates of about a hundred microns per second. Neighboring cells are more likely to be synchronized in their firing than distant pairs. These correlated patterns of activity are present during the same period of time as synaptic modifications in the LGN, and the waves subside as the LGN patterns stabilize. Thus the spatiotemporal structure of these patterns supports the refinement of the topography and eye-specific segregation in the LGN.

5.3 OBJECTIVES OF THE CHAPTER

We will focus our attention in the next two sections on the construction of the multiple constraint model of Fraser and Perkel[19] in which several cooperative and competing constraints are concurrently operative. As is the case for other frustrated systems, there are multiple metastable states, or local minima, many of which may be selected by the biological system under varying sets of experimental (environmental) conditions. Under normal circumstances several mechanisms, each stochastic in character, cooperate to produce the precise set of required connections usually found. Under differing experimental manipulations, some of these interactions can either be disabled or brought into conflict with one another to produce different wiring solutions. We will start in Section 5.4 by reviewing the experimental findings in the frog *Xenopus* together with the pioneering single and dual interaction models inspired by the data. We will then present in Section 5.5 the multiple constraint model that combines many of the features of the earlier models into a single unified approach.

We observe in Fig. 5.2 that there is a transition in the LGN from six-layered lamination representing central vision to four-layered lamination encoding peripheral vision. The gaps appearing in the vicinity of the transition correspond to the blind spot produced by the exit of the optic nerve from the eye through the optic disk. In Section 5.6 we will find that this association is a consequence of a singularity produced by the blind spot in an otherwise smooth potential gradient representing the interactions between retinal and geniculate cells. This finding, due to Lee and Malpeli,[7,8] illustrates in a biological setting that not only can stochastic processes lead to precise results but also that local anomalies can produce global effects.

The physiological basis for the thermodynamic models lies in the stochastic mechanisms that guide and sustain growth cones motility, neurite outgrowth, and synaptogenesis. Our formalism allows us to relate the appearance of precise structural features, observations of the time course of laminar development, regenerative properties, and the results of experimental manipulations of tissues to these physiological processes. The predictions of multiple interactions acting in concert to produce favorable outcomes serves to highlight the importance of signal transduction and the accompanying integrative processes within the neural substrate. In Sections. 5.7 and 5.8 we will discuss the modulatory and integrative functions of G-proteins and calcium channels with this view in mind. We will also briefly discuss recent findings on diffusible and surface-bound molecular agents that promote growth cone motility and neurite outgrowth, and whose existence was predicted on theoretical grounds.

The states found in our modeling efforts using simulated annealing represent static equilibrium patterns of neural connections formed during development. In Section 5.9 we look at some results of experiments that probe the dynamic properties of receptive fields and topographic maps subsequent to their initial formation.

5.4 DEVELOPMENTAL MODELS

There are three classes of wiring solutions in retinotectal systems, namely normal development, regeneration, and forced innervation. In normal development the initial contacts are rather orderly, exhibiting considerable dorsoventral topography and, somewhat later, nasotemporal topography. During growth there is an orderly migration of terminal arbors over the tectal surface that maintains the global topography. As indicated in Fig. 5.1, the terminal fibers span appreciable portions of the tectum, and neighboring ganglion cell arbors largely overlap. These arbors become more restricted in their spread during electrical-activity refinement.

If the optic nerve is severed, the system is able to reestablish a set of topographically ordered projections to the optic tectum. The precision is far lower in regeneration than seen in normal development, and retinal electrical activity is needed to refine these connections. In systems created by grafts, or transplants such as the ones depicted in Fig. 5.3, retinal axons are able to form functional synapses with tectal targets. In the case of tectal graft experiments,

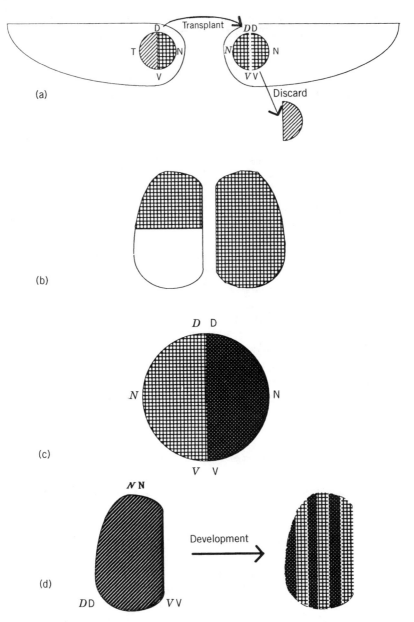

Figure 5.3. Construction of a double nasal eye: (*a*) Nasal half of right eye replaces temporal half of left eye. (*b*) Graft tissue establishes contacts with tectal targets in register with the host's nasal retinal projections and therefore retains its original specificity. (*c*) Double nasal eye shown with different patterns. (*d*) Initial projections from the double nasal eye overlap, then segregate into bands. (From Udin and Fawcett[4]. Reprinted with permission of Annual Reviews, Inc.)

more than one wiring solution is observed.[19] In experiments in which the neural epithelium through which the axons must navigate is rotated and reattached, Harris[20] found that local positional cues embedded in the neural epithelium contribute guidance information.

Experiments have been done by Constantine-Paton and Law[21] in which a third eye rudiment was implanted into *Rana pipiens* embryos (Fig. 1.4). The host and third eye subsequently formed functional retinotopic maps, and the efferents self-organized into alternating ocular dominance clumps or patches in the tectum resembling those which form in afferent layer 4 of mammalian striate cortex. As shown by Cline et al,[22] retinal electrical activity is crucial for the formation of these stripes, and exposure to an NMDA receptor antagonist leads to their desegregation.

5.4.1 Minimalist Marker Models

The term "markers" is used to designate a property distributed across cells or groups of cells that generates either directly or indirectly a distribution of affinities between retinal and tectal elements; it is related to spatial location but is distinct from these locations. Several investigations of marker-induced retinotopy have been reported. In the earliest of these studies, by Prestige and Willshaw,[23] graded distributions of markers were shown to be capable of generating topographic projections. In this study it was discovered that there was a need to supplement graded distributions of markers with a mechanism that prevents large numbers of fibers from synapsing on tectal cells expressing the highest marker concentrations. This was accomplished by imposing saturation conditions that, by limiting the available axonal territory, lead to competitive (repulsive) fiber-fiber interactions. Difficulties in describing the tectal compression and expansion data were noted by the authors.

Fiber-fiber interactions were also studied by Hope, Hammond, and Gaze.[24] In their minimalist "arrow" model, marker gradients simply provided polarity information. A nearest-neighbor exchange kinetics, similar to that used to promote order-disorder transitions in binary alloys, was used to generate bundles of optic fibers that sort themselves out enroute to the optic tectum. This form of nearest-neighbor sorting produces ordered fiber bundles resembling those produced by selective fasciculation in insect systems. However, the mechanism is inconsistent with results of studies of fiber bundling and, as pointed out by the authors, with the tectal graft and rotation data.

In the model of Whitelaw and Cowan,[25] retinotectal connectivity arises from the combined influence of chemospecificity and Hebbian, activity-dependent synaptic strengthening. First, the adhesion of retinal and tectal cells is differentially mediated by chemical markers. Second, the formation of modifiable synapses is modulated by neural activity. To combine these two mechanisms, the synaptic strengths are weighted by marker concentrations. This model is able to describe the establishment of initial retinotectal contacts, generate a local, neighbor-preserving ordering of the connections (conserve retinotopy), and correctly specify polarity (map orientation). The precursors for the Whitelaw-Cowan model include the correlated electrical activity[26] and marker induction[27] models of Willshaw and Malsburg.

5.4.2 Molecular Gradients

Molecular gradients[28] that generate an optimally adhesive site on the tectum for each retinal fiber can be constructed from graded distributions of two or more species of cell surface molecule across the tectal and growth cone surfaces. In its simplest form, two types of molecules distributed in a graded fashion such that their summed density is constant at each location would generate a triangular well-shaped potential for a given direction. Tectal and growth cone distributions do not have to be identical but merely coordinated in their orientations. Although both homophilic and heterophilic mechanisms will produce a unique mapping from retina to optic tectum, only a homophilic mechanism will preserve nearest-neighbor relationships among fibers, and is fully consist with regeneration data[28].

The role of chemical markers in guiding growth cone motility was explored in a series of studies by Gierer.[29–31] The basic question addressed in these studies was that of the form of the gradients needed to properly guide growth cones to specific tectal targets, starting from any of a number of possible locations. As in the multiple constraint model the system seeks a minimum in the interaction energy.

5.5 THE MULTIPLE CONSTRAINT MODEL

In Steinberg's[32] differential adhesion hypothesis a mechanism is proposed by which different cells can sort themselves out to form tissues. The underlying idea is that in systems that cohere while maintaining mobility, the tendency to adhere is related to a preference to minimize the adhesive-free energy. The stable states of the system are states of minimal adhesive-free energy. The relative strengths of adhesion of the mixed population of cell types or phases will then determine the overall morphology of the system. A heirarchy of positions will occur that reflect the hierarchy of cellular adhesivenesses. The differences in adhesion between two species may result from differences in the types of surface molecules and/or from differences in their number.

In the multiple constraint model of Fraser and Perkel, the influences of chemotropic factors, electrical activity, and cell surface molecules are described in terms of adhesive-free energies of fiber-fiber and fiber-tectum interations. These include position-dependent chemoaffinities between retinal fibers and tectal cells, position-independent repulsions between retinal fibers, electrical-activity and position-dependent interactions between retinal fibers, and position-independent adhesion between retinal fibers and cells in the tectum. We will examine in this section the forms required of these stereotypic

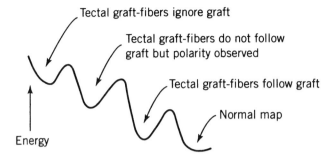

Figure 5.4. Correspondence between metastable states and observed for normal development and tectal grafts. (From Fraser[28]. Reprinted with permission of Academic Press.)

interactions in order to describe the data on plasticity, transplants and grafts, and segregation. In the next section we will discuss the physiological basis for the interactions.

If we arrange the aforementioned interactions into hierarchy of strengths so that the position-dependent chemoaffinity is the weakest of these forces, the repulsions and electrical-activity-driven fiber-fiber interactions are intermediate in strength, and the general adhesion is the strongest, then the retinotectal projection will possess multiple metastable states generated by these several, sometimes conflicting and possibly frustrated constraints. Under these interactions the system would evolve toward a state of minimum adhesive- (interfacial) free energy. It would reach this state by simultaneously maximizing its energetically favorable adhesive contacts while minimizing its energetically unfavorable repulsive interactions. The correspondence between the metastable states generated by the model and the experimentally observed patterns of connectivity for normal development and tectal grafts is depicted in Fig. 5.4.

5.5.1 Position-Independent Affinity

In the model the optic tectum and terminal arbors are represented as rigid discs of fixed size. The diameters of the arbor discs are set at about 10% of the diameter of the tectal disc, in accordance with the data for unrefined arborization. The first term in the model is a position-independent adhesive interaction between target cells in the optic tectum, and the optic fiber growth cones. We can write the contribution to the total free energy of the position-independent affinity of the ith fiber as

$$E_{0i} = -c_0 \phi_i \tag{5.1}$$

where c_0 is a positive, coupling strength, and ϕ_i is the fractional overlap of the ith fiber disc with the optic tectum. Axonal growth cones will be attracted with equal strength to any region of the optic tectum through this interaction, except at the boundaries where the percentage overlap is varying. If we ignore

158 THE PATTERNING OF NEURAL CONNECTIONS

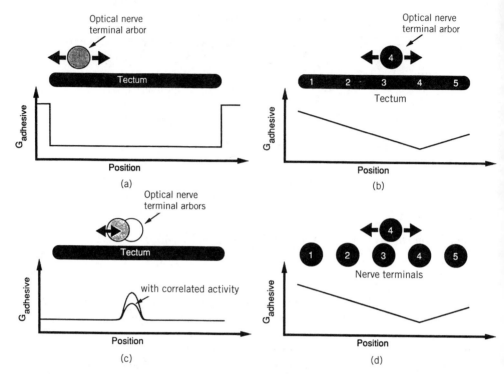

Figure 5.5. Graphical representations of the interaction potentials (adhesive-free energies). *Upper left*: Position-independent affinity. *Lower left*: Short-range fiber-fiber repulsion and nearest-neighbor correlational energy. The nearest-neighbor interaction weakens the repulsion between fibers. *Right-hand panel*: Position-dependent affinities. This potential is decomposed into an interaction between terminals and tectum (*upper right*) and between fibers (*lower right*). The points illustrated in the plots are that each fiber has a preferred site on the tectum, and terminals from the same region of the retina are weakly attracted to one another. (From Fraser and Perkel[19]. © 1990 John Wiley & Sons, Inc.)

these boundary effects, then we may represent the interaction by the attractive square well potential depicted in Fig. 5.5a. The position-independent affinity attracts growth cones to the tectal surface and helps define tectal boundaries.

5.5.2 Fiber-Fiber Repulsion

The next term in the model is a position-independent repulsion between fibers. The repulsion is of short range and is proportional to the amount of overlap between the two arbor discs. The fiber-fiber repulsion due to interactions of the ith fiber with the others may be expressed as

$$E_{1i} = c_1 \sum_j \left(\frac{r_{ij} - a_1}{r_{ij} - a_1 + a_2} \right) \chi_{ij} \tag{5.2}$$

In Eq. (5.2), c_1 is a positive coupling constant smaller in magnitude than c_0, χ_{ij} is the percent overlap between fibers i and j, r_{ij} is the distance between the retinal gangion cells responsible for fibers i and j, and a_1 and a_2 are constants. This repulsive interaction plays a role in the patterning of neural connections analogous to the short-range repulsion in the van der Waals or lattice gas pictures of a liquid. It ensures that there is a uniform distribution of fibers across the tectum, preventing multiple attachments at some points in the tectum and empty areas elsewhere. The form of this free energy generated by this expression is illustrated in Fig. 5.5c.

5.5.3 Nearest-Neighbor Correlated Activity

The fiber-fiber repulsion is modulated by a weak, electrical-activity-dependent interaction that encourages axonal growth cones from neighboring retinal ganglion cells to synapse on neighboring tectal cells. This interaction energy is independent of position in the tectum. It has the same dependence upon the overlap between arbor discs as does the fiber-fiber repulsion and can be either attractive or repulsive depending on whether or not the two fibers are neighbors. This interaction energy represents the experimental observations that correlated spontaneous electrical activity in neighboring retinal cells promotes retinotopy, perhaps by signaling nearest-neighbor information. As shown in Fig. 5.5c, correlated (nearest neighbor) activity will decrease the fiber-fiber repulsion and promote stability.

5.5.4 Position-Dependent Affinity

The final contributions to the adhesive free energy are a set of position-dependent affinities between growth cones and their tectal targets. Following Sperry and others, these attractive interactions may be portrayed as two gradients, one running in the anterior-posterior direction and the other operating in the dorsal-ventral direction. Appropriate gradients can be generated in several ways. One manner in which they can be produced is through homophilic interactions between cell surface molecules distributed in a graded manner on the growth cones and optic tectum. This mechanism will be discussed in greater detail shortly. Homophilic interactions produce a tendency for fibers originating from neighboring sites in the retina to attract one another in the tectum. Thus we treat the position-dependent affinities as being composed of distinct contributions from fiber-fiber and fiber-tectum intractions. The fiber-fiber attractive interaction involving the ith fiber is

$$E_{2i} = -c_2 \sum_j r_{ij}^{AP} \chi_{ij} - c_3 \sum_j r_{ij}^{DV} \chi_{ij} \qquad (5.3)$$

with the usual distance relation for orthogonal directions

$$r_{ij}^2 = (r_{ij}^{DV})^2 + (r_{ij}^{AP})^2 \qquad (5.4)$$

and the corresponding fiber-tectum term is

$$E_{3i} = -c_4\phi_i(1 - t_i^{AP}) - c_5(1 - t_i^{DV}) \tag{5.5}$$

where the t_i denotes the distance between the location in the tectum of the ith fiber and its optimal location. The fiber-tectum and fiber-fiber interactions are illustrated schematically in Fig. 5.5b and d. The total energy is obtained by summing the contributions from each of the fibers:

$$E_{tot} = \sum_i (E_{0i} + E_{1i} + E_{2i} + E_{3i}) \tag{5.6}$$

5.5.5 Multiple Stable States in the Retinotectal Projection

In the multiple constraint model there are a number of stable states that may be exploited by the system. Relationships between the various minima and some of the experimentally manipulated systems are illustrated in Fig. 5.4. In this plot we observe that the deepest global minimum corresponds to a normal retinotectal projection. Progesssively shallower minima, denoting the several possible wiring solutions, are found in retinotectal systems formed by grafting tectal tissues. The plot gives an accurate representation of the depths of the various minima, but the heights of the barriers sparating these minima are path dependent. The barrier heights shown are approximate, and the separations of the minima in the figure are not intended to reflect the distances between the corresponding valleys in the appropriate multidimensional space.

As we have done for the other multiple constraint systems, we use simulated annealing to evolve the system into one of the favored, low free energy states. In the calculations a terminal is selected at random. This terminal is allowed to randomly translate a fraction of its diameter. The change in adhesive free energy is determined, and if the free energy in the new position is lower than that in the old position, the move is allowed. If the move leads to an increase in the free energy, the move is allowed in accordance with the Metropolis algorithm. This procedure is repeated for all terminal arbors until a stable pattern emerges. As before, simulated annealing allows the system to avoid being trapped in shallow local minima. The barriers associated with trapping in shallow energy minima may be interpreted as activation energies.

When neither the retinal input nor the tectal environment has been modified, a normal projection can be generated starting from a completely or partially random ordering. Results of the simulated annealing calculations are displayed in Fig. 5.6. We observe in this figure that a few arbors are incorrectly positioned but most are topographically arranged. In a normal projection all constraints are simultaneously satisfied, and the interations mutually reinforce one another. Elimination of one of the interactions does not have a major effect on the topography. If the synchronous (nearest-neighbor) activity generated interaction is removed, map formation takes a larger number of iterations, and the precision of the topography is reduced.

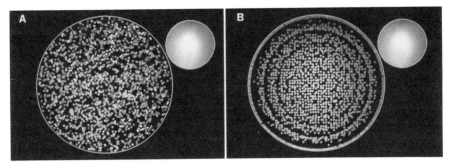

Figure 5.6. Normal development of retinotectal topography: (A) Random initial configuration of the system. The small circle in the upper-right-hand corner represents the retina, and the large circle corresponds to the optic tectum. The spatial origin of terminals in the retina is denoted by color. (B) Emergence of topography. The lack of spatial uniformity near the edge can be eliminated by modifying the position independent adhesion so that it is weaker near the boundary than in the central region. (From Fraser and Perkel[19]. © 1990 John Wiley & Sons, Inc.)

If a portion of the optic tectum is translocated or rotated, there are several possible stable solutions. The growth cones may navigate correctly to the rotated or translocated tissue, or they may ignore the alterations made to the tectum, to establish a topography that is indistinguishable from that of a normal projection. In cases where the graft has been ignored, the stable solution corresponds to one of the several local minima depicted in Fig. 5.4, while the global minimum conforms to the case where the growth cones follow the graft. All solutions have been seen experimentally. In contrast to a normal projection, the constraints cannot be concurrently satisfied in the presence of a graft. In following a grafted tissue, nearest-neighbor relations among retinal fibers are disturbed. In more detail, the synchronous-activity-generated fiber-fiber interactions, and the position-dependent fiber-fiber interactions are in conflict with the position-dependent fiber-tectum interactions, and the system is frustrated. Initial conditions are more important in this situation than they are for a normal projection. The degree of randomness and the size of the tectal graft influence which state is selected.

A type of experiment that is somewhat different from grafting is *ablation*, where a portion of either retina or optic tectum is removed. An example of this type of preparation was depicted in Fig. 5.3. In these situations expansions or compressions occur. If half of the retinal tissue is removed, the remaining fibers expand to fill the entire tectum. Similarly, if a half-tectum is formed, the retinal fibers compress into the remaining tectal tissue. The fiber-fiber repulsion drives the expansion, while the position-independent fiber-tectum attraction promotes the compression. In both instances the position-dependent fiber-tectum interactions partially oppose the direction of change. The multiple constraint model is able to explain both the majority and the minority results. In the analysis the ablation data help establish the scale of interaction strengths.

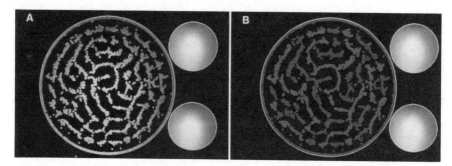

Figure 5.7. Development of ocular dominance stripes in three-eyed preparations. (A) Initial diffuse pattern of connections in a tectum receiving inputs from two retinas. (B) Formation of ocular dominance stripes. Terminals from the normal retina re color coded green and those from the supernumary retina are colored blue. (From Fraser and Perkel[19]. © 1990 John Wiley & Sons, Inc.)

The three-eyed preparations[21] and the resulting appearance of ocular dominance stripes are a particularly dramatic example of the responses possible in frustrated systems. The requirement for ocular dominance stripe formation is input to a tectum from two eyes. The kinetics of the process is that topography emerges first followed by eye-specific segregation. Results of simulations for this system are presented in Fig. 5.7. In the calculations we find that the size of the ocular dominance domains is determined primarily by the spatial extent of a terminal arbor. In this situation the relative strengths of the interactions are not the crucial factor. Instead, the wiring patterns depend on the presence or absence of electrical-activity-dependent fiber-fiber interactions that drive the segregation. The results in Fig. 5.7 illustrate the emergence of well-defined ocular dominance stripes from an initially diffuse set of connections. In this type of preparation, the electrical-activity-dependent interactions are in conflict with the position-dependent fiber-tectum interactions. Both constraints cannot be satisfied, and the result is the creation of small domains, the stripes, that represent a compromise between the nearest-neighbor and position-dependent topography. If the electrical activity in one eye is eliminated, these stripes still form in agreement with the experimental observations.

5.6 MORPHOGENESIS OF THE LATERAL GENICULATE NUCLEUS

We will now examine the emergence of laminar order in the segregation in the lateral geniculate nucleus of afferents from the left and right hemiretinas in the rhesus monkey. The model of Lee and Malpeli[7,8] which we will examine in this section accounts for the precise patterns of connectivity which emerge late in development from an initial diffuse pattern of retinogeniculate connections. In their model the terminals are endowed with an equilibrium dynamics that describes their evolution from a disordered initial state to a final ordered configuration where the afferents are segregated into their appropriate layers and each layer is topographically ordered and registered with respect to the others. The energy function that is minimized contains attractive and repulsive potentials that characterize the retinotopic and laminar properties of the configuration of terminals.

5.6.1 Interaction Potentials

The representation of the visual field by layers 2, 4, and 6 of the LGN is shown in Fig. 5.8. The retina is partitioned by eccentricity and polar angle into roughly equal rectangular patches in this plot. The two-dimensional model of

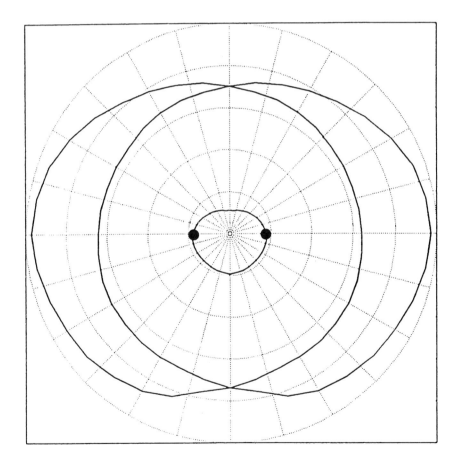

Figure 5.8. Portions of the visual field of the retina of the rhesus monkey represented by layers 6, 4, and 2 of the LGN. The blind spots are denoted by filled circles. Eccentricities are marked in 20-degree increments. (From Lee and Malpeli[8]. Reprinted with permission of Prof. Joseph G. Malpeli.)

the LGN displayed in Fig. 5.2 represents the horizontal meridian. A retinotopic mapping from retina to LGN is described by eccentricity e and a horizontal coordinate x that increases in the posterior to anterior direction. Terminals are grouped into projection columns labeled by a projection column number k, and these are ordered so that the eccentricities increase with increasing projection column number. The contribution to the total interaction energy due to interactions of terminal i with the remaining terminals is given by

$$E_i = E_i(e_i, x_i) + E_i^{corr} + E_i(y_i) \tag{5.7}$$

where the first term in Eq. (5.7) promotes retinotopy and the other two encourage laminar development. The retinotopy-generating term is straightforward and so will not be discussed further. The segregation of the various classes of afferents into layers driven by the last two terms in Eq. (5.7) is more interesting. The first quantity is a correlational energy, and the second is a vertical (y) positioning energy. The correlational energy for a given terminal i is a sum of several types of interactions, each of the general form

$$E_i^a = \sum_{j \in W_a} G^a(d_{ij}) \tag{5.8}$$

where d_{ij} is the distance between terminals i and j, W_a is the set of all terminals participating in interaction a, and G^a is a Gaussian of the form

$$G^a(d_{ij}) = B^a \exp\left(-\frac{d_{ij}^2}{s^a \Phi(k_i)}\right) \tag{5.9}$$

In Eq. (5.9), B^a sets the overall strength of the interaction energy of type a and s^a scales the widths of the Gaussians. These widths represent interaction distances, and $\Phi(k_i)$ established a gradient in these widths with increasing projection number (eccentricity) k_i according to the formula

$$\Phi(k_i) = 0.0015 k_i + 0.4 \tag{5.10}$$

There are six types of correlational interaction energy. These are shown in Fig. 5.9 as a function of the distance between terminals. The first term is a short-range repulsive term anologous to the short-range hard-core potential of Chapter 3. This potential acts among all terminals to promote an even spacing. The next term is a repulsive potential between magnocellular and parvocellular terminals. The third and fourth potentials shown in Fig. 5.9 provide an attractive contribution between terminals for the same eye and repulsive one between dissimilar terminals. The last two contributions to the correlational interaction energy treat center polarities of the parvocellular terminals. If the polarities are the same, the energy is lowered, and if different, the energy is increased.

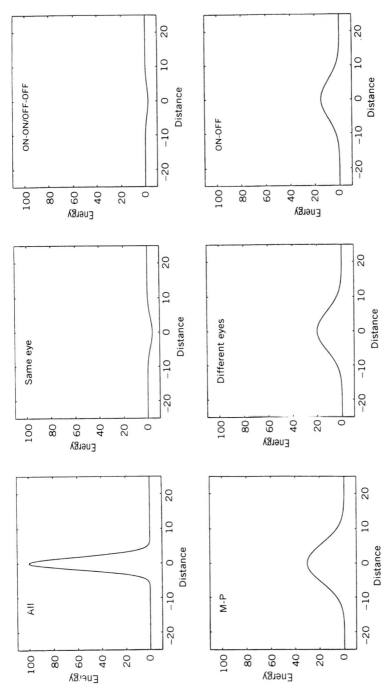

Figure 5.9. Interaction potentials as a function of distance between fibers. *Upper left*: Short-range repulsion between fibers. *Lower left*: Repulsion between magnocellular and parvocellular terminals. *Center*: Attraction between fibers from the same eye and repulsion between terminals from different eyes. *Right*: Attraction between parvocellular terminals having the same center polarity and repulsion between parvocellular terminals having opposite polarities. (From Lee and Malpeli[8]. Reprinted with permission of Prof. Joseph G. Malpeli.)

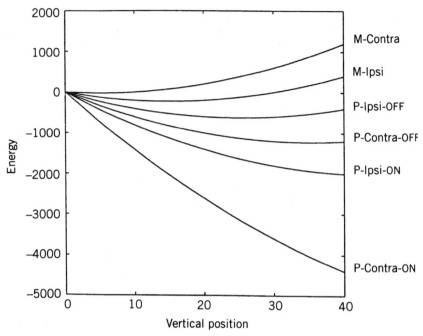

Figure 5.10 Potentials governing the vertical positioning of each of the six terminal types. (From Lee and Malpeli[8]. Reprinted with permission of Prof. Joseph G. Malpeli.)

The final contribution to the total energy is provided by the vertical positional energies $E_i(y_i)$. These are given by the expression

$$E_i(y_i) = ay_i^2 + K_g y_i \qquad (5.11)$$

where y_i is the vertical position of the ith terminal, $a = 1.5$, and K_g are a sequence of progressively more negative constants for the six terminal types. These energies are plotted in Fig. 5.10. We observe that the the location of the minima in the energy progressively shift toward the right in a manner that promotes the empirically observed laminar order.

5.6.2 Induction of the Laminar Transition

In examining the relative strengths of the contributions to the correlational energy, we observe that the ocularity-specific potentials are stronger than the polarity-preserving terms. As a result the correlational energies tend to favor a segregation of the afferents into four geniculate layers, whereas the vertical positioning energy supports a six-layered structure. The gradient in the widths of the Gaussians given by Eq. (5.8) plays a key role in inducing the laminar transition from six posterior layers to four anterior layers. The gradient controls the balance between the correlational and positioning energies. In the posterior region the positional energy dominates. The correlational energies gradually increase in strength relative to the positional energies as the interaction distances increase. The result is a laminar transition in which terminals in layers 4 and 5 exchange positions so that two pairs of eye-specific layers each become merged together anteriorly.

5.6.3 Trapping of the Transition by the Blind Spot

The retinal blind spot is modeled by ghost magnocellular (layer 1) and parvocellular, ON polarity (layer 6) terminals in a small portion of the contralateral eye. These terminal are influenced by the retinotopic hard-core repulsion and

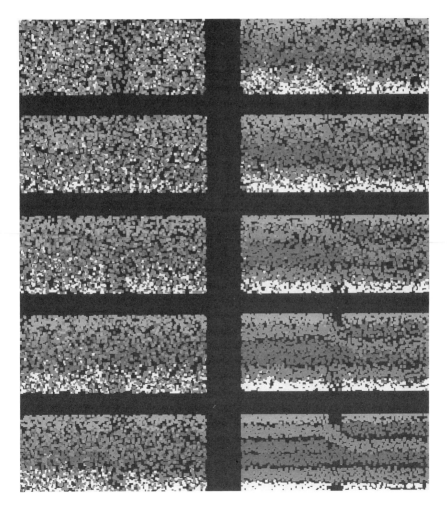

Figure 5.11. Emergence of eye-specific lamination in the LGN at various stages of simulation. The progression from upper left to lower right gives the patterns produced at iterations 1, 4, 10, 20, 40, 60, 70, 90, 120, and 230. (From Lee and Malpeli[8]. Reprinted with permission of Prof. Joseph G. Malpeli.)

vertical positioning energies. However, they are not subject to the M/P, ocularity, or center-polarity correlational energies.

In the calculations, terminals are randomly distributed in the dorsoventral direction, while a rough anteroposterior retinotopy establishes the representation of the fovea in the posterior part of the LGN. A terminal is randomly selected and its energy calculated. A trial move is randomly picked in an arbitrary direction with a Gaussian probability distribution of distances. The energy change ΔE is computed. If ΔE is negative, the new position is accepted; otherwise, the rearrangement is selected with probability given by the Metropolis algorithm. Another terminal is then chosen, and the process is repeated in pseudorandom order until all terminals have been sampled. The temperature is then lowered following a geometric cooling schedule with $T_{n+1} = 0.985 T_n$. The initial temperature is chosen so that at least 60% of the terminals move to new positions.

In the absence of the ghost terminals, the laminar transition occurs in a broad zone over the posterior half of the LGN. The addition of the ghost terminals traps the transition at its stereotypic position by introducing a singularity, due to the absence of the M/P, ocularity, and polarity terms, in the potential gradient-controlling laminar development. Typical results at various stages in a simulation are shown in Fig. 5.11. We observe that the transition is sharply delineated at the proper location in the vicinity of the optic disk gap. As found in earlier chapters and in the last section, the global laminar pattern of connectivity emerges from a sequence of stochastic interactions, each local in character. In this case we have an additional element, the seeding of a transition in the global laminar structure by a local perturbation, the optic disk gap, which otherwise might be thought of as being unrelated to any of the properties of the global structure.

5.7 GROWTH CONE GUIDANCE AND NEURITE OUTGROWTH

The neuronal growth cone is responsible for the sensing and navigational activities of developing axons. Filopodia and the weblike lamellipodia are advance filaments that radiate out from the leading margin of the advancing growth cone to probe the local microenvironment. Filopodia are able to extend in length, move about, differentially adhere to permissive surfaces, and/or retract. Neurite outgrowth can be divided into neurite elongation and growth cone motility. The former refers to the elaboration of new neurite length, and the latter denotes the sampling and forward movements of the growth cone and their filopodia and lamellipodia. These structures contain dense networks and bundles of actin filaments. From a mechanical point of view, neurite elongation depends on microtubules concentrated in the neurite shaft, while growth cone motility depends primarily on cytoskeletal actin and associated proteins.

Actin filaments, actin-associated proteins, and cell surface molecules are distributed on growth cones, and on their filopodia and lamellipodia. Thus we find[33] that the leading margin of chick dorsal root ganglion (DRG) nerve growth cones (including their filopodia and lamellipodia) contain the actin-associated molecules filamin, α-actin, myosin, tropomyosin, talin, and vinculin. These molecules help link actin filament bundles and networks to integrins, stabilize actin filaments against depolymerization, and contribute to the generation of tension forces. Cell surface molecules found include integrins, L1, NCAM, and N-cadherin. Integrins and L1 are found in high concentrations on DRG growth cone filopodia; NCAM and N-cadherin are more widely distributed on growth cones, but NCAM is mostly absent from filopodia.

Local mechanical, chemical, and electrical interactions all play important roles in producing the precise connections that characterize the nervous system. In the last two sections we presented models that included contributions from these multiple interactions in their hamiltonians or free energies. The colocation of several different types of cell surface molecules, and evidence of their acting in concert,[34] provide support for this view. Other important constituents of the plasma membrane of growth cones include G-proteins and voltage-dependent calcium channels. G-proteins and calcium channels serve several important signaling and integrating functions. Recent findings on their physiological function in growth cones provide futher evidence that growth cone motility is guided by the joint actions of a variety of agents including diffusible molecules, action potentials, and depolarization.

In the remainder of this section, we will briefly explore the physiological substrate for neurite outgrowth and growth cone motility. Since we have already discussed electrical-activity-mediated interactions and their effects on the refinement of the patterns of connectivity, we will focus our attention in this discussion on biochemical cell-cell and cell-substratum interactions and their intracellular integration. We will start by examining the findings on molecules that impart directional information through spatial gradients in their concentrations. We will then discuss the nondirectional modulation of motility by cell adhesion molecules. Finally, we will discuss signal transduction mechanisms, concentrating on the regulation of neurite outgrowth and growth cone motility by intracellular calcium, and on the strong cooperativity promoted by G-proteins. Our exploration of the physiological substrate will be resumed in Chapter 8, where we discuss the physiological basis for the BCM theory. In Chapter 8 we will direct out attention to postsynaptic processes and the cooperative and integrative properties of NMDA receptors and second-messenger calcium.

5.7.1 Chemoattractants

The search for chemotropic factors that began with Ramon y Cajal's hypothesis that chemotropic agents guide neurite outgrowth has produced both chemoattractants and chemorepulsants. Netrins, as reported by Serafini et al.[35] and Kennedy et al.,[36] function as circumferential chemoattractants emanating from the floor plate at the ventral midline of the developing spinal cord. These molecules provide guidance signals for the growth cones of the commissural neurons whose axons extend toward and across the floor plate. Estimates of the effective range of the diffusible chemoattractants such as the netrins are 100 to 300 microns. These studies were preceded by an investigation by Tessier-Lavigne et al.[37] of the influences of rat floor-plate cells explants on axon outgrowth. In that earlier set of *in vitro* experiments, evidence was supplied that a diffusible chemotropic factor emanating from the floor plate influenced the growth and orientation of commissural axons. This chemotropic factor was specific and did not affect the behavior of other axons. Those results were extended further by Placzek et al.[38] who showed that the chemotropic agent could diffuse considerable distances through a collagen gel matrix and through the neural epithelium *in vitro*, while remaining effective in providing directional guidance to the commissural axons.

This work that led to the netrins was aided by the development of an *in vitro* assay for the study of chemoattraction by Lumsden and Davies.[39] These authors co-cultured embryonic mouse sensory neurons and their peripheral target tissues at a developmental stage prior to their normal *in vivo* contact. Neurite outgrowth was observed to be exclusively directed toward the peripheral tissue and not toward any inappropriate tissues. Further support for chemoattractants was provided by Heffner et al.[40] in a study of budding and ingrowth of axon collaterals in the mammalian corticopontine pathway. This connection is from layer V of the neocortex to the basilar pons. During development axons initially grow past the basilar pons and only later extend branches to it. Thus the target detection is not by growth cones of primary axons. In their experiments, cells from the cortex and basilar pons were frozen into spatially separate positions within a collagen matrix. Effects of diffusible attractants, while excluding influences from local guidance cues, were clearly visible.

5.7.2 Chemorepulsants

We have seen in the modeling studies that both positive and negative guidance and regeneration cues are important. Luo et al.[41] have discovered that a secreted protein, collapsin, extracted from adult chick brain membrane has the ability to induce the collapse and paralysis of growth cones. Recombinant collapsin is specific to dorsal root ganglion (DRG) cells. It has no effect on retinal ganglion growth cones, and conversely, extracts from liver membrane do not possess any DRG growth cone collapsing ability. This diffusible protein is soluble but also may be able to bind to cell surfaces or to the ECM. As is the case for the netrins, the molecule is highly conserved, being closely related to fasciclin IV, a guidance molecule found in grasshopper.

The study by Luo et al.[41] was preceded by several sets of findings on chick-membrane-derived chemorepulsive activity, including those of Walter et al.,[42] Kampfhammer and Raper,[43] and Raper and Kampfhammer[44] on the formation of distinct territories for growth cone advancement, the avoidance of inappropriate regions, and the development of growth cone collapse *in vitro* assays. Similar findings have also been reported in the developing mammalian central nervous system by Pini et al.[45] In experiments with the olfactory system of embryonic rat, Pini[45] found that the septum releases a repulsive factor that diffuses into the olfactory bulb. In another set of *in vitro* experiments dealing with the developing rat spinal cord, Fitzgerald et al.[46] showed an inhibitory effect of the embryonic ventral horn on neurite outgrowth from the dorsal root ganglion neurites. This inhibitory effect was developmentally regulated, diminishing beyond a certain period of time.

We end our brief overview of chemoattractants and chemorepulsants with two observations. First, the ability of growth cones to read gradients of surface-associated information has been demonstrated through *in vitro* experiments by Baier and Bonhoeffer.[47] These authors found that chick embryo retina growth cones are sensitive to small changes in concentration of molecular gradients. The strength of the spatial gradient, and not the absolute concentration, seems to be the critical factor in directing outgrowth. Second, with regard to specificity, Goodman[48] has argued that that some guidance molecules may be attractive for some growth cones and repulsive for others, depending on the receptor. That is, there may be attraction- and repulsion-flavored receptors. This is analogous to the situation for neurotransmitters where a particular neurotransmitter released at a synapse may be either excitatory or inhibitory depending on the postsynaptic receptor.

5.7.3 Cell Adhesion Molecules

Three classes of cell adhesion promoting receptors are colocated on growth cones. These are integrins;[49] members of the immunoglobulin gene superfamily such as NCAM,[50,51] and members of the cadherin family such as N-cadherin.[52] Integrins promote growth cone motility and neurite outgrowth over the extracellular matrix (the basement membrane located in the extracellular region). NCAM and N-cadherin support these activities too but differ from integrins in that they mostly operate through homophilic rather than heterophilic binding mechanisms. That is, they serve as their own receptors and ligands, and bind to cells of the neural epithelium. All three classes of receptors are transmembrane-spanning glycoproteins with cytoplasmic domains and function as recognition and two-way signaling devices.

Integrins[49] play an important role in cell migration and neural outgrowth, both during central nervous system development and during physiological function such as lymphocyte traffic and wound healing in the mature adult. Most of the ligands bound by integrins, such as laminin, are extracellular matrix proteins that are involved in cell-substratum adhesion. Affinities are usually low with attachment resulting from multiple weak adhesions. Some integrins can mediate cell-cell interactions, binding to ligands (counterreceptors) on the other cell. Integrins can switch between activated and deactivated states and vary their affinity for a specific ligand in response to signals from within the cell. They also relay signals into the cell from without, triggering events such as tyrosine phosphorylation.

Extracellular matrix molecules such as laminin provide a permissive substrate for cellular migration and neurite outgrowth. These molecular agents do not appear to influence motility in a graded, directional manner as do the chemotropic factors. Instead, the binding of laminin modulates the manner in which cells adhere to the substratum. In a study of the migration of embryonic olfactory cells, Calof and Lander[53] found that there was no direct correlation between the migratory actions of the cells and the strength of the adhesion. Instead, motility arises through a modulation of the structure and dynamics of the actin-based cytoskeleton. The cytoskeleton influences cell shape, cell stiffness, and the formation of strongly adhesive focal contacts. Laminin acts by inhibiting the formation of focal contacts by the cells with the substratum. Growth cone motility also depend on the mechanical behavior of the actin-based cytoskeleton. As noted by Reichardt and Tomaselli,[54] integrin-mediated interactions with F-actin can regulate actin polymerization, migration, and subsequent depolymerization.

In studying white blood cells, we find[55] that multiple cell adhesion receptors are involved in the execution of a given task. In general, several different receptors operate cooperatively in most adhesion operations. In many instances cross talk leads to receptor activation. Effective adhesion can be produced by combined effects of several receptors and by coupling weak specific and strong nonspecific receptors. Typically a given cell will have a collection of cell-surface receptors and operate by multiple binding to multiple ligands on the opposing cell or matrix.

There are some differences in the neurite-promoting activities of the various classes of cell adhesion molecules. In a study of neurite outgrowth from rat cerebellar neurons cultured on a substrate of NCAM-expressing cells, Doherty et al.[56] found that the critical element in inducing neurite outgrowth was the relative level of NCAM expression. In particular, there was a critical threshold level of NCAM expression at the cell surface below which there was no outgrowth. Once the threshold concentration was exceeded, there was a rapid rise in the neurite response as measured by several indicators such as neurite length and branches per neurite. Thus NCAM is highly cooperative[50,56] with a clear threshold response. In contrast, N-cadherin is rather linear in its response properties.

5.8 SIGNAL TRANSDUCTION AND INTEGRATION

5.8.1 Dynamic Regulation of Cell Surface Receptors

Transmembrane signaling follows receptor-ligand interactions, and all classes of cell surface receptors are dynamically regulated on neurons. In studies of the neurite outgrowth-promoting properties of cell adhesion molecules, Doherty et al.[57,58] observed a time-dependent switching from neurite outgrowth-promoting activity to a stability-promoting mode. Two regulatory influences were shown to be involved in the regulation of plasticity. Polysialic acid (PSA) was found to be a positive modulator of NCAM-dependent neurite outgrowth, while the NCAM isoform switching (exon VASE) was found to be a negative modulator, downregulating plasticity and promoting adhesive interactions. Loss of PSA and VASE expression produced the observed bahavioral changes.

Integrins operate dynamically in more than one way and in concert with other receptors. They interact directly with cytoskeletal proteins, and they activate second-messenger pathways. In more detail,[59] receptor clustering or aggregation can result in the formation of focal points of contact where integrins link ECM proteins to cytoskeletal complexes and actin filaments. Protein phosphorylation is an early event that follows ligand binding. The linking of integrin receptors to tyrosine kinases can regulate tyrosine phosphorylation activated by integrin receptor clustering. Integrin activation can induce increases in Ca^{2+} signaling, and integrin signaling pathways may be integrated with other receptor pathways through G-protein-coupled receptors. For instance, G-protein-linked receptors may enhance the aforementioned integrin-dependent tyrosine phosphorylation. Finally, integrins can be downregulated upon establishment of functional contacts with their targets. For example, Cohen et al.[60] have shown that the ablation of the optic tectum reduces the loss of functional laminin receptors on retinal ganglion cells. This downregulation appears to be epigenetically triggered by the target tissue.

5.8.2 Intracellular Calcium Signaling

Evidence has accumulated[61] in support of the notion that intracellular calcium regulates neurite outgrowth and growth cone motility. The resulting calcium hypothesis[61] is a statement that if the level of calcium either falls below an optimal range or rises above it, motility and outgrowth are inhibited. In this picture calcium influences are graded, and the regulation is homeostatic—perturbations of the calcium levels are opposed so that the levels are restored to near-basal concentrations. The calcium hypothesis provides one possible mechanism for an electrical-activity-dependent stop signal. The idea is that excitatory neurotransmitters lead to alterations in membrane potential that result in the opening of voltage-sensitive calcium channels, producing an increase in intracellular calcium levels that serves to inhibit further motility and outgrowth. One prediction of this hypothesis is that any effect produced by a stimulus will depend on the calcium level at the time of the stimulation. Thus electrical activity can either increase or decrease motility and outgrowth, dependent on the calcium levels present at the time of stimulation. As already noted, growth cones integrate information from multiple cues, including those provided by bound and soluble (e.g., neurotransmitter) molecules and electrical activity. Neurotransmitters and action potentials each cause changes in the intracellular calcium levels, and the integrated overall level will determine whether to continue to grow or to stop.

In the classification of Tsien and Tsien,[62] there are four types of voltage-gated calcium channels. These are designated as L, T, N, and P. Common characteristics of these channels include a steeply depolarization-dependent activation, selectivity by high affinity binding of Ca^{2+}, and selective modulation by neurotransmitters, G-proteins, and diffusible molecules. Another common feature is their nonuniform spatial distribution; that is, they may be found clustered together in one region and absent in another. These channels differ from one another in their single-channel conductance, voltage- and time-dependent kinetics of their opening and closing, pharmacology, and cellular distribution. The high-voltage-activated L and N-channels are found in neurons.

It is well established that electrical activity influences the pattern of connectivity. In the preceding paragraphs we discussed biochemical mechanisms, especially those involving intracellular calcium that link the promotion and cessation of growth cones motility to electrical activity. Another possible connection of these processes to growth cone turning has been made by Silver et al.[63] The idea is that unstimulated growth cones grow straight forward, but electrical activity, and therefore the marginal growth caused by electrical activity, is asymmetric. Calcium hot spots produced by clustering of L-type Ca^{2+} channels may provide a mechanism for triggering the turning of electrically active neurites. This would happen when Ca^{2+} concentrations in a hot spot rise above the level needed to activate proteins that cut, nucleate, and cap actin filaments. Lastly, we note that Ca^{2+} participates in a number of molecular signaling networks, including the Ras pathway which transduces extracellular signals into changes in gene expression.[64]

5.8.3 G-Proteins

G-protein-linked receptors,[65] mentioned earlier, are a large family of seven-pass, membrane-spanning proteins that respond to extracellular signaling agents such as hormones and neurotransmitters. Included in this family of proteins are receptors for odorants in the neuroepithelium of the nose, for light in rods and cones in the retina, for acetylcholine in heart muscle cells, for endorphins and opioids in neurons, and for mating signals in baker's yeast. This highly conserved family of proteins appears to have evolved from sensory receptors in ancient unicellular organisms.

The heterotrimeric G-proteins, to which these receptors are linked, are loosely bound to the inner surface of the plasma membrane. Their three distinct subunits are termed α, β, and γ. In the resting state the three units form a complex that is bound by guanosine diphosphate (GDP). When a ligand binds to a G-protein-linked receptor, the receptor undergoes a conformational change that allows it to bind to a G-protein. This binding promotes the release of GDP by the alpha subunit. The site left vacant by the release can be occupied by an intracellular guanosine triphosphate (GTP) molecule. When this happens, the alpha subunit becomes activated. The activated alpha unit next dissociates from the $\beta\gamma$ complex (the beta and gamma units remain bound to one another to form a single complex) and diffuses along the inner membrane to the site of an effector molecule, which is then activated. The alpha unit subsequently deactivates itself by the hydrolysis of ATP to ADP and reassociates with a $\beta\gamma$ complex. Thus a G-protein-linked receptor act indirectly to regulate the activity of an effector molecule.

The G-protein effector molecule is a separate plasma-membrane-bound protein. This target molecule can be either an enzyme or an ion channel. The activation of the target protein can result in either an alteration in the concentration of an intracellular mediator (enzyme) or a change in the permeability of the plasma membrane (ion channel). Therefore we have a mechanism by which diffusible molecules that bind to a G-protein-linked receptor in the plasma membrane can influence the response properties of other transmembrane receptors such as a voltage-gated calcium channels. The intracellular mediators may alter the behavior of still other proteins in the cell, and by this means signals received by the G-protein-linked receptors at the cell surface can be relayed to the nucleus where they may influence gene expression.

The GTP-binding proteins, G_0 and G_i, are found in the brain. Actin, tubulin, and the G_0 alpha and beta subunits are the most abundant proteins in the growth cone membrane.[66] Some of the identified targets of neural G-proteins are voltage-dependent calcium channels, several types of potassium channels, and phospholipase C, an enzyme that changes second-messenger calcium levels. Another molecule found in large concentrations in growth cones is the growth-associated protein (GAP) known as GAP-43. Experiments by Strittmatter et al.[66] demonstrate that the binding by G_0 to the GTP analog, GTP-γ-S, is enhanced by GAP-43. The significance of this activity is that it shows that GAP-43 can serve as a link between the local intracelllular environment and the activation of the alpha subunit of G_0. Thus G_0 in the growth cone may be regulated both intracellularly and extracellularly. The transduction and integration[67] of signals by G_0 is depicted in Fig. 5.12.

The integrative actions of G-proteins are strongly cooperative. The state of activity of a G_0-protein at a point in time depends on its state of activity at its immediately preceding time step, on signals received from outside, and on the states of molecular elements in its local intracellular microenvironment (that perhaps relay other extracellular signals). There are many versions of this theme. For example, G-protein $\beta\gamma$ subunits also have regulatory effects. The response of an effector to an α subunit may be enhanced, inhibited, or neither, by a $\beta\gamma$ dimer.[68] In some cases a $\beta\gamma$ complex belonging to a different G-protein type conveys a coincident signal from other molecular elements in the microenvironment (perhaps relayed from outside). The interactions between the two subunits of the two different G-protein may be direct or they may both bind to a third molecule which serves as a coincidence detector.

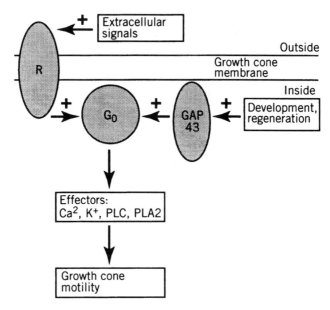

Figure 5.12. Transduction and integration. Schematic view of the role of G_0 in the integration of extracellular and intracellular signals in growth cones. (From Strittmatter and Fishman[67]. Reprinted with permission of ICSU Press.)

5.9 DYNAMICS

In our study of the retinotectal projection, we observe that the multiple interactions and cooperativity support the presence of multiple stable states. Biological systems are adaptive in the sense that they are able to exploit this multiplicity by finding useful minima (optima). We are able to observe this neural plasticity by altering the mix of interactions, disabling some while strengthening others through surgical manipulations. The presence of multiple stable states and the order-disorder transitions we have been exploring in an equilibrium dynamics in this chapter may be regarded as *static* adaptive properties. The full dynamics come into play when we consider the responses of these systems to variations in their parameters. The presence of a variety of stable states over the system parameter ranges, coupled to an ability of the system to alter its parameters as the need arises in response to signaled changes in the internal and external environments, may be viewed as *dynamic* adaptive properties of the system.

5.9.1 Reorganization in the Adult Cortex

The central nervous system of the adult retains its adaptive capabilities. We observe dynamic adaptive changes in adult mammalian receptive fields and topographic organization when we induce injuries (lesions) and when we modify the character of the input stimuli. Plastic changes have been found in a variety of primary cortical areas and thalamic nuclei, and they seem to operate at the synaptic level through mechanisms that are Hebbian. The changes in synaptic efficiency may be promoted by the dense networks of horizontal connections and by the feedforward and feedback circuitry.

The development of new receptive field properties by neurons divested of their normal inputs has been demonstrated in deprivation studies by Pons et al.,[70] in surgical alterations of the sensory input patterns by Allard et al.[71] and in lesion studies by Gilbert and Weisel.[72] Dynamic reorganizations of topographical maps have been observed in thalamic nuclei by Garraghty and Kaas[73] and by Nicholelis et al.,[74] and in the primary cortices by Jacobs and Donoghue.[75] Evidence and arguments in favor of mechanisms involving changes in synaptic efficiency, that is, by unmasking processes involving alterations in the balance between excitation and inhibition, were presented by Gilbert and Weisel,[72] Jacobs and Donoghue[75] and Hess and Donoghue.[76] In these experiments changes in receptive field properties begin to appear within a few minutes to a few hours after altering normal input to the cells.

5.9.2 Time Course of Developmental Events

We end this chapter by noting that the morphogenesis of the primate LGN has been studied in a dynamic model by Tzonev, Schulten and Malpeli.[77] Their findings not only confirm the earlier conclusions of Lee and Malpeli, but also extend the inferences. In the investigation of the time course of laminar development, Tzonev et al.[77] find that local cell-cell interactions generate a wave of development of neuronal receptive fields that propagates through the nucleus and establishes the two distinct laminar patterns. In the dynamic picture the gaps not only trap the transition in its stereotypic position but also induce the change in lamination. At the position of the transition the system is in a metastable state, and perturbations in retinotropy due to the gaps bring on the transition to the favored four-layered pattern.

5.10 REFERENCES

1. Goodman, C. S., and Shatz, C. J. (1993). Developmental mechanisms that generate precise patterns of neuronal connectivity. Cell, **72**/Neuron, **10** (suppl.), 77–98.
2. Shatz, C. J. (1983). The prenatal development of the cat's retinogeniculate pathway. J. Neurosci., **3**, 482–499.
3. Shatz, C. J., and Sretavan, D. W. (1986). Interactions between retinal ganglion cells during the development of the mammalian visual system. Ann. Rev. Neurosci., **9**, 171–207.
4. Udin, S. B., and Fawcett, J. W. (1988). Formation of topographic maps. Ann. Rev. Neurosci., **11**, 288–327.

5. Rakic, P. (1977). Prenatal development of the visual system in rhesus monkey. Proc. Roy. Soc. Lond., **B278**, 245–260.
6. Rakic, P. (1981). Development of visual centers in the primate brain depends on binocular competition before birth. Science, **214**, 928–931.
7. Lee, D., and Malpeli, J. G. (1994). Global form and singularity: Modeling the blind spot's role in lateral geniculate morphogenesis. Science, **263**, 1292–1294.
8. Lee, D., and Malpeli, J. G. (1994). Role of the blind spot in the laminar morphogenesis of the rhesus lateral geniculate nucleus: A thermodynamic model. Soc. Neurosci. Abst., **19**, 525.
9. Shatz, C. J., and Stryker, M. P. (1988). Prenatal tetrodotoxin infusion blocks segregation of retinogeniculate afferents. Science, **242**, 87–89.
10. Cline, H. T., and Constantine-Paton, M. (1989). NMDA receptor antagonists disrupt the retinotectal topographic map. Neuron, **3**, 413–426.
11. Sperry, R. W. (1963). Chemoaffinity in the orderly growth of nerve fiber patterns and connections. Proc. Nat. Acad. Sci. USA, **50**, 703–710.
12. Harris, W. A., Holt, C. E., and Bonhoeffer, F. (1987). Retinal axons with and without somata, growing to and arborizing in the tectum of *Xenopus* embryos: A time-lapse video study of single fibres *in vivo*. Development, **101**, 123–133.
13. Fraser, S. E., and O'Rourke, N. (1990). *In situ* analysis of neuronal dynamics and positional cues in the patterning of nerve connections. J. Exp. Biol., **153**, 61–70.
14. O'Rourke, N. A., Cline, H. T., and Fraser, S. E. (1994). Rapid remodeling of retinal arbors in the tectum with and without blockade of synaptic transmission. Neuron, **12**, 921–934.
15. Galli, L., and Maffei, L. (1988). Spontaneous impulse activity of rat retinal ganglion cells in prenatal life. Science, **242**, 90–91.
16. Maffei, L., and Galli-Resta, L. (1990). Correlation in the discharges of neighboring rat retinal ganglion cells during prenatal life. Proc. Nat. Acad. Sci. USA, **87**, 2861–2864.
17. Meister, M., Wong, R. O. L., Baylor, D. A., and Shatz, C. J. (1991). Synchronous bursts of action potentials in ganglion cells of the developing mammalian retina. Science, **252**, 939–943.
18. Wong, R. O. L., Meister, M., and Shatz, C. J. (1993). Transient period of correlated bursting activity during development of the mammalian retina. Neuron, **11**, 923–938.
19. Fraser, S. E., and Perkel, D. H. (1990). Competitive and positional cues in the patterning of nerve connections. J. Neurobiol., **21**, 51–72.
20. Harris, W. A. (1989). Local positional cues in the neuroepithelium guide retinal axons in embryonic *Xenopus* brain. Nature, **339**, 218–221.
21. Constantine-Paton, M., and Law, M. I. (1978). Eye-specific termination bands in tecta of three-eyed frogs. Science, **202**, 639–641.
22. Cline, H. T., Debski, E. A., and Constantine-Paton, M. (1987). *N*-Methyl-D-Aspartate receptor antagonist desegregates eye-specific stripes. Proc. Nat. Acad. Sci. USA, **84**, 4342–4345.
23. Prestige, M. C., and Willshaw, D. J. (1975). On a role for competition in the formation of patterned connections. Proc. Roy. Soc. Lond., **B190**, 77–93.

24. Hope, R. A., Hammond, B. J., and Gaze, R. M. (1976). The arrow model: retinotectal specificity and map formation in the goldfish visual system. Proc. Roy. Soc. Lond., **B194**, 447–466.
25. Whitelaw, V. A., and Cowan, J. D. (1981). Specificity and plasticity of retinotectal connections: A computational model. J. Neurosci., **1**, 1369–1387.
26. Willshaw, D. J., and Malsburg, C. von der (1976). How patterned neural connections can be set up by self-organization. Proc. Roy. Soc. Lond., **B194**, 431–445.
27. Willshaw, D. J., and Malsburg, C. von der (1979). A marker induction mechanism for the establishment of ordered neural mappings: Its application to the retinotectal problem. Philos. Trans. R. Soc. Lond., **B287**, 203–243.
28. Fraser, S. E. (1980). A differential adhesion approach to the patterning of nerve connections. Dev. Biol., **79**, 453–464.
29. Gierer, A. (1981). Development of projections between areas of the nervous system. Biol. Cybern., **42**, 69–78.
30. Gierer, A. (1983). Model for the retinotectal projection. Proc. Roy. Soc. Lond., **B218**, 77–93.
31. Gierer, A. (1987). Directional cues for growing axons forming the retinotectal projection. Development, **101**, 479–489.
32. Steinberg, M. S. (1970). Does differential adhesion govern self-assembly processes in histogenesis? Equilibrium configurations and the emergence of a hierarchy among populations of embryonic cells. J. Exp. Zool., **173**, 395–434.
33. Letourneau, P. C., and Shattuck, T. A. (1989). Distribution and possible interactions of actin-associated proteins and cell adhesion molecules of nerve growth cones. Development, **105**, 505–519.
34. Bixby, J. L., Pratt, R. S., Lilien, J., and Reichardt, L. F. (1987). Neurite outgrowth on muscle cell surfaces involves extracellular matrix receptors as well as Ca^{2+}-dependent and -independent cell adhesion molecules. Proc. Nat. Acad. Sci. USA, **84**, 2555–2559.
35. Serafini, T., Kennedy, T. E., Galko, M. J., Mizayan, C., Jessel, T. M., and Tessier-Lavigne, M. (1994). The netrins define a family of axon outgrowth-promoting proteins homologous to *C. elegans* UNC-6. Cell, **78**, 409–424.
36. Kennedy, T. E., Serafini, T., de la Torre, J. R., and Tessier-Lavigne, M. (1994). Netrins are diffusible chemotropic factors for commissural axons in the embryonic spinal cord. Cell, **78**, 425–435.
37. Tessier-Lavigne, A., Placzek, M., Lumsden, A. G. S., Dodd, J., and Jessell, T. M. (1988). Chemotropic guidance of developing axons in the mammalian central nervous system. Nature, **336**, 775–778.
38. Placzek, M., Tessier-Lavigne, M., Jessell, T., and Dodd, J. (1990). Orientation of commissural axons *in vitro* in response to a floor plate-derived chemoattractant. Development, **110**, 19–30.
39. Lumsden, A. G. S., and Davies, A. M. (1983). Earliest sensory nerve fibres are guided to peripheral targets by attractants other than nerve growth factor. Nature, **306**, 786–788.
40. Heffner, C. D., Lumsden, A. G. S., and O'Leary, D. D. M. (1990). Target control of collateral extension and directional axon growth in the mammalian brain. Science, **247**, 217–220.

41. Luo, Y., Raible, D., and Raper, J. A. (1993). Collapsin: A protein in brain that induces the collapse and paralysis of neuronal growth cones. Cell., **75**, 217–227.
42. Walter, J., Henke-Fahle, S., and Bonhoeffer, F. (1987). Avoidance of posterior tectal membranes by temporal retinal axons. Development, **101**, 909–913.
43. Kampfhammer, J. P., and Raper, J. A. (1987). Collapse of growth cone structure on contact with specific neurites in culture. J. Neurosci., **7**, 201–212.
44. Raper, J. A., and Kampfhammer, J. P. (1990). The enrichment of a neuronal growth cone collapsing activity from embryonic chick brain. Neuron, **4**, 21–29.
45. Pini, A. (1993). Chemorepulsion of axons in the developing mammalian central nervous system. Science, **261**, 95–98.
46. Fitzgerald, M., Kwiat, G. C., Middleton, J., and Pini, A. (1993). Ventral spinal cord inhibition of neurite outgrowth from embryonic rat dorsal root ganglia. Development, **117**, 1377–1384.
47. Baier, H., and Bonhoeffer, F. (1992). Axon guidance by gradients of a target-derived component. Science, **255**, 472–475.
48. Goodman, C. S. (1994). The likeness of being: Phylogenetically conserved molecular mechanisms of growth cone guidance. Cell, **78**, 353–356.
49. Hynes, R. O. (1992). Integrins: Versatility, modulation and signalling in cell adhesion. Cell, **69**, 11–25.
50. Cunningham, B. A., Hemperly, J. J., Murray, B. A., Prediger, E. A., Brackenbury, R., and Edelman, G. M. (1987). Neural cell adhesion molecule: Structure, immumoglobulin-like domains, cell surface modulation and alternative RNA splicing. Science, **236**, 799–806.
51. Edelman, G. M., and Crossin, K. L. (1991). Cell adhesion molecules: Implications for a molecular histology. Ann. Rev. Biochem., **60**, 155–190.
52. Takeichi, M. (1991). Cadherin cell adhesion receptors as a morphogenetic regulator. Science, **251**, 1451–1455.
53. Calof, A. L., and Lander, A. D. (1991). Relationship between neuronal migration and cell-substratum adhesion: Laminin and merosin promote olfactory neuronal migration but are antiadhesive. J. Cell Biol., **115**, 779–794.
54. Reichardt, L. F., and Tomaselli, K. J. (1991). Extracellular matrix molecules and their receptors: Functions in neural development. Ann. Rev. Neurosci., **14**, 531–570.
55. Hynes, R. O., and Lander, A. D. (1992). Contact and adhesion specificities in the associations, migrations and targeting of cells and axons. Cell, **68**, 303–322.
56. Doherty, P., Fruns, M., Seaton, P., Dichson, G., Barton, C. H., Sears, T. A., and Walsh, F. S. (1990). A threshold effect of the major isoforms of NCAM on neurite outgrowth. Nature, **343**, 464–466.
57. Doherty, P., Skaper, S. D., Moore, S. E., Leon, A., and Walsh, F. S. (1992). A developmentally regulated switch in neuronal responsiveness to NCAM and N-cadherin in the rat hippocampus. Development, **115**, 885–892.
58. Doherty, P., Moolenaar, C. E. C. K., Ashton, S. V., Michalides, R. J. A. M., and Walsh, F. S. (1992). The VASE exon downregulates the neurite growth-promoting activity of NCAM 140. Nature, **356**, 791–793.
59. Clark, E. A., and Brugge, J. S. (1995). Integrins and signal transduction pathways: The road taken. Science, **268**, 233–239.

60. Cohen, J., Nurcombe, V., Jeffrey, P., and Edgar, D. (1989). Developmental loss of functional laminin receptors on retinal ganglion cells is regulated by their target tissue, the optic tectum. Development, **107**, 381–387.
61. Kater, S. B., and Mills, L. R. (1991). Regulation of growth cone behavior by calcium. J. Neurosci., **11**, 891–899.
62. Tsien R. W., and Tsien, R. Y. (1990). Calcium channels, stores, and oscillations. Ann. Rev. Cell Biol., **6**, 715–760.
63. Silver, R. A., Lamb, A. G., and Bolsover, S. R. (1990). Calcium hotspots caused by L-channel clustering promote morphological changes in neuronal growth cones. Nature, **343**, 751–754.
64. Ghosh, A., and Greenberg, M. E. (1995). Calcium signaling in neurons: Molecular mechanisms and cellular consequences. Science, **268**, 239–247.
65. Neer, E. J. (1995). Heterotrimeric G-proteins: Organizers of transmembrane signals. Cell, **80**, 249–257.
66. Strittmatter, S. M., Valenzuela, D., Kennedy, T. E., Neer, E. J., and Fishman, M. C. (1990). G_0 is a major growth cone protein subject to regulation by GAP-43. Nature, **344**, 836–841.
67. Strittmatter, S. M., and Fishman, M. C. (1991). The neuronal growth cone as a specialized transduction system. BioEssays, **13**, 127–134.
68. Tang, W.-J., and Gilman, A. G. (1991). Type-specific regulation of adenylyl cyclase by G protein $\beta\gamma$ subunits. Science, **254**, 1500–1503.
69. Bourne, H. R., and Nicoll, R. (1993). Molecular machines integrate coincident synaptic signals. Cell, **72**/Neuron, **10** (suppl.). 65–75.
70. Pons, T. P., Garraghty, P. E., Ommaya, A. K., Kaas, J. H., Taub, E., and Mishkin, M. (1991). Massive cortical reorganization after sensor deafferentation in adult macaques. Science, **252**, 1857–1860.
71. Allard, T., Clark, S. A., Jenkins, W. M., and Merzenich, M. M. (1991). Reorganization of somatosensory area 3b representations in adult owl monkeys after digital syndactyly. J. Neurophysiol., **66**, 1048–1058.
72. Gilbert, C. D., and Weisel, T. N. (1992). Receptive field dynamics in adult primary visual cortex. Nature, **356**, 150–152.
73. Garraghty, P. E., and Kaas, J. H. (1992). Dynamic features of sensory and motor maps. Curr. Opin. Neurobiol., **2**, 522–527.
74. Nicolelis, M. A. L., Lin, R. C. S., Woodward, D. J., and Chapin, J. K. (1993). Induction of immediate spatiotemporal changes in thalamic networks by peripheral block of ascending cutaneous information. Nature, **361**, 533–536.
75. Jacobs, K. M., and Donoghue, J. P. (1991). Reshaping the cortical motor map by unmasking latent intracortical connections. Science, **251**, 944–947.
76. Hess, G., and Donoghue, J. P. (1994). Long-term potentiation of horizontal connections provides a mechanism to reorganize cortical motor maps. J. Neurophysiol., **71**, 2543–2547.
77. Tzonev, S., Schulten, K., and Malpeli, J. G. (1995). Morphogenesis of the lateral geniculate nucleus: How singularities affect global structure. In G. Tesauro, D. Touretzky and T. Lean (eds.), *Advances in Neural Information Processing Systems*, vol. 7. Cambridge: MIT Press.

6

MARKOV RANDOM FIELDS

6.1 DEFINITIONS AND INTRODUCTORY REMARKS

A Markov random field (MRF) is a mathematical generalization of the notion of a one-dimensional temporal Markov chain to a two-dimensional (or higher) spatial lattice or graph. The mathematical properties of systems endowed with spatial Markovian properties were first elucidated by Lévy[1] and by Dobruschin.[2] Lévy's spatial Markovian process, as refined by Wong[3] and extended to discrete systems by Woods,[4] is known as a Pth-order Markov random field or, alternatively, a Markov P-process.

6.1.1 The Markov P-Process

We can describe a Markov P-process with respect to a two-dimensional space endowed with a metric in the following informal manner. Let us consider a pair of smooth, closed curves forming a band of minimum width P. This band, composed of contiguous lattice points, separates the two-dimensional space into an inside G^+ and an outside G^- so that each point in G^+ is at least a distance P from any point in G^-, and vice versa. The partitioning of our space by the band is illustrated in Fig. 6.1. The points of the lattice are the sites of the elements of interest to us, such as atomic spins or grey-valued pixels. These elements may each occur in one of a number of states in a manner that can be described probabilistically. A particular instance of the system is called a configuration. We now let X_{G^+} denote any particular configuration of lattice elements associated with sites in G^+. Similarly we let X_{G^-} and $X_{\delta G}$ designate configurations restricted to lattice sites in G^- and in the boundary band δG. The spatial Markovian property is expressed as a requirement on the form of

DEFINITIONS AND INTRODUCTORY REMARKS

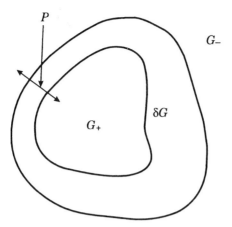

Figure 6.1. Inner, outer, and boundary regions of a Markov P-process. (From Derin and Kelly[7]. Reprinted with permission of IEEE.)

the conditional probabilities for the system. We have a Markov P-process if and only if the conditional probabilities satisfy the relation

$$p(X_{G^+} | X_{G^-}, X_{\delta G}) = p(X_{G^+} | X_{\delta G}) \tag{6.1}$$

for all such bands δG. That is, in a Markov P-process outside and inside lattice elements do not interact directly with one another. If we think of the inside and outside as representing the past and future, then a Markov P-process is one for which the past and future are independent given the present.

6.1.2 Markov Random Fields and Neighborhood Systems

Let us consider a discrete two-dimensional rectangular lattice Λ, and let X_{mn} denote a random variable defined on Λ that takes on the values x_{mn} at lattice site (m, n). We define a configuration of the lattice system to be the set of values of the random variables, one for each lattice site. This definition of a configuration can be written

$$X = \{x_{mn}\} = \{x_{mn}; (m, n) \in \Lambda\} \tag{6.2}$$

If we now define a joint probability measure p on the set of all possible configurations Ω of the random variable over the lattice or any sublattice, then the tuple (Ω, Λ, p) is a *random field*.

A Markov random field can be thought of best as a generalization of the Ising model studied in Chapter 3. Recall that in the Ising model a given spin element interacted with its four nearest neighbors. These four sites are said to form a nearest-neighbor, or neighborhood, system. If, for the two-dimensional

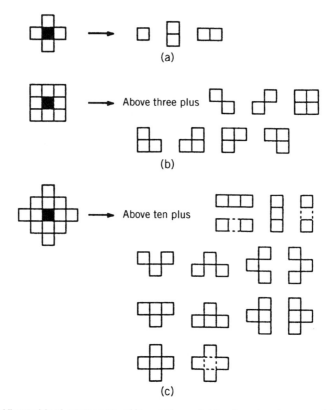

Figure 6.2. Hierarchical sequence of low-order neighborhood systems and their associated cliques: First-order (a), second-order (b), and third-order (c) neighborhood systems shown to the left of the arrows. The cliques belonging to these systems are given to the right of the arrows. (From Derin and Kelly[79]. Reprinted with permission of IEEE.)

Ising spin system, (i,j) is the site that labels the neighborhood, then the set of sites $(i, j - 1)$, $(i, j + 1)$, $(i - 1, j)$, and $(i + 1, j)$ constitutes a first-order neighborhood system. Neighborhood systems of several different orders are depicted in Fig. 6.2. It may be noted that in contrast to the inside-outside ordering of a Markov P-process and the temporal ordering of a Markov chain, there is no notion of causality in these neighborhood systems. In addition we observe in Fig. 6.2 that the labeling site is not part of the neighborhood system. We are now ready to formalize the notion of a neighborhood in a lattice. A *neighborhood system* N_{ij} associated with a lattice Λ is defined as

$$N_{ij} = \{\eta_{ij} \subset \Lambda; (i, j) \in \Lambda\} \tag{6.3}$$

such that

$$(i, j) \notin \eta_{ij}$$
$$(k, l) \in \eta_{ij} \Rightarrow (i, j) \in \eta_{kl} \tag{6.4}$$

DEFINITIONS AND INTRODUCTORY REMARKS   185

A *Markov random field* is a random field for which:

1. The joint probability distribution has associated conditional probabilities that are local in character; that is, they obey the spatial Markovian relationship

$$p(X_{mn} = x_{mn} | X_{rs} = x_{rs}, rs \neq mn) = p(X_{mn} = x_{mn} | X_{rs} = x_{rs}, rs \in \eta_{mn}) \quad (6.5)$$

   for all random variables defined on the lattice, and where η_{mn} is the neighborhood system for lattice site (m, n).
2. The probability distribution is positive definite for all values of the random variable.
3. The conditional probabilities are invariant with respect to neighborhood translations.

6.1.3 Gibbs Random Fields and Clique Potentials

We now introduce a second way of defining a random field by adopting the Gibbs distribution discussed in earlier chapters as our probability measure. We define the tuple (Ω, Λ, p) to be a *Gibbs random field* if and only if its joint probability distribution is of the form

$$p(X = \mathbf{x}) \equiv p(\mathbf{x}) = \frac{1}{Z} \exp\left(-\frac{1}{T} U(\mathbf{x})\right) \quad (6.6)$$

where we use a lowercase vector (bold) to designate a configuration of the lattice system. The quantity $U(\mathbf{x})$ is a potential that encapsulates the global properties of the system, and in the absence of an explicit kinetic term, $U(\mathbf{x})$ is formally identical to the hamiltonian discussed in our earlier chapters. As before, Z is the partition function, the normalization constant representing the sum over all configurations of the system:

$$Z = \sum_{\mathbf{x} \in \Omega} \exp\left(-\frac{1}{T} U(\mathbf{x})\right) \quad (6.7)$$

The parameter $\lambda = 1/T$ is our familiar inverse "temperature," which will be used when we discuss simulated annealing. The temperature parameter has no constructive role in establishing the Gibbs-Markov equivalence, and it will be set to unity until needed.

Guided by our study of the Ising model, we consider the following class of potentials: First, we want the total potential to be a sum of individual contributions from potentials V_i for each lattice site

$$U(\mathbf{x}) = \sum_{i \in \Lambda} V_i(x_i) \quad (6.8)$$

Second, we incorporate the key idea of nearest-neighbor interactions by decomposing the contribution from each lattice site into a sum of clique potentials V_c:

$$V_i(x_i) = \sum_{c \in C_i} V_c(x_c) \qquad (6.9)$$

In an Ising hamiltonian, we have contributions from single sites denoting interactions of a spin element with an external field and from two-body terms representing interactions between adjacent spin elements. Cliques and clique potentials are a generalization of the nearest-neighbor interactions of the Ising model. They provide us with a means of portraying a sequence of progressively more elaborate sets of interactions involving a site element and its neighbors, starting with single-element potentials and including pair interactions, three-body interactions, and so on.

We formally define a *clique* to be either a single site or a collection of site plets in which every site is a neighbor of every other site. The quantity C_i appearing in the summation in Eq. (6.9) is the set of all cliques associated with the site i. Shown in Fig. 6.2 are the cliques that can be built from elements of the first-, second-, and third-order neighborhood systems. As their name indicates, *clique potentials* describe the interactions between clique elements. We will use them to construct a polynomial expansion of the site potential in Section 6.3, and we will employ them to solve a variety of image-processing problems in several of the following sections of this chapter.

6.1.4 The Gibbs-Markov Equivalence

The conditional probability structure must satisfy consistency conditions in order for there to be an associated joint probability distribution, and there is more than one way of endowing a nearest-neighbor system with a probabilistic structure. These issues were the subject of studies by Whittle,[5] Brook,[6] and Besag.[7] The most general form for the conditional probabilities of a nearest-neighbor system that are consistent with a joint probability emerges from the Hammersley-Clifford theorem.

The most important aspect of this theorem is that it asserts that the global character of a Gibbs random field (defined through local interactions) is equivalent to the purely local character of a Markov random field. The equivalence between the two ways of defining a random field is not surprising when considering that a Gibbs distribution with a potential composed of a sum of local contributions leads to a Markov random field. However, the converse that any Markov random field has an equivalent Gibbs random field expressed in terms of clique potentials, quantities that arise in the physics of the Ising model, is striking.

The Hammersley-Clifford[8] theorem establishes the equivalence between Gibbs and Markov random fields for finite lattices and graphs. A proof of the

Gibbs-Markov equivalence based on a polynomial expansion of the potential has been given for lattice systems by Besag.[9] Another proof of the equivalence based on Möbius inversion has been presented for finite graphs by Grimmett.[10] Additional details pertaining to the positivity condition on the probabilities, which serve to complete the exposition of the Gibbs-Markov equivalence, were subsequently provided by Moussouris,[11] and insightful discussions into the Gibbs-Markov equivalence have been given by Spitzer.[12]

The Hammersley-Clifford theorem also establishes the functional form of the conditional probability distribution at each lattice site imposed by the requirement of a mathematically consistent joint probability structure. We begin in Section 6.3 with Besag's[9] proof of the Hammersley-Clifford theorem and the construction of potential functions for lattice systems. We follow this discussion in Section 6.4 with Grimmett's proof of the Gibbs-Markov equivalence for finite graphs and a sketch of its recent application to neural systems.

6.2 RANDOM FIELDS AND IMAGE PROCESSES

Our focus in this chapter is primarily on Markov random fields and their use in image processing. Two-dimensional images are highly organized spatial systems. From the time of the earliest work on Markov random fields, their utility in the image-processing domain was recognized. Some of the data-processing tasks performed using Markov random fields are texture analysis and synthesis, and the reconstruction, segmentation, and fusion of images. Data processed in this manner include visual images, complex-valued synthetic aperture radar data, and registered range and reflectance images acquired using imaging laser radar operating at near-infrared wavelengths. In the remainder of this chapter we will consider two-dimensional images as our problem domain.

6.2.1 Random Field Models

There are several different kinds of random field model. Some are causal such as the Markov mesh random field of Abend, Harley, and Kanal[13] and the Pickard random field,[14] while others are noncausal. In causal models, a random variable X_{ij}, conditioned on random variables defined over a portion of the lattice above and to its left, will depend only on random variables at sites immediately above and to the left. If we denote the larger segment of the lattice above and to the left of a site (i,j) by Φ_{ij}, and we represent the small subset of that segment called the *support set* by S_{ij}, then we define a *causal random field* by the relation

$$p(x_{ij}|x_{kl};(k,l)\in\Phi_{ij}) = p(x_{ij}|x_{kl};(k,l)\in S_{ij}) \qquad (6.10)$$

The utility of the Markov mesh and Pickard causal models for image reconstruction processes has been discussed by Derin et al.[15] At this early stage in the evolution of a theory of cooperative processes, the tendency is to favor noncausal models in which there is no preferred directionality.

One of the most widely utilized classes of noncausal models are the Gauss-Markov random fields introduced by Woods[4] and elaborated upon by Moran,[16,17] Besag,[9] Chellappa,[18] and Sharma and Chellappa.[19] A second major class of model, which can be formulated as either a causal or noncausal random field, are the simultaneous autoregressive (SAR) models first described by Whittle[5] and then by Besag,[9] and utilized in the image-processing domain by Chellappa and Kashyap,[20] Kashyap and Chellappa,[21] Kashyap,[22] and Sharma and Chellappa.[19] We follow the discussions of Sections 6.3 and 6.4 with an overview of the major classes of random field models. We start our tour in Section 6.5 with Besag's[9] auto models. We next introduce the notion of a wide sense Markov process and begin a discussion of Gauss-Markov random fields that will be continued in a later section. We then continue in Section 6.6 with an exploration of some of the properties and uses of SAR random fields, and we characterize the image reconstruction process.

6.2.2 Gibbs Distributions and Simulated Annealing

One implication of the Gibbs-Markov equivalence is that we may exploit the global properties to model the system or problem of interest and then use the local characteristics to design a method or algorithm to evaluate the consequences. In particular, we may use simulated annealing to endow the system with an equilibrium dynamics and carry out a pixel-by-pixel iterative reconstruction of the image. In the method of Geman and Geman,[23] Markov random fields, Bayesian inferencing, and simulated annealing are combined to process two-dimensional images. The energy minimum sought in this stochastic approach represents a maximum a posteriori estimate of the true solution given the degraded data. We will examine this synergistic joining of Markov random fields with simulated annealing in Sections 6.7 and 6.8.

Several complementary approximations and alternatives to the method of Geman and Geman have been developed. We will study two of these methods, Marroquin's[24] maximizer of the posterior marginals and Besag's[25] iterated conditional modes of the posterior distribution, in Sections 6.9 and 6.10. Both of these methods provide us with algorithms with convergence properties that are more rapid than simulated annealing.

A central element in Geman and Geman's modeling of complex spatial structures is the use of multiple, coupled Markov random fields. For example, to reconstruct a digital image, at least two random field models are needed. One random field, the pixel process, treats the smoothly varying surface properties, and the second, a line process, handles the attendant surface discontinuities. Other random fields may be needed to assign physical or logical labels to elements of the image. Several examples of multiple random field models will be encountered in this chapter. We will examine their appearance in the Gauss-Markov models developed by Jeng and Woods[26] in Section 6.11.

6.2.3 Potentials and Deterministic Approaches

The interactions between and among pixels, line elements, and labels express the physical constraints of the system. In maximum a posteriori approaches, the potentials introduced through the Bayesian priors are used to generate physically meaningful reconstructions. One of our main objectives in studying image-processing applications is to explore ways of constructing useful potential functions. We will find that there are several sources of inspiration for the construction of prior potentials including regularization theory and a number of closely related mechanical models.

We will construct a number of useful potentials using mean-field annealing and a related mechanical approach in the next three sections. In the mean-field annealing models of Bilbro and Snyder[27] and Geiger and Girosi,[28] we have two deterministic approximations to the stochastic method of Geman and Geman that can be applied to continuous-valued pixel data. The mean-field annealing methods are closely related to another deterministic alternative, graduated nonconvexity. Developed by Blake and Zisserman,[29] graduated nonconvexity is a mechanical approach that provides insight into the cooperativity encapsulated by a Markov random field. The potential functions generated in these deterministic approaches are closely related to one another. In Sections 6.12–6.14 we will develop the mean-field formalism and resulting potentials and examine the corresponding graduated nonconvexity potential functions.

The final technical section of the chapter, Section 6.15, illustrates the breadth of the marriage of MRF and constrained optimization techniques. This section is devoted to a brief look at Gidas's[30] renormalization group method. In this approach Markov random fields, simulated annealing, and the renormalization group methods introduced in Chapter 3 are combined into a hierarchical method of performing multilevel-multiresolution image reconstruction tasks.

6.3 THE HAMMERSLEY-CLIFFORD THEOREM FOR FINITE LATTICES

We begin our proof of the Hammersley-Clifford theorem by defining a probability measure for the configurations, assuming that if $p(x_m) > 0$ at each site m, then the probability for the configuration, $p(x_1, \ldots, x_m, \ldots, x_r)$, exists and is strictly positive. We denote the joint probability of a configuration as $p(\mathbf{x})$ and designate the sample space of all possible configurations with positive probability as Ω:

$$\Omega = \{\mathbf{x}: p(\mathbf{x}) > 0\} \tag{6.11}$$

The joint and conditional probabilities are related to one another through Bayes's theorem. Specifically, we observe that

$$p(\mathbf{x}) = p(x_m | x_1, \ldots, x_{m-1}, x_{m+1}, \ldots, x_r) p(x_1, \ldots, x_{m-1}, x_{m+1}, \ldots, x_r) \quad (6.12)$$

We now assume that the value 0 is available at every site. By the positivity condition, we have $p(\mathbf{0}) > 0$, and we may define the quantity $Q(\mathbf{x})$ as

$$Q(\mathbf{x}) = \log\left(\frac{p(\mathbf{x})}{p(\mathbf{0})}\right) \quad (6.13)$$

for any configuration $\mathbf{x}$ in the sample space. We now designate by $\mathbf{x}_i$ any configuration in which the ith site takes on the value 0:

$$\mathbf{x}_i \equiv (x_1, \ldots, x_{i-1}, 0, x_{i+1}, \ldots, x_r) \quad (6.14)$$

We observe that it follows from Eqs. (6.12) and (6.13) that

$$\exp(Q(\mathbf{x}) - Q(\mathbf{x}_i)) = \frac{p(x_i | x_1, \ldots, x_{i-1}, x_{i+1}, \ldots, x_r)}{p(0 | x_1, \ldots, x_{i-1}, x_{i+1}, \ldots, x_r)} \quad (6.15)$$

We then introduce the mathematical generalization of the nearest neighbor interactions by expanding $Q(\mathbf{x})$ in the series

$$Q(\mathbf{x}) = \sum_{i=1}^{r} x_i G_i(x_i) + \sum_{i<j=1}^{r} x_i x_j G_{ij}(x_i, x_j) + \sum_{i<j<k=1}^{r} x_i x_j x_k G_{ijk}(x_i, x_j, x_k) + \cdots$$

$$(6.16)$$

We will now establish our main results in a form of a theorem that relates the G-functions to the Markovian properties of the conditional probabilities: *For any $i < j < \cdots < s$ in the range $1, 2, \ldots, r$, the functions, $G_{i,j,\ldots,s}$ are nonzero if and only if the sites form a clique.* The proof is as follows: From Eq. (6.15) we have that for any configuration $\mathbf{x}$, $Q(\mathbf{x}) - Q(\mathbf{x}_i)$ can only depend on x_i and the values at neighboring sites. For notational convenience, let us choose site i to be site 1. Then we observe that

$$Q(\mathbf{x}) - Q(\mathbf{x}_1) = x_1 \left[G_1(x_1) + \sum_{j=2}^{r} x_j G_{1j}(x_1, x_j) + \sum_{j<k=2}^{r} x_j x_k G_{1jk}(x_1, x_j, x_k) | \cdots \right]$$

$$(6.17)$$

can only depend on x_1, and the values at those sites that are neighbors of site 1. Suppose that site m is neither site 1 nor a neighbor of site 1. Then, the difference $Q(\mathbf{x}) - Q(\mathbf{x}_1)$ must be independent of site m for all configurations $\mathbf{x}$. Let us choose a configuration $\mathbf{x}$ for which $x_i = 0$ at all sites except at sites 1 and m. We find that since

$$Q(\mathbf{x}) - Q(\mathbf{x}_1) = x_1 G_1(x_1) + x_1 x_m G_{1m}(x_1, x_m) \qquad (6.18)$$

can only depend on site 1 and values at its neighbors, the quantity G_{1m} must vanish. A similar construction gives an identical result for G_{1mn} for a suitably chosen configuration, and so on. We may therefore conclude that the G-functions are nonzero only if the sites $i, j, \ldots,$ form a clique.

Conversely, any group of G-functions will generate probability distributions that are positive definite. Since the differences $Q(\mathbf{x}) - Q(\mathbf{x}_i)$ depend upon x_m only if there is a nonzero G-function linking x_i to x_m, we see that the conditional probabilities generated by the G-functions are Markovian by construction.

To conclude this section, we observe that the Hammersley-Clifford theorem is a statement of the equivalence between two ways of describing a system. One description is in terms of a system's local Markovian properties, while the second makes use of the statistical mechanics notion of a global Gibbsian energy composed of a sum of clique potentials.

6.4 RANDOM FIELDS ON GRAPHS

Up to this point the emphasis in our discussions of Markov random fields has been exclusively on spatial lattice systems. However, the notion of a Markov random field and its equivalence to a Gibbs distribution defined over a set of clique potentials is far more general. It can be applied to systems composed of interacting elements irregardless of their spatial organization. A particularly informative example of this universality is the application by Martignon et al.[31] of random field methods to the identification of cooperative assemblies of spiking neurons. The key observation in using Markov random fields to detect temporal correlations among spiking neurons, the problem being addressed by Martignon et al., is the Hammersley-Clifford theorem which allows us to construct a potential function of a random field defined on a graph. Before we discuss the construction of the potential function, we will describe what is meant by a cooperative neural assembly. We will then define Markov random fields on graphs and establish a Markov-Gibbs equivalence for these systems following the approaches of Grimmett,[10] Moussouris,[11] and Kindermann and Snell.[32]

6.4.1 Cooperative Assemblies

The concept of a neural assembly has its origins in the work of Sherrington[33] and Hebb.[34] In a Hebbian assembly the criterion for membership is coactivation of synaptic input. Interactions between members of the assembly are not necessary, since shared input may be sufficient to produce coactivation. A more modern and interesting way of defining a neural assembly is that it is a collection neural elements able to self-organize and function cooperatively to solve a given task.[35] With the development of simultaneous multicellular recording techniques, it is possible to detect assemblies of cells in operation. We can use this stronger definition of cooperativity as a means of recognizing assemblies as they form and reform.[35] The underlying interactions among the elements of the assembly will then manifest themselves through their (delayed) correlated spike trains.

The detection, analysis, and interpretation of correlations among neural spike trains has been the subject of several investigations. In an earlier study by Gerstein, Perkel, and Dayhoff,[36] a "gravity" method of analysis of temporally related firing patterns was developed. This approach provided a conceptual picture of assembly dynamics by mapping the correlated activities of neurons into the motions of particles in a multidimensional euclidean space. Viscous forces were introduced and particles that fired together formed cooperative clusters. In the approach of Martignon et al., these ideas together with concepts from graph and information theory are formalized using Gibbs potentials. The result is a method of detecting higher-order correlations among spiking events in populations of neurons.

6.4.2 Gibbs-Markov Equivalence for Random Fields on Graphs

Our primary goal in this section is to establish the Gibbs-Markov equivalence for the case where the systems is a set of sites of an arbitrary finite graph and where each site of a graph has a finite number of states. Following Grimmett, and others, we will exhibit the equivalence by using the Möbius inversion theorem to construct a preferred potential for the random field. We begin by defining a Markov random field over of finite graph.

A *graph* G on Λ is a set of vertices or sites and a set of edges connecting pairs of elements of Λ. If sites i and j are connected by an edge, they are said to be neighbors of one another. The set of neighbors of a site k defines a neighborhood of that site, denoted as N_k. A clique is a subset of Λ whose sites are all neighbors of one another. We can consider such a graph as representing sets of neurons (sites or vertices) and their interactions (edges). The interactions in turn may be regarded as representing those processes underlying the synchronous spiking. As illustrated in Fig. 6.3, we are given a collection of neurons, labeled 1 to N. Each of these neurons can be in one of a finite number of states, $1, 2, \ldots$. A configuration x of the system is an array of states of the form $(x_1, x_2, \ldots, x_N)$. That is, a configuration is an assignment of a label or color to each of the neurons of the graph. The interactions represented by edges (i, j) in the graph appear in the hamiltonian as nonzero terms containing the labels i and j as subindexes.

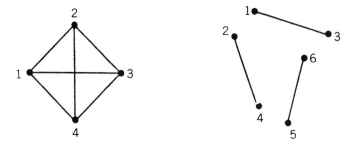

Figure 6.3. Two examples of graphs. (From Martignon et al.[31] Reprinted with permission of Springer-Verlag.)

As was the case for lattice systems, we describe the local characteristics of our system through conditional probabilities associated with the configurational joint probabilities. We again let Ω denote the space of all configurations of the system. If we introduce a probability measure p on (Ω, Λ), then the tuple (Ω, Λ, p) defines a random field. This random field is a Markov random field if the joint probabilties are positive definite for all configurations and if the conditional probabilities satisfy the Markovian relationship

$$p(X_i = x_i | X_j = x_j, i \neq j) = p(X_i = x_i | X_j = x_j, j \in N_i) \qquad (6.19)$$

Let us relabel our states to be $0, 1, 2, \ldots$. The state labeled 0 will assume the role of a preferred state, to be discussed further below. We denote by x_A the configuration that coincides with x on a subset A of Λ and assigns the label 0 to all sites outside of A. Also we use the notation $A + B$ and $A - B$ to indicate the union and difference A and B. We now have the following alternative way of defining a Markov random field. For any $x \in \Omega$, $A \subset \Lambda$, and nonneighboring pair of sites a and b, such that $a \in A, b \notin A$, a random field (Ω, Λ, p) is a Markov random field if the joint probabilties are positive definite for all configurations and satisfy the nearest-neighbor relationship

$$\frac{p(x_{A+b})}{p(x_A) + p(x_{A+b})} = \frac{p(x_{A+b-a})}{p(x_{A-a}) + p(x_{A+b-a})} \qquad (6.20)$$

The key element is the numerator which is a statement that the state at a given site b, given the states at all other sites, depends only on the states at neighboring sites, namely on the states at sites that interact with the given site. The denominator handles normalization issues. This nearest-neighbor relationship may be rewritten as

$$\frac{p(x_{A+b})}{p(x_A)} = \frac{p(x_{A+b-a})}{p(x_{A-a})} \qquad (6.21)$$

We next define a Gibbs distribution within this framework. A random field (Ω, Λ, p) is a Gibbs distribution if there exists a potential function V such that

$$p(x) = Z^{-1} \exp\left\{\sum_{K \subseteq L(x)} V(x_K)\right\} \tag{6.22}$$

for all $x \in \Omega$, where $L(x)$ is the set of sites not labeled 0 in the configuration, the summation is over all cliques in $L(x)$, and Z is the partition function

$$Z = \sum_{x \in \Omega} \exp\left\{\sum_{K \subseteq L(x)} V(x_K)\right\} \tag{6.23}$$

Now suppose that $f(A)$ and $g(A)$ are two real-valued functions defined on the subsets of a finite set such that

$$f(A) = \sum_{B \subseteq A} g(B) \tag{6.24}$$

Then the Möbius inversion formula tells us that the $g(B)$ can be recovered from $f(A)$ by forming the sum

$$g(B) = \sum_{C \subseteq B} (-1)^{|B-C|} f(C) \tag{6.25}$$

where $|C|$ denotes the number of elements in the set C. We can apply this inversion procedure to our potential. From our definition of a Gibbs distribution we have

$$\ln p(x_B) = \sum_{K \subseteq B} V(x_K) \tag{6.26}$$

and by Möbius inversion we obtain

$$V(x) = \sum_{B \subseteq L(x)} (-1)^{|L(x) - B|} \ln p(x_B) \tag{6.27}$$

We now have the following theorem (Grimmett): A random field (Ω, Λ, p) is a Gibbs distribution if and only if it is a Markov random field. If (Ω, Λ, p) is a Markov random field, then its potential function is given by Eq. (6.27) for all $x \in \Omega'$, where Ω' is restricted to those configurations of Λ that assign 0 to all sites of Λ except those belonging to cliques, and the summation is over all subsets of the clique $L(x)$. The proof of this theorem is as follows: Suppose that (Ω, Λ, p) is a Markov random field, and assume that the potential V is defined

RANDOM FIELDS ON GRAPHS 195

by Eq. (6.27) for all $x \in \Omega$. Now let $x \in \Omega$ and $A \subseteq L(x)$. Then we have

$$V(x_A) = \sum_{B \subseteq A} (-1)^{|A-B|} \ln p(x_B) \tag{6.28}$$

We now prove that $V(x_A)$ vanishes unless A is a clique. Suppose that A contains two sites, a and b, that are not neighbors. Then

$$V(x_A) = \sum_{\substack{B \subseteq A \\ x, y \in B}} (-1)^{|A-B|} \ln p(x_B) + \sum_{\substack{B \subseteq A \\ x \in B, y \notin B}} (-1)^{|A-B|} \ln p(x_B)$$

$$+ \sum_{\substack{B \subseteq A \\ x \notin B, y \in B}} (-1)^{|A-B|} \ln p(x_B) + \sum_{\substack{B \subseteq A \\ x, y \notin B}} (-1)^{|A-B|} \ln p(x_B) \tag{6.29}$$

$$= \sum_{B \subseteq A-a-b} (-1)^{|A-B|} \ln \left(\frac{p(x_{B+a+b}) p(x_B)}{p(x_{B+b}) p(x_{B+a})} \right) = 0$$

We thus have

$$p(x) = p(\beta) \exp \left\{ \sum_{K \subseteq L(x)} V(x_K) \right\}, \qquad \forall x \in \Omega \tag{6.30}$$

since

$$V(\beta) = \ln p(\beta) \tag{6.31}$$

and thus (Ω, Λ, p) is a Gibbs random field with a potential given by Eq. (6.27).

We now demonstrate that the converse is true. Suppose that (Ω, Λ, p) is a Gibbs random field with a potential function V. By construction, the probabilities are positive definite

$$p(x) > 0, \qquad \forall x \in \Omega \tag{6.32}$$

We then have

$$\begin{aligned}
\ln p(x_{A+a}) - \ln p(x_A) &= \sum_{K \subseteq A+a} V(x_K) - \sum_{K \subseteq A} V(x_K) \\
&= \sum_{\substack{K \subseteq A+a \\ a \in K}} V(x_K) \\
&= \sum_{\substack{K \subseteq A-b+a \\ a \in K}} V(x_K) \\
&= \sum_{K \subseteq A-b+a} V(x_K) - \sum_{K \subseteq A-b} V(x_K) \\
&= \ln p(x_{A-b+a}) - \ln p(x_{A-b})
\end{aligned} \tag{6.33}$$

and therefore by Eq. (6.21) we deduce that (Ω, Λ, p) is a Markov random field.

6.4.3 Higher-Order Interactions

Gibbs potentials defined as above within a graph-theoretic framework were used by Martignon et al. to represent the distribution of firing frequencies obtained from multiunit recording data. A central element of their modeling effort was the recognition that higher-order interactions are required to describe the correlations among spiking neurons. In the next section we will study linear random field models in which terms up to second order in the polynomial expansion are retained. In a linear model there is a one-to-one correspondence between the weights assigned to edges and the coefficients of the interaction potentials. This mapping fails when higher-order terms are included. To address this issue, Martignon et al. introduced constellations and assembly diagrams. These generalizations of a graph were used to model their data.

An alternative approach to the treatment of higher-order interactions was given by Miller and Goodman.[37] In their work the coefficients appearing in the polynomial expansion of the potential function were used to model joint probabilities of fields (attributes) in databases. These authors explored ways of correcting the model when the polynomial expansion is truncated, thereby neglecting higher-order terms. The solution found to be appropriate for their database investigation was to renormalize the potentials appearing in the Möbius inversion relation. The renormalized potentials satisfy the relations

$$\ln p(x_A) = \sum_{b \subseteq A} \binom{|A| - 1}{|b| - 1}^{-1} V(x_b) \qquad (6.34)$$

and

$$V(x_A) = \ln p(x_A) - (|A| - 1)^{-1} \sum_{\substack{b \subset A \\ |b| = |A| - 1}} \ln p(x_b) \qquad (6.35)$$

Proofs and lemmas associated with these findings are given by Miller and Goodman.[37]

6.5 RANDOM FIELD MODELS

One of the most important categories of random field models is the auto (linear) model mentioned above in which only the first two terms in the polynomial expansion of $Q(x)$ are retained. This class of models, which was introduced by Besag[9] and includes the autobinary, autonormal (Gaussian), autobinomial, and auto-Poisson models, is widely used in image-processing tasks such as texture analysis and image reconstruction. In this section we briefly explore the autobinomial model, the autonormal model, commonly termed the Gauss-Markov random field model, and the related simultaneous autoregressive model. In doing so, we will introduce the notion of a wide-sense Markov process. Another simple model used to study a variety of image-processing problems, including texture analysis and segmentation, is the Derin-Elliott[38] model. The potential for the Derin-Elliott model will be presented in Section 6.7.3.

6.5.1 Besag's Auto Models

As mentioned above, we are interested in examining random field model for which $Q(\mathbf{x})$ only has contributions from single-site and pair interactions and thus takes the form

$$Q(\mathbf{x}) = \sum_{i=1}^{n} x_i G_i(x_i) + \sum_{i<j=1}^{n} x_i x_j G_{ij}(x_i, x_j) \qquad (6.36)$$

We further restrict our attention to conditional probabilities for each site that are of an exponential form. These probabilities can be written

$$\ln p_i(x_i | x_j, j \in N_i) = a_i(x_j, j \in N_i) b_i(x_i) + c_i(x_i) + d_i(x_j, j \in N_i) \qquad (6.37)$$

It then follows[9] that α_i must satisfy the equation

$$a_i(x_j, j \in N_i) = \alpha_i + \sum_{j=1}^{n} \beta_{i,j} b_i(x_i) \qquad (6.38)$$

where $\beta_{i,j} = \beta_{j,i}$ and $\beta_{i,j} = 0$ unless i and j are neighboring sites. The quantities G_{ij} appearing in Eq. (6.36) can be cast into a product form

$$G_{ij} = \beta_{i,j} h_i(x_i) h_j(x_j) \qquad (6.39)$$

with

$$x_i h_i(x_i) = b_i(x_i) - b_i(0) \qquad (6.40)$$

Besag's auto models follow from the additional requirement that b_i is linear in x_i so that $Q(\mathbf{x})$ simplifies to

$$Q(\mathbf{x}) = \sum_{i=1}^{n} x_i G_i(x_i) + \sum_{i<j=1}^{n} \beta_{ij} x_i x_j \qquad (6.41)$$

Random fields of this form are called *auto* or *linear models*. Upon evaluating the quantity $Q(\mathbf{x}) - Q(\mathbf{0})$, we find that these models have the conditional probability structure

$$\frac{p(x_i | x_1, \ldots, x_{i-1}, x_{i+1}, \ldots, x_r)}{p(0 | x_1, \ldots, x_{i-1}, x_{i+1}, \ldots, x_r)} = \exp\left\{x_i\left[G_i(x_i) + \sum_{j=1}^{n} \beta_{ij} x_j\right]\right\} \quad (6.42)$$

Three classes of auto models are widely used in image processing. These are the autologistic, autobinomial, and autonormal, or Gauss-Markov models. Autologistic models are auto models for which the random variables may assume one of two values at each site. Typically the binary values are chosen as 0 and 1 (or equivalently -1 and $+1$). The potential function for an autologistic model can be written

$$Q(\mathbf{x}) = \sum_{i=1}^{n} \alpha_i x_i + \sum_{i<j=1}^{n} \beta_{ij} x_i x_j \quad (6.43)$$

If the pair interactions are restricted to nearest neighbors, then we have our familiar Ising hamiltonian. Hassner and Sklansky[39] used binary random field models of this form to synthesize textures. In their approach normalized conditional probabilities of the form

$$p(x_i | x_j, j \in N_i) = \frac{\exp\{x_i(\alpha_i + \sum_{j \in N_i} \beta_{ij} x_j)\}}{1 + \exp\{\alpha_i + \sum_{j \in N_i} \beta_{ij} x_j\}} \quad (6.44)$$

were used with a first-order neighborhood system. Their work was extended to the synthesis of natural textures using an autobinomial random field model by Cross and Jain.[40]

6.5.2 The Autobinomial Model

In texture generation (synthesis) and analysis we take a textured image and extract a set of statistical parameters that describes its spatial properties using a random field model. We then use the random field model and parameter values to generate a textured image that resembles the input image as closely as possible. In modeling textures, the brightness of each pixel of the lattice is regarded as a random variable that may assume one of G possible values in the range 0 to $G - 1$. In the autobinomial model the brightness values for a given pixel in the lattice are binomially distributed with parameters determined by the neighboring pixels. The conditional probabilities that define the autobinomial model are of the form

$$p(X_i = k | X_j = x_j, j \in N_i) = \binom{G-1}{k} \theta_i^k (1 - \theta_i)^{G-1-k} \quad (6.45)$$

In this expression the binomial coefficient $(G - 1)!/k!(G - 1 - k)!$ represents the number of ways of selecting k objects from a bin containing $(G - 1)$ objects, and the probability of success θ_i is dependent on the values of the neighboring pixels:

$$\theta_i = \frac{\exp(T_i)}{1 + \exp(T_i)} \qquad (6.46)$$

and

$$T_i = \alpha_i + \sum_{j \in N_i} \beta_{ij} x_j \qquad (6.47)$$

When $G = 2$, this model reduces to the autobinomial model described above.

We can use this model to synthesize textured images by means of a Metropolis sampling procedure. We proceed by generating a Markov chain of texture states that converges to an autobinomial limiting distribution characterized by a given set of parameters. This is accomplished by noting that for any two texture realizations $\mathbf{x}$ and $\mathbf{y}$, we have

$$\frac{p(\mathbf{y})}{p(\mathbf{x})} = \prod_{i=1}^{n} \frac{p(y_i | x_1, \ldots, x_{i-1}, y_{i+1}, \ldots, y_n)}{p(x_i | x_1, \ldots, x_{i-1}, y_{i+1}, \ldots, y_n)} \qquad (6.48)$$

The expression was used by Cross and Jain,[40] together with a sampling procedure closely modeled on the exchange algorithm used to study binary alloys described in Section 4.4.2. In modeling textured images, the configurations $\mathbf{y}$ and $\mathbf{x}$ are the same except for two sites selected at random whose pixel grey values are exchanged.

6.5.3 Wide-Sense Markov Processes

We begin our discussion of the next two classes of random field model by broadening our definition of a spatial Markov process. Up to now we have defined Markov random fields through the requirement that conditional probabilities associated with the joint probability distribution of the configurations of a system possess a neighborhood property. Models satisfying this type of Markovian requirement are termed *strict-sense Markov random field models*. Alternatively, we may define spatial Markovianess in a broader sense either through a neighborhood condition on the low-order moments of the distribution or though recurrence relations. *Wide-sense Markov processes* may be defined in several equivalent ways. They may be defined in terms of the linear minimum mean square error estimate $\hat{E}$ or through the first- and second-order moments, the mean and autocorrelation. Thus we have a wide-sense Markov process if the random variables X_{ij} satisfy the conditional error relation

$$\hat{E}(X_{ij} | X_{kl}, (k, l) \in \Lambda) = \hat{E}(X_{ij} | X_{kl}, (k, l) \in \eta_{ij}) \qquad (6.49)$$

for $(i, j) \in \Lambda$. As is the case throughout this chapter, η_{ij} in the above expression denotes a noncausal neighborhood of the lattice point (i, j). Alternatively, we may define a wide-sense Markov process though the autoregressive (AR) family of linear equations

$$f_{mn} = \sum_{(k,l) \in \eta_{mn}} h_{kl} f_{m-k, n-l} + u_{mn} \qquad (6.50)$$

In this expression h_{kl} are a set of coefficients and u_{mn} is a random process that is othogonal to f_{mn}:

$$\langle f_{mn} u_{kl} \rangle = \sigma_u^2 \delta_{mk} \delta_{nl} \qquad (6.51)$$

The random process is not uncorrelated. Instead, we find that the autocorrelation manifests itself through the coefficients h_{kl} as

$$\sigma_u^{-2} \langle u_{mn} u_{kl} \rangle = \begin{cases} 1, & (m, n) = (k, l) \\ -h_{m-k, n-l}, & (m, n) \in \eta_{kl} \\ 0 & \text{otherwise} \end{cases} \qquad (6.52)$$

The quantities h_{kl} are the coefficients of the linear minimum mean square error estimate of f_{mn} given its neighbors. The coefficients for f_{mn} vanish for sites (m, n) not in the neighborhood η_{kl}, and

$$\sigma_u^2 = \left\langle \left[f_{mn} - \sum_{(k,l) \in \eta_{mn}} h_{m-k, n-l} f_{kl} \right]^2 \right\rangle \qquad (6.53)$$

In using the above representation as an image model, we treat the pixel grey values at a given site f_{mn} as a linear combination of the pixel grey values at the neighboring sites plus an additive noise term. It is worthwhile to examine this image model in more detail. This may be done most easily using Hunt's[41] lexicographic notation in which we stack the rows of our $N \times N$ image one after the other to form a vector $\mathbf{f}$ of length N^2. In this notation Eq. (6.50) takes the form

$$\mathbf{B(h)f = u} \qquad (6.54)$$

where $\mathbf{B}$ is a $N^2 \times N^2$ matrix of a special form. In a circulant matrix the rows are related to each other by a circular shift to the right. A matrix element shifted out to the right reappears to the left, and the first row is a circular shift of the last row. A block circulant matrix is a circulant martix partitioned into N^2 blocks, each of which is a circulant matrix of dimension $N \times N$. The matrix

B(h) is block circulant and is written

$$\mathbf{B(h)} = \begin{bmatrix} \mathbf{B}_0 & \mathbf{B}_1 & \cdots & \mathbf{B}_{M-2} & \mathbf{B}_{M-1} \\ \mathbf{B}_{M-1} & \mathbf{B}_0 & \cdots & & \mathbf{B}_{M-2} \\ \mathbf{B}_{M-2} & \mathbf{B}_{M-1} & \mathbf{B}_0 & \cdots & \\ \vdots & \vdots & \vdots & \vdots & \vdots \\ \mathbf{B}_1 & \mathbf{B}_2 & \cdots & \mathbf{B}_{M-1} & \mathbf{B}_0 \end{bmatrix} \quad (6.55)$$

where each partition $\mathbf{B}_j$ is circulant. We will explore the characteristics of block circulant matrices in the next section.

6.5.4 Gauss-Markov Random Fields

Our next class of models is the Gauss-Markov random fields. These models are described by wide-sense AR representations of the form given by Eq. (6.50) in which the random variables u_{mn} are Gaussian distributed. We may cast the joint and conditional probabilities into the compact forms:

$$p(\mathbf{f}) = (2\pi\sigma_u^2)^{-(1/2)n} |\mathbf{B}|^{1/2} \exp\left\{-\frac{\mathbf{f}^T \mathbf{B} \mathbf{f}}{2\sigma_u^2}\right\} \quad (6.56)$$

and

$$p(f_{mn} | f_{kl}, (k, l) \in \eta_{nm}) = (2\pi\sigma_u^2)^{-1/2} \exp\left\{\frac{-(f_{mn} - \Sigma_{(k,l)\in\eta_{kl}} h_{m-k, n-l} f_{kl})^2}{2\sigma_u^2}\right\} \quad (6.57)$$

In the above, $\sigma_u^2 \mathbf{B}^{-1}$ is the covariance matrix for $\mathbf{f}$. These models are strict-sense Markov random fields by construction and can be represented by a Gibbs distribution with its associated clique potentials.

Examples of textures synthesized in a Gauss-Markov random field model are shown in Fig. 6.4. The original texture is that of wood. It is illustrated in the upper left panel. Textured images synthesized by Chellappa[18] using a fourth-order Gauss-Markov random field model with 12 parameters are presented in the other three panels. Results derived using the full model are given in the lower left panel, while those obtained in simplified calculations are given in the two panels on the right. We observe that the vertical streaking and periodicity of the wood pattern are captured by the model and its parameters.

We will discuss Gauss-Markov models again in Section 6.11 where we examine a compound Gibbs representation developed by Jeng and Woods.[26] In this approach the authors construct a model built from the line process of Geman and Geman and a Gauss-Markov random field. The resulting com-

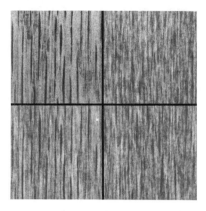

Figure 6.4. Example of wood texture synthesis using a Gauss-Markov random field model. *Upper left*: original wood texture. The information contained in the texture is split between the model parameters and the residuals. *Lower left*: Synthesized using a fourth-order model and residuals quantized to one bit each. *Upper right*: Synthesized using the model parameters and a pseudorandom number array. *Lower right*: Synthesized using the model parameters and half the residuals. (From Chellappa[18]. Reprinted with permission of Elsevier Science, the Netherlands.)

pound representation can be used for the discontinuity-preserving reconstruction of images composed of continuous-valued pixel grey levels.

6.6 SIMULTANEOUS AUTOREGRESSIVE MODELS

6.6.1 Simultaneous Autoregressive Random Fields

The final class of models in this brief overview are the simultaneous autoregressive models defined through AR representations with finite support for the autoregression and independent driving noise. In more detail, the values of the pixel grey levels in an SAR model are determined by a linear combination of the values at neighboring pixels plus a noise that varies independently from pixel to pixel. Let us define a set of zero-mean independent (noise) random variables u_{ij} and a set of real-valued coefficients $b_{ij,kl}$ over our lattice, namely for all $(i, j) \in \Lambda$ and $(k, l) \in \Lambda$. The random variables X defined over the lattice by

$$\sum_{(k,l) \in S \cup (i,j)} b_{ij,kl} x_{ij} = u_{ij} \tag{6.58}$$

form a SAR random field, with $(i, j) \in \Lambda$ and support $S = \{(k, l) : (k, l) \neq (i, j)$ and $b_{ij,kl} \neq 0\}$. Let us introduce another set $A_{ij} \subset \Lambda$ defined for all $(i, j) \in \Lambda$ by the relation $A_{ij} = \{(k, l) : (k, l) \neq (i, j)$, where for some $(m, n) \in \Lambda, b_{mn,ij} \neq 0$ and $b_{mn,kl} \neq 0\}$. The set A_{ij} is termed the *set of acquaintances* of (i, j). The SAR

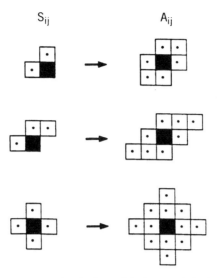

Figure 6.5. Autoregressive support sets S_{ij} and their associated sets of acquaintances A_{ij} for simultaneous autoregressive models. (From Derin and Kelly[79]. Reprinted with permission of IEEE.)

random field defined by Eq. (6.58) is both a Markov random field and a wide-sense Markov process, with neighborhoods that are subsets of A_{ij}. Several examples of support sets and the associated acquaintance sets are illustrated in Fig. 6.5.

To show that a SAR random field as defined above is a Markov random field, we proceed by casting the joint probability distribution $p(\mathbf{x})$ into the form

$$p(\mathbf{x}) = |\mathbf{B}| \prod_{i=1}^{M} p_i(\mathbf{b}_i^T \mathbf{x}) \qquad (6.59)$$

where $\mathbf{b}_i^T$ denotes the ith row of the matrix $\mathbf{B}$, and the det $\mathbf{B}$ factor preserves the overall normalization of the probability distribution. The product form is a direct consequence of the independence assumption. The proof that we have a Markov random field is completed by using Eq. (6.14). We then find that

$$p(x_{ij} | x_{kl}, (i, j) \neq (k, l)) = p(x_{ij} | x_{kl}, (k, l) \in A_{ij}) \qquad (6.60)$$

One of the most widely used types of SAR model is the (finite) toroidal lattice SAR field. In this model we fold the lattice into a torus connecting the left boundary to the right boundary and the top to the bottom. We assume that the support set contains a fixed collection of lattice points, and we partition the lattice into interior and boundary regions. The boundary region contains all lattice points (i, j) having a support set S that is wrapped around the lattice. The toroidal SAR field is defined by a pair of equations, one for the interior region and one for the boundary, of the form

$$x_{ij} + \sum_{(k,l) \in S} b_{kl} x_{i-k, j-l} = u_{ij} \tag{6.61}$$

This SAR field serves as an image model in the image reconstruction process. An approach frequently taken[19-22] is to exploit the properties of the block circulant matrices appearing in both image and degradation models, resulting in the design of fast Fourier-transform-based algorithms for image reconstruction. Before proceeding with this construction, let us formalize our image reconstruction problem. We will then examine the connection between block circulant matrices and Fourier transforms.

6.6.2 Image Reconstruction

We formulate our image reconstruction problem by establishing a relationship between the ideal image $\mathbf{f}$ and the observed image $\mathbf{g}$. This relationship is characterized by a degradation process and a noise process. The degradation process is represented symbolically by an operation of the form $s[D(\mathbf{f})]$, where s is the detector-response function and D is the point-spread function of the imaging system. The noise process $\mathbf{n}$ may or may not contain detector-dependent contributions, and it assumes an additive, multiplicative, or some other form. Our image reconstruction or restoration goal is to form an estimate of the image $\mathbf{f}$ given the data $\mathbf{g}$. To restrict the many possible reconstructions to physically meaningful ones, we use random field models to encode some general properties of the desired image into an image model. The image model together with a model of the degradation and noise processes plus a method for finding the optimal solution is used to reconstruct the image. In the remainder of this chapter, we will explore a number of different ways of achieving this objective.

As is customary, we will simplify our problem by assuming that the detector-response function s is linear and can be set equal to unity. Also we assume that our point-spread function D is linear and space invariant. We further assume that our noise process is independent of the imaging process. We represent the relationship between $D\{f(x, y)\}$, $g(x, y)$, and $n(x, y)$ as a discrete convolution, which can be expressed in one dimension as

$$g_m = \sum_{k=0}^{M-1} d_{m-k} f_m + n_m \tag{6.62}$$

and in two dimensions by

$$g_{mn} = \sum_{k=0}^{M-1} \sum_{l=0}^{N-1} d_{m-k, n-l} f_{mn} + n_{mn} \tag{6.63}$$

In these expressions f and d are arrays that are periodic of length M in one dimension, and M and N in two dimensions, with values in the range $m = 0, \ldots, M - 1$ and $n = 0, \ldots, N - 1$. The alignment of these arrays is accomplished by appending zeros wherever needed. The convolutions of these arrays may be written in stacked notation as

$$\mathbf{g} = \mathbf{Df} + \mathbf{n} \qquad (6.64)$$

The quantity d appearing in Eqs. (6.62) and (6.63) is the degredation-describing, point-spread function. Its matrix representation $\mathbf{D}$ is circulant in one dimension and block circulant in two dimensions.

There are a number of image reconstruction methods. One of the ways of reconstructing images is *constrained least-squares estimation* (LSE). In this optimization approach we minimize $\|\mathbf{Qf}\|^2$ subject to the (soft) constraint $\|\mathbf{g} - \mathbf{Df}\|^2 = \|\mathbf{n}\|^2$. As is customary, we treat the constraint using the method of Lagrange multipliers. The minimization of $\|\mathbf{Qf}\|^2 + \lambda \|\mathbf{g} - \mathbf{Df}\|^2$ with respect to $\mathbf{f}$ yields the optimal estimate $\mathbf{f}_{LSE} = [\mathbf{D}^T\mathbf{D} + (1/\lambda)\mathbf{Q}^T\mathbf{Q}]\mathbf{D}^{-1}\mathbf{g}$, where λ is our Lagrange multiplier. We have to select an appropriate form for the linear operator $\mathbf{Q}$. A typical choice for $\mathbf{Q}$ is $\mathbf{Q}^T\mathbf{Q} = \mathbf{R}_f^{-1}\mathbf{R}_n$, where $\mathbf{R}_f = \langle \mathbf{ff}^T \rangle$ is the correlation matrix for the image $\mathbf{f}$, and $\mathbf{R}_n = \langle \mathbf{nn}^T \rangle$ is the correlation matrix of the noise $\mathbf{n}$. This method leads to the well-known Wiener filter.

A second approach is *linear minimum mean square error (MMSE) estimation* of the image $\mathbf{f}$. In this method we define an error vector $\mathbf{e}$ as the difference between the true image and our estimate $\mathbf{f}_{MMSE}$. We then minimize the positive quantity $\langle \mathbf{e}^T\mathbf{e} \rangle = \langle \text{Tr}\,\mathbf{ee}^T \rangle$, assuming that $\mathbf{f}$ transform into $\mathbf{g}$ under a linear operator. In other words, we define $\mathbf{e} = \mathbf{f} - \mathbf{f}_{MMSE}$, and introduce the linear operator $\mathbf{L}$ through the relation $\mathbf{f}_{MMSE} = \mathbf{Lg}$. We then find that

$$\begin{aligned} \langle \mathbf{e}^T\mathbf{e} \rangle &= \langle \text{Tr}(\mathbf{f} - \mathbf{Lg})(\mathbf{f} - \mathbf{Lg})^T \rangle \\ &= \langle \text{Tr}(\mathbf{f} - \mathbf{L}(\mathbf{Df} + \mathbf{n}))(\mathbf{f} - \mathbf{L}(\mathbf{Df} + \mathbf{n}))^T \rangle \\ &= \text{Tr}[\mathbf{R}_f - 2\mathbf{LDR}_f + \mathbf{LDR}_f\mathbf{D}^T\mathbf{L}^T + \mathbf{LR}_n\mathbf{L}^T] \end{aligned} \qquad (6.65)$$

In the last step we have assumed signal-independent noise so that $\langle \mathbf{f}^T\mathbf{n} \rangle = \langle \mathbf{n}^T\mathbf{f} \rangle = 0$; the autocorralation matrices $\mathbf{R}_f$ and $\mathbf{R}_n$ are as defined above, and we have exploited the linearity of the trace to interchange the order of trace and expectation operations. The result of minimizing this expression with respect to $\mathbf{L}$ is the linear minimum mean square error estimate

$$\mathbf{f}_{MMSE} = \mathbf{R}_f\mathbf{D}^T(\mathbf{DR}_f\mathbf{D}^T + \mathbf{R}_n)^{-1}\mathbf{g} \qquad (6.66)$$

We now assume that $\mathbf{f}$ is stationary and approximate $\mathbf{R}_f$ by the block circulant covariance matrix $\mathbf{Q}_f$. If we further assume a white noise process with common variance v so that $\mathbf{R}_n = v\mathbf{I}$, then the linear minimum mean square error estimate simplifies to

$$\mathbf{f}_{MMSE} = \mathbf{Q}_f \mathbf{D}^T(\mathbf{D}\mathbf{Q}_f\mathbf{D}^T + \nu\mathbf{I})^{-1}\mathbf{g} \qquad (6.67)$$

The MMSE estimate of the image given by Eq. (6.67) was used, together with the aforementioned SAR random field image model defined over a toroidal lattice, to restore images by Chellappa and Kashyap.[20] In further developments of the approach, simultaneous autoregressive and Gaussian image models were used for 2-D spectral estimation by Sharma and Chellappa,[19] and issues related to estimation and selection of appropriate neighborhood systems were investigated by Kashyap and Chellappa.[21] MMSE estimation methods implemented in the Fourier domain were discussed earlier by Andrews and Hunt.[42] The eigenvalues and eigenvectors of circulant and block circulant matrices, which we will now examine, were also discussed in the text by Gonzales and Wintz.[43] Additional characteristics of the Fourier spectrum were given by Sharma and Chellappa.[19]

6.6.3 Fourier Computation and Block Circulant Matrices

We start by considering a circulant matrix $\mathbf{R}$:

$$\mathbf{R} = \begin{bmatrix} r_0 & r_1 & \cdots & r_{M-2} & r_{M-1} \\ r_{M-1} & r_0 & \cdots & \cdots & r_{M-2} \\ r_{M-2} & \cdots & \cdots & \cdots & \cdots \\ \vdots & \vdots & \vdots & \vdots & \vdots \\ r_1 & r_2 & \cdots & r_{M-1} & r_0 \end{bmatrix} \qquad (6.68)$$

We proceed by forming the sums λ_k given by

$$\lambda_k = r_0 + r_1 e^{2\pi i(k/M)} + r_2 e^{2\pi i(2k/M)} + \cdots + r_{M-1} e^{2\pi i([M-1]k/M)} = \sum_{j=0}^{M-1} r_j e^{2\pi i(jk/M)} \qquad (6.69)$$

If we next build the column vectors $\mathbf{w}_k$,

$$\mathbf{w}_k = \begin{bmatrix} 1 \\ e^{2\pi i(k/M)} \\ e^{2\pi i(2k/M)} \\ \vdots \\ e^{2\pi i([M-1]k/m)} \end{bmatrix} \qquad (6.70)$$

then we have the equation

$$\mathbf{R}\mathbf{w}_k = \lambda_k \mathbf{w}_k \tag{6.71}$$

and thus the $\mathbf{w}_k$ are the eigenvectors of the circulant matrix $\mathbf{R}$ and the λ_k are its eigenvalues. We may use the eigenvectors $\mathbf{w}_k$ to construct a matrix $\mathbf{W}$ that can be used to transform $\mathbf{R}$ to a diagonal form. We carry out the construction by using the eigenvectors $\mathbf{w}_k$ as column vectors of $\mathbf{W}$:

$$\mathbf{W} = [\mathbf{w}_0, \mathbf{w}_1, \ldots, \mathbf{w}_{M-1}] \tag{6.72}$$

The kjth element of this matrix is

$$W_{kj} = e^{2\pi i(kj/M)} \tag{6.73}$$

The inverse of the matrix $\mathbf{W}$ exists and has matrix elements

$$W_{kj}^{-1} = \frac{1}{M} e^{-2\pi i(kj/M)} \tag{6.74}$$

We now use this matrix to transform $\mathbf{R}$ to the diagonal matrix $\mathbf{T}$:

$$\mathbf{T} = \mathbf{W}^{-1}\mathbf{R}\mathbf{W} \tag{6.75}$$

whose diagonal elements T_{kk} are the eigenvalues λ_k of the matrix $\mathbf{R}$.

Let us now return to the degredation model represented in one dimension by the convolution given in Eq. (6.62). Neglecting the noise term temporarily, we have

$$\mathbf{W}^{-1}\mathbf{g} = \mathbf{W}^{-1}\mathbf{D}\mathbf{f} = \mathbf{W}^{-1}\mathbf{D}\mathbf{W}\mathbf{W}^{-1}\mathbf{f} = \mathbf{T}\mathbf{W}^{-1}\mathbf{f} \tag{6.76}$$

We now observe that the quantity $\mathbf{W}^{-1}\mathbf{f}$ is a column vector with elements F_k given by

$$F_k = \frac{1}{M} \sum_{j=0}^{M-1} f_j e^{-2\pi i(kj/M)} \tag{6.77}$$

for $k = 0, \ldots, M - 1$. These elements are the discrete Fourier transforms of the components f_j of $\mathbf{f}$. Similar results hold for $\mathbf{g}$ and $\mathbf{n}$. That is, multiplication of these vectors by $\mathbf{W}^{-1}$ produces vectors whose components are the Fourier transforms of the elements of the original vectors. Furthermore we see that

$$T_{kk} = \lambda_k = \sum_{j=0}^{M-1} d_j e^{-2\pi i(kj/M)} = MD_k \tag{6.78}$$

where D_k is the discrete Fourier transform of the components of the point-spread function d. Combining these observations yields the result that

$$G_k = M D_k F_k \tag{6.79}$$

This result may be recognized as the discrete convolution theorem which relates convolution of a pair of arrays in the spatial domain to element-by-element multiplication in the frequency domain, and vice versa.

Similar results can be derived for the two-dimensional image model. By proceeding as before, we construct a matrix **W** that can be used to diagonalize the block circulant matrix **B**. The matrix **W** is of dimension $MN \times MN$ and contains M^2 blocks of size $N \times N$, where **f**, **g** and **n** are now stacked vectors of length MN. The quantity $\mathbf{W}^-\mathbf{f}$ is a MN-dimensional column vector with components

$$F_{kl} = \frac{1}{MN} \sum_{m=0}^{M-1} \sum_{n=0}^{N-1} f_{mn} z_M^{mk} z_N^{nl} \tag{6.80}$$

with $z_M = \exp(-2\pi i/M)$ and $z_N = \exp(-2\pi i/N)$. The elements F_{kl} are the two-dimensional discrete Fourier transforms of the components f_{mn} of **f**. Again, analogous expressions hold for **g** and **n**, and we find that in the Fourier domain

$$G_{kl} = M N D_{kl} F_{kl} + N_{kl} \tag{6.81}$$

where we have now included the noise term N_{kl}.

In implementing the MMSE reconstruction formula, Eq. (6.67), Chellappa and Kashyap[20] addressed two problems. The first was the need to have an efficient means of performing the indicated matrix inversion. This issue was resolved by using fast Fourier transforms. The second problem was the search for a reliable way of evaluating the covariance matrix $\mathbf{Q}_f$ given only the degraded image **g**. This goal was achieved by estimating the covariance matrix from the degraded image **g** using the SAR and Gaussian random field image models. Since $\mathbf{Q}_f$ and $\mathbf{Q}_g$ are block circulant, Fourier computation was exploited in their evaluation to yield MMSE estimates of the image **f**.

6.7 THE METHOD OF GEMAN AND GEMAN

Another approach to image reconstruction is *maximum a posteriori* (MAP) estimation. In the MAP method of Geman and Geman,[23] we have three main ingredients. The first of these is the use of a Bayesian formalism that allows us to incorporate a variety of constraints into the image reconstruction process. These constraints encode general properties of the physical world and serve as our image model. The second component of the method of Geman and

Geman[23] is the full exploitation of the Gibbs-Markov equivalence in constructing the prior and posterior probability distributions. As a consequence we are able to describe these probabilities in terms of Gibbs distributions and then use simulated annealing to find a desired low-energy state corresponding to the MAP estimate. The third element of the method is the introduction of a second, dual lattice system containing discontinuity-preserving information. The associated line process, constraints, and coupling interactions permit us to suspend smoothing and other pixel-related processes at surface boundaries and to eliminate weakly defined boundary elements while reinforcing and strengthening others.

6.7.1 Maximum A posteriori (MAP) Estimation

The formal structure of our MAP solution to the constrained restoration problem is provided by Bayes's theorem. If we put $\mathbf{x} = \mathbf{f}$ and $\mathbf{y} = \mathbf{g}$, in the notation of Section 6.1.2, then Bayes's theorem becomes a statement that the posterior distribution, $p(\mathbf{f}|\mathbf{g})$ is proportional to the product of the likelihood $p(\mathbf{g}|\mathbf{f})$ and the prior distribution $p(\mathbf{f})$:

$$p(X = \mathbf{f}|Y = \mathbf{g}) = \frac{p(Y = \mathbf{g}|X = \mathbf{f})p(X = \mathbf{f})}{p(Y = \mathbf{g})} \quad (6.82)$$

Our goal is to obtain a maximum a posteriori estimate of the true image $\mathbf{f}$ given the degraded noisy data $\mathbf{g}$ and prior information we have about the general properties of the physical solution. In Bayesian approaches the degradation model is expressed by the likelihood. We now present a simple model for the degradation process and then discuss the construction of the prior distribution and its incorporation of constraints. We begin by adopting the model of the degradation process presented in Eq. (6.64) containing additive noise, which we describe by the multivariate Gaussian

$$p(\mathbf{n}) = [(2\pi)^{M/2}|R_n|^{1/2}]^{-1} \exp\left[-\left(\frac{1}{2}\right)(\mathbf{n} - \langle\mathbf{n}\rangle)^T R_n^{-1}(\mathbf{n} - \langle\mathbf{n}\rangle)\right] \quad (6.83)$$

In the above, R_n is the covariance matrix and $\langle n \rangle$ is the mean value of the noise. We next assume that the noise is uncorrelated and stationary with zero mean. We then have

$$p(\mathbf{n}) = (2\pi\sigma^2)^{-M/2} \exp\left[-\left(\frac{1}{2\sigma^2}\right)(\mathbf{g} - \mathbf{Df})^T(\mathbf{g} - \mathbf{Df})\right] \quad (6.84)$$

where σ^2 is the noise variance. The likelihood assumes its simplest form in the absence of blurring and other effects taken into account by the point-spread function. We then observe that this likelihood has the property that it can be decomposed into a product of terms, one for each element in the image array:

$$p(\mathbf{g}|\mathbf{f}) \propto \prod_{i \in \Lambda} p(g_i|f_i) = \prod_{i \in \Lambda} \exp\left(-\frac{1}{2\sigma^2}(g_i - f_i)^2\right) \quad (6.85)$$

This simplified degradation model will be used for illustrative purposes throughout most the remainder of this chapter. More general image formation models that include blurring, artifacts, and other degradation-producing processes can be treated within this formalism. The crucial feature of interest to us is that the likelihood assumes the form of a product. We will consider a Gauss-Markov generalization in Section 6.11.

As was done previously, we model the image as a Markov random field. The quantities X which may assume any of a number of possible grey values f_i at each lattice site are our random variables. The configurations of the system are the possible image restorations $\mathbf{f} = \{f_i\} = \{f_i, i \in \Lambda\}$. These are sets of grey values, one for each lattice site. These random variables have a joint probability distribution $p(X = \mathbf{f})$. We assert that the associated conditional probabilities are Markovian for an appropriate neighborhood system which we must still define. Thus we have defined a Markov random field. We then invoke the Hammersley-Clifford theorem to cast the prior distribution into the form of a Gibbs distribution

$$p(X = \mathbf{f}) = \frac{1}{Z} \exp(-\alpha U(\mathbf{f})) \quad (6.86)$$

with an energy $U(\mathbf{f})$ composed of a sum of local contributions from the individual lattice sites i,

$$U(\mathbf{f}) = \sum_{i \in \Lambda} V_i(f_i) \quad (6.87)$$

and with a partition function Z calculated by summing over all allowed configurations $\{\mathbf{f}\}$ of the system:

$$Z = \sum_{\{\mathbf{f}\}} \exp\left(-\alpha \sum_{i \in \Lambda} V_i(f_i)\right) \quad (6.88)$$

The parameter α is a strength parameter that plays the role of the inverse of the temperature in thermodynamic applications, and it will now assume values other than unity. The site potentials V_i are composed of sums of clique potentials V_c

$$V_i(f_i) = \sum_{c \in C_i} V_c(f_c) \quad (6.89)$$

where, as done in Section 6.1, the quanitity C_i denotes that we are adding contributions from the cliques associated with the neighborhood system for the site.

6.7.2 Clique Potentials

There are two types of clique potentials. The first are those potentials that express the constraints on the reconstruction process. The second are the interaction energies that allow us to integrate multiple cues. These quantities endow the formalism with a measure of biological content, especially with regard to early vision. This aspect will be discussed further in Section 6.10. Our primary focus in the present section is on the smoothness-enforcing constraints that guide the image reconstruction, and on the structure of the process itself. Digital images are highly organized spatial lattice systems. The general property of these lattice systems that we wish to capture with the prior distribution, called *piecewise smoothness*, is a local one — neighboring pixels tend to have similar grey values, while distant ones may or may not share common grey values. More restricted versions of the property are piecewise constancy and piecewise linearity.

There are a number of widely used potentials that encode piecewise-smoothness-enforcing constraints appropriate for discrete grey-valued pixel elements. In the Derin-Elliott[38] model we have a second-order neighborhood system with associated cliques. For two-, three- and four-element cliques, the clique potentials are of the form

$$V_{c_l}(f_i) = \begin{cases} -\xi_l, & f_j = f_i, \ \forall f_j \in c_l \\ \xi_l & \text{otherwise} \end{cases} \quad (6.90)$$

and for single-site cliques, the clique potentials are defined as

$$V_c(f_i) = \alpha_k \quad \text{for } \forall f_i = q_k \quad (6.91)$$

The first type of clique potential encourages all elements within a given clique to have the same grey level. The strength ξ of the potential may be individually set for each type of clique as labeled by the strength subscript l. The second (single-site) term controls the percentage of pixels of each region. This type of interaction is particularly useful when the image regions are textured. In textured images the α parameters control the marginal distributions while the ξ parameters control clustering.

Examples of texture realizations in the Derin-Elliott[38] model are presented in Fig. 6.6. These patterns were generated in a four-level model with parameters ξ_1 to ξ_4 corresponding to the four, two-element cliques (i.e., horizontal, vertical, upward right diagonal, and upward left diagonal) illustrated in Fig. 6.2. Results presented in Fig. 6.6 were obtained using the Gibbs sampler, which

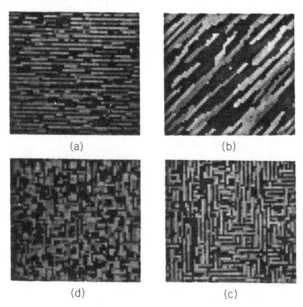

Figure 6.6. Examples of texture realizations in the Derin-Elliott model. These patterns were generated in a four-level model with parameters: (a) $\xi_1 = 1.0$, $\xi_2 = \xi_3 = \xi_4 = -1.0$; (b) $\xi_1 = \xi_2 = \xi_3 = -1.0$, $\xi_4 = 1.0$; (c) $\xi_1 = \xi_2 = 1.0$; $\xi_3 = \xi_4 = -1.0$; (d) $\xi_1 = \xi_2 = 1.0$; $\xi_3 = \xi_4 = -0.5$. (From Derin and Elliott[38]. Reprinted with permission of IEEE.)

we will discuss shortly, for 50 iterations where each iteration represents one raster scan through the image.

A second class of widely used clique potentials are the quadratic difference potentials suggested by regularization theory.[44] These potentials, used in conjunction with an Ising-type first-order neighborhood system, contain pairwise terms of the form

$$V_c(f_i) = (f_i - f_j)^2, \qquad f_j \in \eta_i \tag{6.92}$$

We observe that this potential selectively penalizes, in a progressively stronger manner, any departures of pixel grey values within a neighborhood from that of the central pixel element. We will encounter potentials of this form when we discuss the method of maximation of posterior marginals and when we examine analytically some of the deterministic approximations and alternatives to the method of Geman and Geman.

6.7.3 The Posterior Distribution

We may now combine the likelihood and prior to formally yield the posterior distribution. The results can be expressed in the form

$$p(\mathbf{f} \mid \mathbf{g}) = \frac{1}{Z_p} \exp(-U_p(\mathbf{f}, \mathbf{g})) \tag{6.93}$$

where the several constant terms have been collected together into the partition function Z_p and the energy $U_p(f, g)$ is

$$U_p(\mathbf{f}, \mathbf{g}) = \alpha U(\mathbf{f}) + \left(\frac{1}{2\sigma^2}\right) \sum_{i \in \Lambda} (f_i - g_i)^2 \qquad (6.94)$$

The posterior potential contains the two terms for each lattice site. If we take the quadratic form for the prior potential given by Eq. (6.92), then we have a posterior energy composed of a sum of contributions

$$U_i(f_i, g_i) = \frac{1}{2\sigma^2}(f_i - g_i)^2 + \alpha \sum_{j \in N_i} (f_i - f_j)^2 \qquad (6.95)$$

The posterior distribution represented in general by Eq. (6.94) can be cast into the form of a product of contributions from the individual lattice sites as was done for the likelihood in Eq. (6.85). This parallel form supports the annealing algorithm to be discussed shortly. The first term in $U_i(f_i, g_i)$ encourages restorations that are not too different from the data, discouraging large single-step excursions from the original pixel values. The second term favors maximally smooth restorations, penalizing large pairwise contrasts in pixel values within a local neighborhood.

We now observe that the posterior distribution is Markovian by construction, and therefore we again have defined a Markov random field. We associate a Gibbs distribution with this random field, having the same potential as the posterior distribution:

$$w(\mathbf{f}) = \frac{1}{Z_w} \exp\left(-\frac{1}{T} U_p(\mathbf{f}, \mathbf{g})\right) \qquad (6.96)$$

The parameter T that we have now reinstated in the Gibbs distribution is a second temperature, or control parameter. The configurations of the system that minimize the potential $U_p(\mathbf{f}, \mathbf{g})$ maximize the posterior distribution, Eq. (6.93), and therefore correspond to MAP estimates that we seek.

6.7.4 The Gibbs Sampler

We can find these desired low-energy states of our lattice system using the simulated annealing approach discussed in our earlier chapters. We have chosen the posterior energy as our hamiltonian, and we now endow the image to be reconstructed with an equilibrium dynamics. We then let the system evolve into a desired low-energy state, as was done for Glauber spin flip dynamics, binary alloys, and the traveling salesman's tour. The sampling method we use to guide the evolution of the system is built on the Gibbs distribution of Eq. (6.96) and is known as the Gibbs sampler. This is a generalization of the Glauber heat bath to lattice systems composed of grey-valued pixels. Before discussing the Gibbs sampler, it may be useful for us to summarize our MAP formalism at this point: The prior distribution is constructed from clique potentials representing physical constraints such as piecewise smoothness. In Section 6.10 we will append to the hamiltonian potentials representing interactions between different MRFs. The posterior distribution contains a total energy, composed of contributions from the likelihood and prior, whose minimization yields our MAP estimate. The argument of the exponential appearing in the posterior Gibbs distribution

$$\frac{1}{T}[U(f,g) + \alpha V(f)] \tag{6.97}$$

contains two Lagrange multipliers. One of these multipliers, the α-parameter, controls the relative hardness of the constraints. The other, the T-parameter, plays the role of an annealing temperature. If we fix the α-parameter at a moderate value throughout the restoration, then our constraint is soft. One compromise between maintaining a hard constraint such as the TSP tour requirement, and a soft one is to gradually increase the α-parameter as the temperature is lowered. This strategy[45,46] may be symbolized by the dual limits

$$\alpha \to \infty, \quad T \to 0 \tag{6.98}$$

Our image lattice has a site replacement kinetics that is a grey-value analogue of the Glauber spin kinetics discussed in Chapter 4. In the case of images, we use the Gibbs distribution associated with the posterior distribution to calculate the transition probabilities. The result, known as the Gibbs sampler, takes the form:

$$p_i(\mathbf{f}) = \frac{w_i(\mathbf{f})}{\sum_{j=1}^{n} w_j(\mathbf{f})}, \quad i = 1,\ldots,n \tag{6.99}$$

In this expression the index is used to indicate that we are considering a fixed set of allowed states for the transition. We have already noted that the Markov-Gibbs character of the system allows us to consider the joint distribution as a product of single-site probabilities. We generate new configurations by considering one site at a time. Contributions to the joint probability distribution from the other sites can be neglected, since the terms for the other sites in the numerator cancel those in the denominator of the sampler. In computing the transition probabilities, we need only to consider the allowed values at that site and at those sites in its neighborhood.

As in Chapter 4, our sampler generates an inhomogeneous Markov chain of configurations of the system that converges in distribution to those of minimum energy. These minimum energy equilibrium configurations represent our maximum a posteriori estimate of the image. To ensure convergence, the temperature must be lowered gradually. In general, convergence is assured[23,47] if the temperature T_k used in the kth iteration satisfies the condition

$$T_k \geq \frac{\kappa}{\log(2 + k)}, \quad k = 0, 1, 2, \ldots \qquad (6.100)$$

where κ is an overall scale constant.

The Gibbs sampler algorithm is of the general form:

1. Choose an initial temperature, cooling schedule, and α-parameter dependence.
2. Initialize by selecting a pixel value at each lattice site that maximizes the likelihood $p(\mathbf{g}|\mathbf{f})$.
3. Visit each site in the lattice with equal frequency either in the manner of a raster scan or asynchronously.
4. At each site compute the single site probabilities $p_i(f_i)$ for a range of pixel values.
5. Accept a new pixel value with probability $p_i(f_i)$.
6. Repeat N_{iter} times.
7. Lower the temperature by an amount given by the cooling schedule.
8. Repeat steps 3 through 7.

6.8 MAXIMIZER OF THE POSTERIOR MARGINALS

6.8.1 Optimal Bayesian Estimation

In this section we present a method of approximating the simulated annealing algorithm. Called the maximizer of the posterior marginals by Marroquin,[24,48] this approximation is based on the idea of using the expected value of an error functional as a hamiltonian. We begin by defining the posterior marginal distribution of the ith element of the lattice as the sum of all posterior distributions for which the element f_i has the value r, namely

$$p_i(r|g) = \sum_{f:f_i=r} p(f|g) \qquad (6.101)$$

We next define an energy function that may be regarded as providing a measure of the error in assigning the correct label, or color, f^a, at each lattice site. This energy function, obtained by summing the individual contributions from each site,

$$E(f, f^a) = \sum_i E_i(f_i, f_i^a) \qquad (6.102)$$

has the property that the energy, cost, or penalty is zero if the identification is correct, $f_i = f_i^a$, and is assigned some positive number c otherwise:

$$E(f_i, f_i^a) = \begin{cases} 0, & f_i = f_i^a \\ c & \text{otherwise} \end{cases} \tag{6.103}$$

We now define our optimal Bayesian estimator f^{a*} with respect to the energy E as the global minimizer of the expected value of E for all possible f and g. Since E is positive definite, and the prior density is constant for a set of observations, this estimator may be defined in terms of the posterior distribution as

$$\sum_f E(f, f^{a*}) p(f \mid g) = \inf_{f^a} \sum_f E(f, f^a) p(f \mid g) \tag{6.104}$$

We may now state the key element in the MPM approach, namely that the optimal estimate of f with respect to the energy E can be found by minimizing independently the expected marginal energy for each element in the lattice:

$$f_i^{a*} = q \in Q_i : \sum_{r \in Q_i} E_i(r, q) p_i(r \mid g) \leq \sum_{r \in Q_i} E_i(r, s) p_i(r \mid g) \tag{6.105}$$

for all $s \neq q$ over the finite set of possible values Q_i. This assertion can be show to be correct by rearranging the terms on the right-hand side of Eq. (6.104).

$$\inf_{f^a} \sum_f E(f, f^a) p(f \mid g) = \inf_{f^a} \sum_f \sum_i E_i(f_i, f_i^a) p(f \mid g)$$

$$= \inf_{f^a} \sum_i \sum_{r \in Q_i} \sum_{f : f_i = r} E_i(r, f_i^a) p(f \mid g) \tag{6.106}$$

$$\sum_i \inf_{f^a} \sum_{r \in Q_i} E_i(r, f_i^a) p_i(r \mid g)$$

In the above sequence of steps, we employed Eq. (6.102), interchanged the order of summation, and expressed the sum over f as a double sum. We then reexpressed the double sum by collecting all instances where $f_i = r$, and added together the contributions from the different r values. In the last step, we used the definition of the marginals, Eq. (6.101), and interchanged the first two operations, noting that the energies are positive definite.

6.8.2 Segmentation and Reconstruction

Let us consider segmentation and reconstruction tasks as typical examples. For segmentation we may take as our error criterion the number of elements that are classified incorrectly. This can be accomplished by defining a segmentation energy as

$$E_i(f_i, f_i^a) = 1 - \delta(f_i - f_i^a) \tag{6.107}$$

where

$$\delta(x) = \begin{cases} 1, & x = 0 \\ 0 & \text{otherwise} \end{cases} \tag{6.108}$$

Similarly we may define a smoothness enforcing energy through the quadratic form

$$E_i(f_i, f_i^a) = (f_i - f_i^a)^2 \tag{6.109}$$

In the former case we have

$$\sum_{r \in Q_i} (1 - \delta(r - f_i^{a*})) p_i(r \mid g) = 1 - p_i(f_i^{a*} \mid g) \tag{6.110}$$

which yields the optimal segmentation estimate

$$f_i^{a*} = q \in Q_i : p_i(q \mid g) \geqslant p_i(s \mid g) \tag{6.111}$$

for all $q \neq s$. In the latter instance we may define the mean

$$\bar{r} = \sum_{r \in Q_i} r p_i(r \mid g) \tag{6.112}$$

and then, noting that

$$\sum_{r \in Q_i} (r - q)^2 p_i(r \mid g) \leqslant \sum_{r \in Q_i} (r - s)^2 p_i(r \mid g) \tag{6.113}$$

implies the inequality

$$(\bar{r} - q)^2 \leqslant (\bar{r} - s)^2 \tag{6.114}$$

we obtain the optimal restoration estimate

$$f_i^{a*} = q \in Q_i : (\bar{f}_i - q)^2 \leqslant (\bar{f}_i - s)^2 \tag{6.115}$$

for all $q \neq s$. This second estimate is sometimes called the *thresholded posterior mean*.

6.8.3 The MPM Algorithm

The segmentation solution, that is, maximizing the marginal posterior probabilities for each pixel in the lattice, may be viewed as an approximation to the MAP approach of Geman and Geman. To obtain the MPM estimate, we generate a sequence of states at a single temperature, $T = 1$. This can be done using the Metropolis or equivalent importance sampler to generate a homogeneous Markov chain of states that converges to the equilibrium posterior distribution $p(f|g)$. Once the equilibrium distribution is generated, we may approximate the posterior marginals by counting the number of times each label occurs at a site:

$$p_i(q|g) = \frac{1}{n-k} \sum_{t=k+1}^{n} \delta(f_i^{(t)} - q) \tag{6.116}$$

Similarly the mean field is found by evaluating the sum

$$\bar{f} = \frac{1}{n-k} \sum_{t=k+1}^{n} f_i^{(t)} \tag{6.117}$$

In the above expressions $f_i^{(t)}$ denotes the label occurring at time (iteration) t when the equilibrium state is achieved, k is the number of iterations required to reach equilibrium, and n is the total number of iterations needed to obtain a useful estimate. The resulting marginal distribution and mean field are then used to select the labels and pixel values satistfying Eqs. (6.111) and (6.115).

The following is a Metropolis-based MPM algorithm:

1. Initialize the lattice system by choosing the pixel value at each site that maximizes the likelihood.
2. For each site in the lattice choose a pixel value at random subject to the symmetry requirement.
3. Calculate the energy change ΔU.
4. If $\Delta U < 0$, accept the move.
5. Or else accept the move with probability $\exp\{-\Delta U\}$.
6. Repeat steps 2 through 5 n times, saving configurations $\mathbf{f}^{(k+1)}$ to $\mathbf{f}^{(n)}$.
7. Form $p_i(q|g)$ according to Eq. (6.116).
8. For each site in the lattice, choose the pixel value q satisfying the condition

$$p_i(q|g) \geq p_i(s|g)$$

for all pixel values s.

6.9 ITERATED CONDITIONAL MODES OF THE POSTERIOR DISTRIBUTION

6.9.1 Annealing and Quenching

In metallurgy, annealing is the process of heating and subsequent cooling of a metal or metal alloy to modify its microstructure and endow the finished product with a set of desired mechanical and physical characteristics. The several parameters that govern annealing—the annealing temperature, annealing time, and cooling schedule—are strongly task and context dependent. For example, there are a variety of procedures for heat treating iron and its alloys. These methods differ considerably from those used for treating cooper and its alloys.

Pure iron is allotropic—it exists in several crystalline forms having different physical properties. At temperatures below 910°C (alpha) iron takes the form of a body-centered cubic lattice. At the phase transformation temperature, the lattice structure changes from body-centered alpha iron to face-centered gamma iron. Carbon is soluble in gamma iron. When a small amount of carbon ($<1.7\%$) is added to gamma iron, and the resulting solid solution, called *austenite*, is cooled, the result is carbon steel. During cooling, austenite undergoes a phase transition to become a mixture of alpha iron (ferrite) and iron carbide (cementite). The grain size and structure of the ferrite-cementite mixture depends on austenite formation temperature and downward gamma-to-alpha transformation point. Annealing at temperatures just above the gamma transformation value enhances the formation of spheroidal particles, while austenizing at more elevated temperatures produces a laminar steel. (It should be noted that in either instance, the melting point is well above the annealing temperature.) The downward phase transition can be controlled by varying the cooling rate. Rapid cooling, or quenching, can lower this transformation point substantially to produce less coarse, stronger steels.

The task and context are different for copper and its alloys. Pure copper is monophasic and is usually annealed to promote recrystallization, grain growth and the migration of defects to grain boundaries. The cooling rate is not crucial as long as it is slow. Copper alloys, on the other hand, are sometimes precipitation hardened. In this procedure copper is supersaturated with an alloying metal and then quenched to freeze-in the alloy. Precipitation at room temperatures followed perhaps by working and/or reheating gives rise to a fine-grained and hardened finished product.

6.9.2 Quenching and ICM

One simple way of generating an iterative algorithm for reconstructing an image is to choose pixel values that maximizes the posterior distribution rather then selecting pixel values based on the Gibbs sampler or Metropolis algorithm. This is the approach taken in Besag's[25] iterated conditional modes of the posterior distribution (ICM). In describing this method, it is customary to make explicit the dependence of the prior distribution upon the pixel values in the neighborhood of the site of interest. This can be done by the replacing the prior probabilities $p(f_i)$ with the more precise characterization in terms of the conditional probabilities $p(f_i | f_{\partial i})$. In the preceding expression for the conditional probabilities, we use $f_{\partial i}$ as a shorthand notation for $f_j, j \in N_i$. Once this replacement is made, we may describe ICM as the use of the proportionality

$$p(f_i | g, f_{\partial i}) \propto p(g | f_i) p(f_i | f_{\partial i}) \qquad (6.118)$$

to find the mode of the posterior distribution at each pixel site, given the current site value and neighboring site pixel values. It is clear that this can be accomplished by simply finding the pixel value that minimizes the posterior energy. Thus the ICM algorithm is deterministic. We choose the pixel value at each site that gives the greatest reduction in energy, and moves that increase the posterior energy do not occur. The ICM method may be recognized as the $T = 0$, or *quenched*, limit of the simulated annealing method of Geman and Geman. This can perhaps be seen best by noting that as the temperature is lowered, the Gibbs distribution given by Eq. (6.96) becomes progressively more sharply focused at its mode or peak value.

There are a number of benefits to taking the quenched limit. By rapidly freezing the lattice system, unwanted phase transitions can be avoided. This property is used to an advantage in metallurgical settings. Another benefit of quenching is that the slow, logarithmic time schedule followed in annealing can be bypassed in those situations where states corresponding to local minima in the energy are sufficiently optimal to be acceptable. An example of the time differences involved in annealing, maximizing the posterior marginals, and locating the iterated conditional modes has been given by Dubes and Jain.[49] The quality of the segmentation solutions obtained using these methods was comparable. Image segmentation times for ICM were several order of magnitude shorter than those for simulated annealing, with times for MPM intermediate between the two extremes. Somewhat different results were reported by Konrad and Dubois.[50] In their investigation of SA and ICM approaches to making MAP estimates of two-dimensional motion fields, substantially poorer results were produced by quenching as compared to those obtained by annealing. Time differences between the two methods were about one order of magnitude.

Still other results have been reported by Rignot and Chellappa[51] for segmentation of synthetic aperature radar imagery using simulated annealing (MAP), MPM, and ICM algorithms. One way of assessing the quality of a segmentation algorithm is by counting the classification errors. This method of evaluation was applied to images composed of two region types characterized by differences in contrast by Rignot and Chellappa[51]. Their results, presented in Fig. 6.7 as a function of the difference in contrast between the two regions, show that the probabilistic ICM, MPM, and MAP methods are superior to those obtained from either a minimim distance (MD) classifier or a maximum

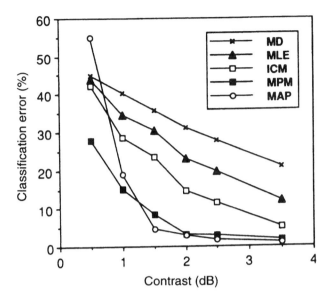

Figure 6.7. Classification error versus contrast for a two-region test image. Errors shown for minimum distance classifier (MD), maximum likelihood estimate (MLE), maximum a posteriori estimate (MAP), maximizer of the posterior marginals (MPM), and iterated conditional modes (ICM). (From Rignot and Chellappa[51]. Reprinted with permission of the Optical Society of America.)

likelihood estimate (MLE). The MPM and MAP results are quite close to one another, and the algorithms are able to separate regions with less than a 10% error in instances where there is a greater than 1 dB difference in contrast.

An example of an MAP segmentation map is presented in Fig. 6.8. Displayed in this figure are the imput image of a forest scene corresponding to a single-look intensity field of synthetic aperture radar complex data, a three-class segmentation map obtained using a minimum distance method, and a segmentation map derived from an MAP estimate. We see that the MAP segmentation is superior to the MD computed map. Results of studies of polarimetric data,[52] and multifrequency, multilook data,[53] have been reported by Rignot and Challappa as well. In their segmentation of L-band polarimetric data,[52] regions were classified into one of 13 different area types. There are more region types than in the earlier study, and the energy function is more complicated. We now find that the ICM segmentation maps are significantly poorer than the MAP maps. It appears that the ICM algorithm gets stuck in poorer local minima and exibits some dependence on the initial configuration.

As in metallurgical annealing and quenching, the selection of a method is dependent on the character of the material (image), especially the properties of its energy surface, and upon the goal of the process. In some applications it may be possible to combine methods starting the reconstruction by annealing

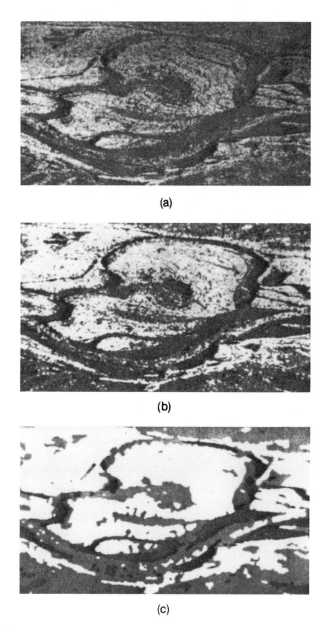

Figure 6.8. Segmentation of single-look synthetic aperture radar complex data: (a) Original intensity field of an Alaska forest scene; (b) MD segmentation map showing results for three classes; (c) MAP segmentation map. In decreasing order of brightness shown are white spruce and balsam poplar, black spruce, and river and clear-cut areas. (From Rignot and Chellappa[51]. Reprinted with permission of the Optical Society of America.)

the image at a high temperature and then after some number of iterations switching to a quench (ICM). This method may be termed *quenched annealing*.

6.9.3 The ICM Algorithm

If we consider potentials of the form given in Eq. (6.92), then the mode of the posterior distribution corresponds to making the substitution

$$f_i^* = \frac{g_i + \kappa \sum_{j \in N_i} f_j}{1 + \kappa |N_i|} \tag{6.119}$$

at each site in the lattice, where $\kappa = 2\alpha\sigma^2$ and $|N_i|$ denotes the number of elements in the set N_i, namely the number of neighbors of pixel i. More generally, ICM algorithms are of the form:

1. Initialize the lattice by choosing the values at each lattice site that maximize the likelihood.
2. For each site in the lattice update the pixel value by selecting the quantity that maximizes the right-hand side of Eq. (6.95), by finding the minimum energy according to Eq. (6.119) or by using an equivalent expression appropriate for a different potential, or in some other manner.
3. Iterate, by repeating step 2 until finished.

6.10 COUPLED MARKOV RANDOM FIELDS

We now introduce the third key ingredient in the method of Geman and Geman, a discontinuity-preserving and boundary-strengthening line process that operates concurrently and interactively with the pixel process. In the pixel process the balance between preserving the information content of the original data and removing noise can be a delicate one. For certain combinations of control parameters, there is a tendency for the lattice system to undergo phase transitions resulting in images dominated by a few colors. In these highly monochromatic images, boundaries delineating the different surfaces are lost. There are several ways of preventing these unwanted phase transitions. One way, which we just discussed in Section 6.8, is to quench, rather than anneal, the image. Another way, which we will now examine in some detail, is to suspend the smoothness constraint at surface boundaries by introducing a line process.

6.10.1 The Line Process

The construction of a line process parallels that of the pixel process. We model the line process as a second Markov random field and employ the Gibbs-

Markov equivalence to produce a simple description in terms of clique potentials. This second Markov random field is defined over a dual lattice and encodes discontinuity information. The dual lattice may be visualized as a two-dimensional array of $2N^2$ line elements, one horizontal element and one vertical line element for each pixel element in the primary pixel lattice. These line elements are positioned between pixel elements, and are binary-valued indicating the presence or absence of a boundary. The line elements associated with first-order vertical and horizontal pixel differences satisfy the relations

$$\|f_{mn} - f_{m+1n}\| > \tau_1, l_{mn}^v = 1; \quad \|f_{mn} - f_{mn+1}\| > \tau_1, l_{mn}^h = 1;$$
$$\|f_{mn} - f_{m+1n}\| \leqslant \tau_1, l_{mn}^v = 0; \quad \|f_{mn} - f_{mn+1}\| \leqslant \tau_1, l_{mn}^h = 0 \quad (6.120)$$

We may consider higher-order pixel differences, and encode that information in additional dual lattices. Although the line elements defined through Eq. (6.120) represent thresholded differences in pixel grey levels, the line elements participate in the equilibrium dynamics co-equally with the pixels.

The suspension of smoothing at surface boundaries is accomplished by introducing a coupling between the two lattices. Let us consider the energy corresponding to a quadratic prior given in Eq. (6.95). We can modify this term so that the interaction energy becomes

$$U_i(f, g) = \frac{1}{2\sigma^2}(f_i - g_i)^2 + \alpha \sum_{j \in N_i} (f_i - f_j)^2 (1 - l_{ij}) \quad (6.121)$$

We now have a situation where the contribution to the energy from the second term is zero whenever a line element is ON. Thus the inclusion of a lattice-lattice coupling prevents the unwanted phase transitions by suspending the smoothness constraint at boundaries. We can then address our second goal, that of strengthening and preserving these boundaries. This is done by discouraging the formation of small islands of pixels separated from their neighbors by line elements. These islands represent local minima in the energy surface. They can be prevented from forming by the addition of the appropriate line element clique potentials to yield a total interaction energy of the form

$$U_i(f, l, g) = \frac{1}{2\sigma^2}(f_i - g_i)^2 + \alpha \sum_{j \in N_i} \left[(f_i - f_j)^2 (1 - l_{ij}) + \sum_{C_l} V_{C_l}(l_{ij}) \right] \quad (6.122)$$

One widely used set of cliques is the off–on system of line elements depicted in Fig. 6.9. The potentials for these six cliques and their rotational equivalents reflect our expectations concerning the relative frequencies of each type of clique. We call the six elements empty, isolated singleton, corner, straight line, tee, and crossing. We then observe that the most likely elements are the empty sites and straight lines, while isolated singletons and crossings are least likely

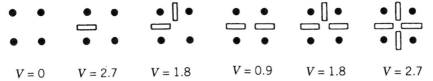

| V = 0 | V = 2.7 | V = 1.8 | V = 0.9 | V = 1.8 | V = 2.7 |

Figure 6.9. Set of cliques for the line process. Filled circles denote pixel elements. The dual lattice for the line process is registered with respect to the pixel lattice so that each line element is located between a pair of pixel elements. The weight associated each clique is shown below the clique. (From Geman and Geman[23]. Reprinted with permission of IEEE.)

and should be discouraged. The potentials or weights assigned to these cliques reflect these expectations.

These clique potentials may be regarded as generalized Ising potentials. We may introduce a variety of clique potentials for the line elements. For example, we may add diagonally oriented cliques to the basic set displayed in Fig. 6.9. However, these potentials are each added to the total interaction energy with a strength that must be empirically determined. The selection of appropriate coupling parameters for these potentials is often difficult. This difficulty places a practical limit on the number of clique potentials that can be utilized in most situations.

Perhaps the simplest choice of interaction potential for the line elements is an Ising hamiltonian. An appropriate Ising hamiltonian can be constructed from two triads of line elements oriented either horizontally or vertically. That is, we consider a central line element and its immediate neighbors to the left and right (horizontal) or above and below (vertical). For notational simplicity let us employ a single index notation to designate a line element associated with pixel i. Then we observe that an Ising hamiltonian for these triads

$$V_i(s) = \sum_{j \in Q_i} J_{ij} s_i s_j = -\sum_{j \in Q_i} s_i s_j = -s_i(s_{i-1} + s_{i+1}) \qquad (6.123)$$

provides a simple way of strengthening boundaries while eliminating isolated line singletons. We observe in Eq. (6.123) that the lowest energies are assigned to triads whose elements are either all ON or all OFF. In this Ising form, spins and line elements related to one another in the standard manner, namely

$$s_i = 2l_i - 1 \qquad (6.124)$$

We will examine several deterministic approaches later in this chapter where we integrate out the line process. As a prelude to these discussions, we note that if we neglect the clique potentials $V_{C_l}(l_{ij})$ and only consider the problem of how to preserve discontinuities while smoothing the data, then we may use a constraint of the form $\phi(D^k X)$, where $\phi(u) = -(1 + u)^{-1}$ in place of the quadratic interaction potential appearing in Eq. (6.121). In this expression[54] X

denotes the pixel intensities and D^k is their corresponding kth order derivative. The treatment of the line processes is implicit in approaches of this type rather than explicit.

6.10.2 Coupled Markov Random Fields and Integration

In our MAP method for image reconstruction, we remove noise, blur, and any other distortions and artifacts from image data while preserving surface boundaries. Our objective in segmenting images is to both reconstruct and identify the individual surfaces in the scene. We achieve these goals by removing noise and artifacts while preserving and strengthening surface boundaries. In integration we have the additional task of combining two-dimensional image data from a number of perceptual domains into an internally self-consistent representation. In many instances we would like to attach labels to the surfaces that describe their physical characteristics. We execute the integration tasks concurrently with the reconstruction and segmentation goals using coupled Markov random fields.

There are several underlying assumptions in the way we formulate image reconstruction problems. The first is the essential notion that 3-D physical scenes are composed of surfaces. Thus, to achieve our goal, we have to design a process, or algorithm, that takes as its imput a 2-D projection of such a 3-D scene and produces as its output an image description in terms of surfaces, their attendant discontinuities, and their stable physical properties. The centrality of surfaces in visual perception has its origin in the work of J. J. Gibson.[55,56] In his investigations of human visual perception, Gibson found that what we perceive are patterned surfaces in relation to one another and in relation to the ground (surface). Gibson asserted that the invariant properties of surfaces were perceived directly from the spatiotemporal patterns of light reaching the observer as contained in the ambient optic array. Barrow and Tenenbaum[57] and Marr,[58] together with Ullman[59] and others, replaced the concept of direct perception with a computational theory for the reconstruction of the surfaces, their discontinuities, and their intrinsic properties.

In the computational theory of perception, the reconstruction process proceeded through key intermediate representations called *intrinsic images*. These images are arrays of elements that are in precise registration with respect to the original image. Each array, or intrinsic image, describes an invariant physical characteristic of the underlying physical surfaces and also contains information regarding the discontinuities associated with that property. Examples of intrinsic characteristics encoded in these arrays are distance (depth), albedo (the ratio of total reflected to total incident illumination), local surface orientation, illumination (the integrated total illumination from all sources), hue, and specularity. In Marr's formulation the local surface orientation, depth, and attendant discontinuities in local surface orientation and depth form the so-called $2\frac{1}{2}$-D sketch. This intermediate representation lies above the primal sketch, which contains more primitive surface information, and convergent higher representations where the disparate information is combined to form object descriptions.

A major component in the computational theme is the realization that constraints are needed to restrict the many possible surface reconstructions to physically meaningful ones. The problem being addressed is as follows: The light convergent onto the retina, or equivalently, the intensity encoded as a pixel value at each location in a digital image, is influenced by many environmental factors. These factors include the character and geometry of the illumination source(s), the orientation and reflectance properties of the surfaces, and the relative position and functional behavior (e.g., of the optics) of the imaging system. The solution to this inverse, ill-posed problem of recovering properties of three-dimensional surfaces from two-dimensional images is to use prior knowledge about the physical properties of surfaces, namely how they are spatially organized. For instance, away from boundaries, neighboring portions of a surface tend to have the same physical characteristics. This prior knowledge is modeled as constraints of piecewise smoothness, statiotemporal continuity, and so on.

For some time it was felt that it is desirable to (1) simultaneously extract surface and boundary information, since they are two aspects of the same phenomenon, and (2) consider the integration of the information contained in the various intrinsic images at the lowest levels possible as part of the surface reconstruction process. Both objectives are achieved by coupling the Markov random fields to one another though the use of the appropriate potentials. In this approach two coupled spatial lattices are introduced for each modality — one describing the intrinsic image and one representing the associated discontinuities. As a result of this dual modeling, surface and boundary aspects not only can be treated simultaneously, but the smoothness requirements can be suspended at the boundaries in a natural way. The integration of the information contained in the different intrinsic images, as well as a variety of labeling operations, can be accommodated easily within the coupled Markov random field methodology. In carrying out the integration in this cooperative approach, we are able to impose additional constraints while giving a precise characterization of the surface properties.

The coupling of Markov random fields representing different perceptual units permits us to treat a set of mutually reinforcing cues through multiple interaction potentials. The emerging integration strategy is depicted in Fig. 6.10 for a perceptual system composed of several early vision modules. Each of the perceptual units is represented by a pair of MRFs, one for the pixel process and one for the line process. We couple together pixel and discontinuity MRFs for a given perceptual unit in the manner described in the previous subsection, and we incorporate interactions between line processes belonging to different modules. As noted by Poggio et al.[60] and Gamble et al.,[61] the resulting mathematical model can be used to classify the discontinuities according to their physical origin such as albedo depth and surface orientation. Line processes belonging to different units mutually reinforce one another and can

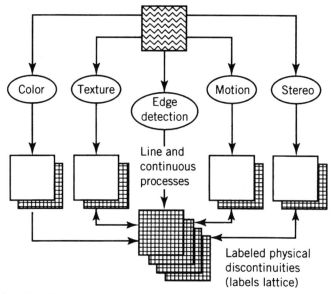

Figure 6.10. Integration of multiple visual cues. Several early vision modules are each represented by a pair of coupled Markov random fields encoding pixel and discontinuity information. The dual lattice of each module is coupled to itself and to the others. The cooperative integration yields a labeling of the underlying physical surfaces. (From Gamble et al.[61]. Reprinted with permission of IEEE.)

be designed to promote the filling-in of sparse data intrinsic to modalities such as motion and stereo depth.

An example of the integration of data from multiple modules is presented in Fig. 6.11. In this study by Beckerman and Sweeney, information from three modalities — depth, reflectance (albedo), and surface orientation — were integrated using an ICM algorithm to produce a segmentation map of the surfaces in the underlying indoor physical scene. The data were acquired using a laser range finder, namely by an imaging the laser radar operating at a near-infrared wavelength of 835 nm. This device produces dense depth maps along with a registered reflectance image of the scene. The input data are shown in the leftmost two panels. The surface orientation information presented in the center panels is derived from the depth map and is treated as a separate modality. Line processes associated with each module are modeled using the formalism of Eqs. (6.120)–(6.124), and the ICM algorithm was implemented using Eq. (6.119). The results are presented in the rightmost panels.

The ICM algorithm for the coupled Markov random fields is as follows:

1. Initialize the pixel lattice by choosing the value at each site that maximizes the likelihood.

2. Initialize all dual lattices using the initial pixel values according to Eq. (6.120).

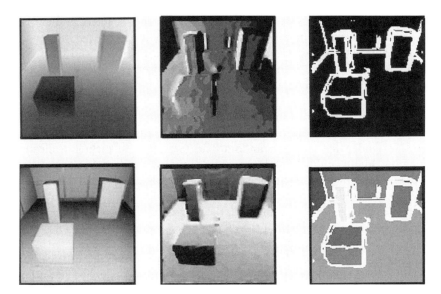

Figure 6.11. Fusion of range, reflectance, and surface orientation modalities. *Left*: Input range (*upper*) and reflectance (*lower*) images. *Center*: Horizontal and vertical components of the surface normals. *Right*: Edge and segmentation maps. (From Beckerman and Sweeney[62]. Reprinted with permission of the Society of Photo-Optical Instrumentation Engineers.)

3. Select a primary dual lattice pair.
4. Select a lattice site in the manner of a raster scan.
5. Replace the pixel grey values with the mode for the neighborhood.
6. Go back to step 4 and continue the raster scan.
7. Initialize the dual lattice using the updated pixel values.
8. Select a dual lattice site in the manner of a raster scan.
9. If flipping the spin lowers the energy, then flip the spin; otherwise do not flip.
10. Go back to step 8 and continue the raster scan.
11. Repeat steps 3 to 10 for the next pair of coupled lattices.
12. Iterate five or six times, repeating steps 3 to 10 for the three lattice pairs.

Another coupled MRF strategy for integrating information from different sensory modules has been utilized in the segmentation of synthetic aperature radar complex imagery. The approach adopted by Derin et al.[63] and by Rignot and Chellappa[51] is similar to the hierarchical coupling of line and pixel processes. Again, two random field models are coupled together, in this case, a lower-level speckle process and a higher-level region-labeling process. The results presented in Fig. 6.8 were obtained using this coupled speckle and label lattice approach.

Coupled random fields have been used to reconstruct, segment and integrate a variety of early vision modules. Coupled optical flow (2-D visual motion) and flow discontinuity fields were used by Murray and Buxton[64] and by Konrad and Dubois[50,65] to segment flow fields. Line processes were coupled by Wright[66] in order to integrate and segment RGB color imagery, and albedo and depth fields were integrated by Nadabar and Jain[67] in order to detect and classify edges. It is sometimes convenient to forbid certain highly unlikely configurations of an image lattice. In a study of texture segmentation, Geman et al.[45] extended the coupled random field formalism to include a means of forbidding highly unlikely states. Finally, coupled random fields have been used by Modestino and Zhang[68] for image interpretation, a higher-level task.

We may summarize our emerging coupled random field picture by noting that it is characterized by parallelism, the cooperative low-level integration of multiple sources of information through simple coupling interactions, and the encoding of general information about the physical world through constraints. This coupled MRF approach has considerable appeal from a biological point of view. The theme of cooperative low-level integration of multimodal information will reappear in a neural setting in Chapter 9 when we discuss integration through temporal tagging.

We have encountered the integration of multiple reinforcing cues previously in our exploration in Chapter 5 of growth cone sensing and navigation. There we observed a scheme where multiple and sometimes redundant molecular agents such as diffusible chemotropic factors and different families of cell surface molecules combined to promote neurite outgrowth and guide developing axons to the neural targets. The integrative character of these developmental processes were brought out dramatically in the tissue manipulation experiments where the molecular agents were brought into conflict with one another.

A body of data pertaining to early vision has been acquired through a variety of psychophysical experiments. In the case of early vision, psychophysical and physiological data appear to be consistent with parallelism, constraints that encode general properties of the environment, multiple cues, and interacting modalities.[69] To our earlier emphasis on physical principles and processes we now impart an engineering flavor exemplified by coupled Markov random fields and their attendant stochastic algorithms.

6.11 COMPOUND GAUSS-MARKOV RANDOM FIELDS

In this section we consider another important class of integrative models, namely the compound Gauss-Markov random fields.[26,70] These models extend the method of Geman and Geman[23] to noncompact, continuous-valued range spaces. In these methods we combine a Gauss-Markov model as developed by Woods[4] and by Kashyap and Chellappa[21] with a Geman and Geman line process. The resulting model is termed compound because it contains several

COMPOUND GAUSS-MARKOV RANDOM FIELDS 231

simple image models, each having different characateristics. Transitions between the simple models are governed by a structure model, a two-dimensional random field with a finite number of discrete values.

Once again, we wish reconstruct an image **f** given the degraded observed image **g**. The point-spread function is d_{kl} with support region η_d. We assume that we have a zero-mean i.i.d. Guassian noise process v with variance σ_v^2. The aforementioned quantities are related to each other by the family of equations

$$g_{mn} = \sum_{(k,l) \in \eta_d} d_{kl} f_{m-k,n-l} + v_{mn} \qquad (6.125)$$

From Section. 6.5.3 we see that there are two types of clique potentials, representing the diagonal and the off-diagonal, neighborhood terms in the probability distributions. These potentials may be written

$$V_{C_{mn}}(f_{mn}) = \frac{f_{mn}^2}{2\sigma_u^2} \qquad (6.126)$$

and

$$V_{C_d}(f_{mn}) = -\frac{h_{m-k,n-l} f_{mn} f_{kl}}{\sigma_u^2} \qquad (6.127)$$

We now extend the formalism presented in Section. 6.5.3 by appending a line or structure process. In place of Eq. (6.50), we have

$$f_{mn} = \sum_{(k,l) \in \eta_{mn}} h_{kl}^{l_{mn}} f_{m-k,n-l} + u_{mn}^{l_{mn}} \qquad (6.128)$$

We see that the covariance and noise is now controlled by the line process l_{mn}. As before, the line process is binary valued and is encoded in a dual lattice. Thus we have a model consisting of a continuous-valued pixel process plus a discrete-valued line process. The posterior hamiltonian $U_p(\mathbf{f}, \mathbf{g})$ for the pixel process consists of a sum over all sites of the site potentials $U_m(f_{mn}, g_{mn})$. These in turn each contain a sum of two quadratic terms:

$$U_{mn}(f_{mn}, g_{mn}) = V_{mn}^{(h)}(f_{mn}) + V_{mn}^{(d)}(f_{mn}, g_{mn}) \qquad (6.129)$$

The first term,

$$V_{mn}^{(h)}(f_{mn}) = \frac{1}{2\sigma_{u^l_{mn}}^2} \left(f_{mn} - \sum_{kl \in \eta_{mn}} h_{kl}^{l_{mn}} f_{m-k,n-l} \right)^2 \qquad (6.130)$$

contains the contribution from the image model, Eq. (6.128). We observe that it contains the two types of clique potential mentioned above. The second term

$$V^{(d)}_{mn}(f_{mn}, g_{mn}) = \frac{1}{2\sigma_v^2} \sum_{ij \in \eta_d} \left(g_{m+i,n+j} - \sum_{kl \in \eta_d} d_{kl} f_{m+i-k,n+j-l} \right)^2 \quad (6.131)$$

is the contribution from the degredation or restoration model as represented by Eq. (6.125).

The posterior hamiltonian for the line process is given by a sum over all dual lattice sites of single-site potentials of the form

$$U^l_{mn}(f_{ij}, f_{i'j'}, l_{mn}) = \frac{f_{ij}^2}{2\sigma_{u^l_{ij}}^2} - \frac{h^{l_{i'j'}}_{i-i',j-j'} f_{ij} f_{i'j'}}{\sigma_{u^l_{ij}}^2} + \frac{f_{i'j'}^2}{2\sigma_{u^l_{i'j'}}^2} \sum_{mn \in C_l} V_{C_l}(l_{mn}) \quad (6.132)$$

In this expression we consider the dual lattice line elements l_{mn} located between pairs of pixel sites f_{ij} and $f_{i'j'}$. The clique system represented by the last term in Eq. (6.132) is the same as that shown in Fig. 6.9.

6.12 MEAN-FIELD ANNEALING

6.12.1 Mechanical Models, Graduated Nonconvexity, and Mean-Field Theory

Mechanical models can provide insight into the design of constraints and interaction energies that are useful in image reconstruction. For example, in Barnard's[71] spring-loaded lever arm model, each state of the system encodes a disparity in the correspondence between a pair of stereo images. Energies associated with these states represent the degree of disparity. The ground-state of the system gives the optimally matched pair of stereo images; that is, the lowest-energy states are configurations of minimum disparity. Two constraints are introduced into the formulation of the stereo-matching problem, namely that matched points should have similar intensities and that the disparity map should vary smoothly. As in Julesz's[72] spring-coupled dipole model, the constraints appear as spring energies. The minimum disparity states of this system were found stochastically by Barnard using the Creutz microcanonical annealing algorithm presented in Chapter 4.

In the next three sections we will explore mechanical and mean-field formulations of the image reconstruction problem modeled by the MAP hamiltonians of Sections. 6.7 and 6.11. As was the case for Geman and Geman's method, the goal is to simultaneously remove noise, and strengthen and preserve boundaries. The resulting reconstruction algorithms are deterministic in character, and they allow for the treatment of continuously valued pixel grey levels. In the graduated nonconvexity approach of Blake and Zisserman,[29] local minima are avoided by constructing convex hamiltonians that approximate the exact nonconvex posterior hamiltonians of the problem.

The effective potentials generated in this mechanical method resemble the mean-field potentials derived by Geiger and Girosi[28] and by Bilbro and Snyder.[27] The deterministic, graduated nonconvexity algorithm may be viewed as a low-temperature limit of the mean-field approaches.

6.12.2 Mean-Field Annealing

In mean-field annealing[27,73] we approximate the posterior hamiltonian by the mean-field hamiltonian. The mean-field that minimizes the posterior hamiltonian corresponds to the equilibrium posterior distribution of the Markov random field. For illustrative purposes we employ a piecewise constant image model in the derivation of the algorithm. The method can be generalized to piecewise linear and other more complex image models. Our starting point in mean-field annealing is the posterior hamiltonian

$$E(\mathbf{f}, \mathbf{g}) = \sum_i \left[\frac{(f_i - g_i)^2}{2\sigma^2} \right] + \alpha \sum_{j \in N_i} b_{ij} V(f_i, f_j) \quad (6.133)$$

where, as before, $\mathbf{g}$ is the observed image, $\mathbf{f}$ is the true image, and α is the parameter governing the strength of the interaction energy that embodies the smoothness-enforcing constraint. The quantities b_{ij} appearing in the equation denote the coupling strengths between the pixel intensities at lattice sites i and j, and V is the prior potential. One chioce for a prior term that enforces piecewise constancy is

$$V(f_i, f_j) = 1 - \delta(f_i - f_j) = \begin{cases} 1, & f_i \neq f_j \\ 0, & f_i = f_j \end{cases} \quad (6.134)$$

This term produces an increase in the energy whenever the pixel intensities at the two lattice sites differ from one another. In order to generate a potential appropriate for continuous-valued variables, we represent the delta function as the limit of a progressively sharper Gaussian:

$$\delta(x) = \lim_{\tau \to 0} \frac{1}{\sqrt{2\pi\tau}} \exp\left(-\frac{x^2}{2\tau}\right) \quad (6.135)$$

The posterior hamiltonian then becomes

$$E(\mathbf{f}, \mathbf{g}) = \sum_i \left[\frac{(f_i - g_i)^2}{2\sigma^2} + \alpha \sum_{j \in N_i} b_{ij} \left(1 - \lim_{\tau \to 0} \frac{1}{\sqrt{2\pi\tau}} \exp\left[\frac{-(f_i - f_j)^2}{2\tau} \right] \right) \right]$$

(6.136)

The first part of the prior contains a sum over the b_{ij}'s. This term gives a constant contribution to the posterior hamiltonian, independent of the pixel values, and can be neglected. Thus

$$E(\mathbf{f}, \mathbf{g}) = \sum_i \left[\frac{(f_i - g_i)^2}{2\sigma^2} - \alpha \lim_{\tau \to 0} \sum_{j \in N_i} \frac{b_{ij}}{\sqrt{2\pi\tau}} \exp\left(\frac{-(f_i - f_j)^2}{2\tau} \right) \right] \quad (6.137)$$

We are now ready to introduce the mean-field $\mathbf{x}$. This quantity is defined through the mean-field hamiltonian $E_{mf}(\mathbf{f}, \mathbf{x})$ as

$$E_{mf}(\mathbf{f}, \mathbf{x}) = \sum_i (f_i - x_i)^2 \quad (6.138)$$

and the accompanying posterior mean-field distribution

$$p_{mf}(\mathbf{f}, \mathbf{x}) = \frac{1}{Z} \exp\left(-\frac{E_{mf}(\mathbf{f}, \mathbf{x})}{T} \right) \quad (6.139)$$

with

$$Z = \int_{R_N} \exp\left(-\frac{E_{mf}(\mathbf{f}, \mathbf{x})}{T} \right) df \quad (6.140)$$

As was the case for a disordered spin glass, the quantities x_i represent the cooperative effect of the interactions at each individual lattice site. The mean-field probability distribution, which is of the form of a Gibbs distribution with a Gaussian normalization, is an approximation to the true posterior distribution expressed in terms of $E(\mathbf{f}, \mathbf{g})$. To determine the form of the mean-field $\mathbf{x}$, we demand that $E_{mf}(\mathbf{f}, \mathbf{x})$ best approximate the true hamiltonian $E(\mathbf{f}, \mathbf{g})$. This best approximation can be found through a self-consistency procedure based on Jensen's inequality.

6.12.3 Convex Functions and Jensen's Inequality

Loosely speaking, a convex function is a smoothly varying function that has a single minimum that can be found starting from any point by a gradient descent algorithm. More formally, a function $f(x)$ defined on an interval I is said to be *convex* if and only if for all $x_1, x_2 \in I$ and $0 \leq a \leq 1$,

$$af(x_1) + (1 - a)f(x_2) \geq f(ax_1 + (1 - a)x_2) \quad (6.141)$$

Generalizing this relation to many elements on the interval gives

$$\sum_i a_i f(x_i) \geq f\left(\sum_i a_i x_i \right) \quad (6.142)$$

where

$$\sum_i a_i = 1, \quad 0 < a_i < 1 \qquad (6.143)$$

If we identify a_i with the discrete probabilities $p_i(x_i)$, then we have

$$\sum_i a_i f(x_i) = \sum_i p_i(x_i) f(x_i) = \langle f(x) \rangle \qquad (6.144)$$

and

$$f\left(\sum_i a_i x_i\right) = f\left(\sum_i p_i(x_i) x_i\right) = f(\langle x \rangle) \qquad (6.145)$$

Therefore we have the inequality

$$\langle f(x) \rangle \geq f(\langle x \rangle) \qquad (6.146)$$

This result applies to continuous functions as well. We define the average with respect to $\mathbf{x}$ of a convex function $F(\mathbf{f})$ by the integral

$$\langle F \rangle_\mathbf{x} \equiv \int_{R_N} F(\mathbf{f}) p_{mf}(\mathbf{f}, \mathbf{x}) d\mathbf{f} \qquad (6.147)$$

Jensen's inequality, when applied to the convex function $F(Q) = \exp(Q)$, is

$$\langle \exp(Q) \rangle_\mathbf{x} \geq \exp(\langle Q \rangle_\mathbf{x}) \qquad (6.148)$$

Taking

$$Q = \frac{E - E_{mf}}{T} \qquad (6.149)$$

yields

$$\ln \frac{\int_{R_N} \exp\{-(E - E_{mf})/T\} \exp\{-E_{mf}/T\} d\mathbf{f}}{\int_{R_N} \exp\{-E_{mf}/T\} d\mathbf{f}} \geq \langle Q \rangle_\mathbf{x} \qquad (6.150)$$

or quivalently

$$\ln \int_{R_N} \exp\left\{-\frac{E}{T}\right\} d\mathbf{f} \geq \ln \int_{R_N} \exp\left\{-\frac{E_{mf}}{T}\right\} d\mathbf{f} - \frac{\langle E - E_{mf} \rangle_\mathbf{x}}{T} \qquad (6.151)$$

and therefore

$$-T \ln \int_{R_N} \exp\left\{-\frac{E}{T}\right\} d\mathbf{f} \leq -T \ln \int_{R_N} \exp\left\{-\frac{E_{mf}}{T}\right\} d\mathbf{f} + \langle E - E_{mf}\rangle_\mathbf{x}$$
(6.152)

Recalling our definition from Chapter 2 of the Helmholtz free energy A as $-T$ times the logarithm of the partition function, we have the inequality

$$A \leq A_{mf} + \langle E - E_{mf}\rangle_\mathbf{x}$$
(6.153)

where A represents the free energy of the lattice with respect to the hamiltonian E and A_{mf} is the corresponding free energy for the hamiltonian E_{mf}.

6.12.4 The Mean-Field Annealing Algorithm

We now use this inequality to find the mean-field $\mathbf{x}$. We observe that the right-hand side of Eq. (6.153) serves as an upper bound to the free energy. By minimizing this upper bound with respect to the mean-field $\mathbf{x}$, we obtain a mean-field hamiltonian E_{mf} that best approximates the hamiltonian E. Therefore the objective of the mean-field annealing algorithm is the minimization of the upper bound; that is, the desired algorithm is a procedure for solving the equation

$$\nabla_\mathbf{x}\left(\langle E - E_{mf}\rangle_\mathbf{x} - T \ln \int_{R_N} \exp\left(-\frac{E_{mf}}{T}\right) d\mathbf{f}\right) = 0$$
(6.154)

To solve this equation, we first evaluate the integral in the last term. We have

$$\int_{R_N} \exp\left\{-\frac{E_{mf}}{T}\right\} d\mathbf{f} = \int_{R_N} \exp\left\{-\frac{\|\mathbf{f} - \mathbf{x}\|^2}{T}\right\} d\mathbf{f}$$

$$= \int_{R_N} \exp\left\{-\frac{\|\mathbf{f}'\|}{T}\right\} d\mathbf{f}' = (\pi T)^{N/2}$$
(6.155)

Second, we evaluate the averaged mean-field hamiltonian

$$\langle E_{mf}\rangle_\mathbf{x} = \int_{R_N} E_{mf} \exp\left\{-\frac{E_{mf}}{T}\right\} d\mathbf{f}$$

$$= \int_{R_N} \|\mathbf{f} - \mathbf{x}\|^2 \exp\left\{-\frac{\|\mathbf{f} - \mathbf{x}\|^2}{T}\right\} d\mathbf{f} = \frac{NT}{2}$$
(6.156)

Third, we find that

$$\langle E \rangle_x = \int_{R_N} \sum_i \left[(f_i - g_i)^2 / 2\sigma^2 \right.$$
$$\left. - \alpha \lim_{\tau \to 0} \sum_{j \in N_i} \frac{b_{ij}}{\sqrt{2\pi\tau}} \exp\left(\frac{-(f_i - f_j)^2}{2\tau}\right) \right] \exp\left\{-\frac{\|\mathbf{f} - \mathbf{x}\|^2}{T}\right\} d\mathbf{f} \quad (6.157)$$
$$= \frac{NT}{4\sigma^2} + \sum_i \left[\frac{(g_i - x_i)^2}{2\sigma^2} - \alpha \sum_{j \in N_i} \frac{b_{ij}}{\sqrt{2\pi T}} \exp\left(\frac{-(x_i - x_j)^2}{2T}\right) \right]$$

This is the crucial term for finding the mean-field. In evaluating the multiple intergrals, we made use of the delta function to carry out one of the integration steps. Combining our results gives

$$\frac{\partial}{\partial x_i}(A_{mf} + \langle E - E_{mf} \rangle_x) = \frac{x_i - g_i}{\sigma^2} + \alpha \sum_{j \in N_i} \frac{\sqrt{2} b_{ij}(x_i - x_j)}{\sqrt{\pi} T^{3/2}} \exp\left(\frac{-(x_i - x_j)^2}{2T}\right)$$

(6.158)

The quantity on the right-hand side of Eq. (6.158) is used to find the minimum of Eq. (6.157). The mean-field annealing algorithm is then:

1. Initialize $\mathbf{x} = \mathbf{g}$.
2. Initialize the temperature T.
3. Use Eqs. (6.157) and (6.158) to find the minimum of Eq. (6.157) using gradient descent or some other deterministic method.
4. Update $\mathbf{x}$ to the value found in step 3, and lower the temperature.
5. Repeat steps 3 and 4 until the stop criterion is met.

6.13 GRADUATED NONCONVEXITY

We now consider the mechanical approach to solving optimization problems in visual surface reconstruction. The mechanical method developed by Blake and Zisserman,[29,74] and called graduated nonconvexity contains several key elements. First, we have the cooperativity intrinsic to mechanical systems built from elements such as rods, springs, and flexible plates. These systems develop global order from local interactions and possess all the properties expected of cooperative systems. For example, in Julesz's spring-loaded dipole model, we find that the system can flip from one state to another and that there are many local minima. A stable state consists of a collection of regions, or domains, each containing magnets of similar orientation. These orientations change abruptly across domain boundaries.

Second, we have piecewise continuity analogous to piecewise smoothness of curves and surfaces in discrete systems. In graduated nonconvexity, smoothness properties are modeled as weak continuity constraints. This means that we are able to suspend continuity constraints at boundaries, as was the case in the method of Geman and Geman. Third, we have nonconvex optimization, in which a nonconvex function is approximated by one that is piecewise convex.

6.13.1 Weak Continuity Constraints

The simplest of the weak continuity problems is the detection of discontinuities in piecewise-smooth lines and surfaces. The mechanical models used to characterize the interaction energies and continuity constraints for these problems are the elastic string and weak membrane. The elastic string is capable of detecting and locating discontinuities in one dimension. The weak membrane is its two-dimensional generalization. The hamiltonians for these models are composed of three terms:

$$E(\mathbf{f}, \mathbf{g}, \mathbf{l}) = D(\mathbf{f}, \mathbf{g}) + S(\mathbf{f}, \mathbf{l}) + P(\mathbf{l}) \tag{6.159}$$

where, as before, $\mathbf{g}$ is the data, $\mathbf{f}$ is the reconstruction, and $\mathbf{l}$ is the line process. The first term, D, measures departures of the string or membrane from the data. The second term assesses how badly the string or membrane is deformed, that is, this quantity represents the elastic energy. The third term assigns a penalty each time a discontinuity is introduced and promotes keeping the number of discontinuities as small as possible. In discretized form the three terms for the elastic string are as follows: For the departure of the data from the surface,

$$D(\mathbf{f}, \mathbf{g}) = \sum_i (f_i - g_i)^2 \tag{6.160}$$

which we recognize as our familiar noise term. For the elastic energy we have

$$S(\mathbf{f}, \mathbf{l}) = \lambda^2 \sum_i (f_i - f_{i-1})^2 (1 - l_i) \tag{6.161}$$

which corresponds to the Bayesian prior in the stochastic approaches. This energy is defined over an Ising-like neighborhood system, and it involves a coupling of the pixel and line processes familiar to us from Section. 6.10. For the penalty term we adopt the simplest possible choice, namely

$$P(\mathbf{l}) = \alpha \sum_i l_i \tag{6.162}$$

6.13.2 Elimination of the Line Process

We now eliminate the discrete-valued line process so that only the real-valued variables **f** remain. To assist in the elimination, we combine the line-process-dependent terms into a potential H:

$$H(f_i - f_{i-1}, l_i) = \lambda^2 (f_i - f_{i-1})^2 (1 - l_i) + \alpha l_i \qquad (6.163)$$

We next notice that our minimization problem is of the form

$$\min_{\{f\},\{l\}} E(\mathbf{f}, \mathbf{g}, \mathbf{l}) = \min_{\{f\}} \left(D(\mathbf{f}, \mathbf{g}) + \min_{\{l\}} \sum_i H(f_i - f_{i-1}, l_i) \right)$$
$$= \min_{\{f\}} \left(D(\mathbf{f}, \mathbf{g}) + \sum_i G(f_i - f_{i-1}) \right) \qquad (6.164)$$

where the prior potential G is

$$G(f_i - f_{i-1}) = \min_{l \in \{0,1\}} H(f_i - f_{i-1}, l_i) \qquad (6.165)$$

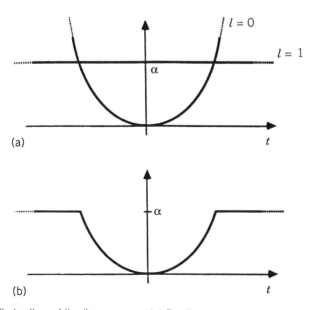

Figure 6.12. Elimination of the line process. (a) The line process dependent potential H as a function of $t = f_i - f_{i-1}$, as given by Eq. (6.163). (b) The line process minimized potential G given by Eq. (6.166). (From Blake and Zisserman[29]. Reprinted with permission of MIT Press.)

240 MARKOV RANDOM FIELDS

We next carry out the minimization over **l** in order to eliminate the line process. This is accomplished by noting that the desired potential is the minimum of the two curves illustrated in Fig. 6.12. The line-process-minimized potential G is

$$G(t) = \begin{cases} \lambda^2 t^2, & |t| < \frac{\sqrt{\alpha}}{\lambda} \\ \alpha, & \text{otherwise} \end{cases} \qquad (6.166)$$

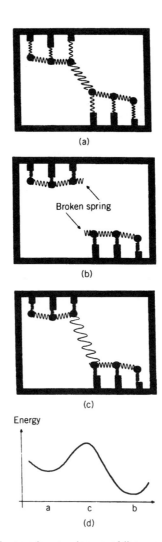

Figure 6.13. Nonconvexity in a spring analogue of the weak string. States (*a*) and (*b*) are stable, while the intermediate state (*c*) has a higher energy than either stable state. (From Blake and Zisserman[29]. Reprinted with permission of MIT Press.)

6.13.3 Nonconvexity

The nonconvexity of the hamiltonian with respect to weak continuity, or the line process, can be understood easily if we think of the system as a set of coupled, breakable springs. In this picture, shown in Fig. 6.13, we use springs to connect data points g_i, to the nodes f_i. These springs represent the D term in the hamiltonian. We now connect a set of lateral springs between nodes. If these springs are deformed too far, they break. Their deformation energy is represented by the G term. The relationship between stable states and spring configurations is shown in Fig. 6.13. We see that the hamiltonians for mechanical systems such as the spring-coupled dipoles, the elastic string and weak membrane will possess a multitude of local minima.

6.13.4 Convex Approximation

The next step in our procedure is to generate a family of convex functions that approximates the prior potential G. This family of potentials $G^{(p)}$ is labeled by a parameter p that varies from 1 to 0:

$$G^{(p)}(t) = d \begin{cases} \lambda^2 t^2, & |t| < q \\ \alpha - c \dfrac{(|t| - r)^2}{2p}, & q \leqslant |t| < r \\ \alpha, & |t| \geqslant r \end{cases} \qquad (6.167)$$

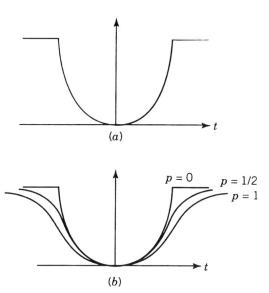

Figure 6.14. Effective potentials in the GNC algorithm as a function of $t = f_i - f_{i-1}$. (a) The original nonconvex function generated by eliminating the line process. (b) Family of convex potentials given by Eqs. (6.167)–(6.169). (From Blake[74]. Reprinted with permission of IEEE.)

where

$$r^2 = \alpha\left(\frac{2p}{c} + \frac{1}{\lambda^2}\right) \tag{6.168}$$

and

$$q = \frac{\alpha}{\lambda^2 r} \tag{6.169}$$

The constant c assumes the value $1/2$ for the elastic string and takes the value $1/4$ for the weak membrane. A general procedure for generating the families of convex approximations has been described by Blake and Zisserman. The family of effective potentials for the elastic string is plotted in Fig. 6.14 for various values of the parameter p. Reducing this parameter from 1 to 0 steadily changes $G^{(p)}$ until it becomes equal to the clipped parabola G.

6.14 EFFECTIVE POTENTIALS

The stochastic Markov random field and deterministic string and membrane models share a common conceptual framework. Both introduce cooperative interactions of the same form, which encode constraints that guide the image reconstruction process, and both generate similar effective potentials. In this section we will explore this last interrelationship, first noted by Geiger and Girosi.[28,75] In doing so, we will find that the statistical mechanics of the Markov random field formalism can be used to generate effective potentials of the graduated nonconvexity form.

We begin by writing the interaction energy for the two-dimensional weak membrane, adopting an i, j notation to explicitly denote the horizontal and vertical coordinates. In this notation we have the MRF posterior hamiltonian

$$E(\mathbf{f}, \mathbf{g}, \mathbf{l}^h, \mathbf{l}^v) = \sum_{i,j} \left\{ \frac{1}{2\sigma^2}(f_{i,j} - g_{i,j})^2 + \lambda((f_{i,j} - f_{i,j-1})^2(1 - l^v_{i,j}) \right.$$
$$\left. + \lambda((f_{i,j} - f_{i-1,j})^2(1 - l^h) + \gamma^v_{i,j} l^v_{i,j} + \gamma^h_{i,j} l^h_{i,j} \right\} \tag{6.170}$$

where the superscripts h and v denote horizontal and vertical line elements, and λ and γ are strength constants for the interaction and penalty terms, respectively. The partition function obtained by summing over all configurations of the weak membrane is

$$Z(\mathbf{f}, \mathbf{g}, \mathbf{l}^v, \mathbf{l}^h) = \sum_{\{\mathbf{f}\},\{\mathbf{l}^v\},\{\mathbf{l}^h\}} \exp\{-\beta E(\mathbf{f}, \mathbf{g}, \mathbf{l}^v, \mathbf{l}^h)\} \tag{6.171}$$

As we did before, we group together the terms involving the summations over the line elements and write the partition function as

$$Z = \sum_{\{f\}} \exp\left\{-\beta \sum_{i,j}\left[\frac{(f_{i,j} - g_{i,j})^2}{2\sigma^2} + \gamma^v_{i,j} + \gamma^h_{i,j}\right]\right\}$$
$$\times \sum_{\{l^v\},\{l^h\}} \exp\left\{-\beta \sum_{i,j}[(1 - l^v_{i,j})G^v_{i,j} + (1 - l^h_{i,j})G^h_{i,j}]\right\} \quad (6.172)$$

where $G^v_{ij} = \lambda(f_{ij} - f_{ij-1})^2 - \gamma^v_{ij}$, and similarly for G^h_{ij}. The above equation is that of a pair of one-dimensional systems of Ising-like spin elements in external fields G^v and G^h. We can carry out the summations over the two values of the line elements at each lattice site. The result is

$$Z = \sum_{\{f\}} \exp\left\{-\beta \sum_{i,j}\left[\frac{(f_{i,j} - g_{i,j})^2}{2\sigma^2} + \gamma^v_{i,j} + \gamma^h_{i,j}\right]\right\}$$
$$\times \prod_{i,j}(1 + \exp\{-\beta G^v_{i,j}\})(1 + \exp\{-\beta G^h_{i,j}\}) \quad (6.173)$$

We may now define the effective potential $G(f)$ for the weak membrane through the expression

$$Z = \sum_{\{f\}} \exp\{-\beta[D(\mathbf{f}, \mathbf{g}) + G(\mathbf{f})]\} \quad (6.174)$$

In the above, the first quantity

$$D(\mathbf{f}, \mathbf{g}) = \sum_{i,j} \frac{(f_{i,j} - g_{i,j})^2}{2\sigma^2} \quad (6.175)$$

is the noise term, and the second

$$G(\mathbf{f}) = \sum_{i,j}\left(\gamma^v_{i,j} + \gamma^h_{i,j} - \frac{\ln[(1 + \exp\{-\beta G^v_{i,j}\})(1 + \exp\{-\beta G^h_{i,j}\})]}{\beta}\right) \quad (6.176)$$

is the effective potential. This family of potentials is parameterized by the temperature $1/\beta$. In the zero temperature limit, the effective potential becomes identical to the graduated nonconvexity form.

6.15 RENORMALIZATION GROUP SIMULATED ANNEALING

We recall from Chapter 3 that renormalization group (RG) methods were introduced to provide a way of treating systems in which multiple degrees of freedom spanning many scales of length are locally coupled. In these strongly cooperative systems, the local coupling of degrees of freedom generates a cascade effect in the whole system leading to scaling phenomena and universality. In the renormalization group approach we generate new potentials (hamiltonians) from old ones by progressively integrating out degrees of freedom at each length scale. At the end of each step, we are left with a hamiltonian representing the interactions over length scales not yet treated. The main step is the identification of a RG transformation that relates the old and new hamiltonians while leaving the form of the partition function alone, collecting the effect of the summation into an effective coupling constant.

Grey-scale images possess many of the same features as other strongly cooperative systems. In particular, we find that changes in grey-scale levels occur over the many scales of length for which the image is organized. In local processing methods such as simulated annealing convergence in the frequency domain of the long wavelength components of the grey-scale pixel values is slow. The basic idea of the renormalization group approach is to combine local processing of information at different length scales, the RG cascade, with an interscale transfer of information, the RG transformation. The Gidas[30] renormalization group algorithm is a procedure for global optimization that generates iteratively a cascade of restored images corresponding to different levels of resolution. It achieves these objectives by merging renormalization group, Markov random field, and simulated annealing procedures into a common formalism. In this section we briefly sketch the real space RG approach to solving the two-dimensional Ising model, and we then explore the Gidas renormalization group algorithm as applied to image restoration.

6.15.1 Kadanoff Transformations

Real-space RG methods are block spin techniques developed for handling lattice systems of low dimensionality such as the two-dimensional Ising model. These methods were introduced by Neimeijer and van Leeuwen.[76] The simplest real space transformation is the decimation transformation of Kadanoff and Houghton.[77] In these approximation methods some spins are held fixed while others are summed over, producing an effective interaction over the fixed spins. In more detail, in decimation the new lattice is chosen to include half the sites of the old one. On the new lattice the new spins are chosen to be the same as the old ones at the corresponding site, and the remaining elements are summed over. Since the summations only involve fixed spins, the summations can be done independently at each site. This is not the only way to generate a scaling transformation. A second method of generating a sequence of progressively sparser lattices is by block transformations. As illustrated in Fig. 3.12, each new spin element appears in the middle of a block of four spins that are its nearest neighbors in the old lattice and that it replaces.

The Kadanoff scaling transformations allow us to go from hamiltonians and partition functions defined in terms of lattice of spin variables to hamiltonians and partition functions defined over a decimated lattices of new spin variables.

The Kadanoff transformation can be written

$$\exp\{-H^{(1)}(\mathbf{t})\} = \sum_{\{\mathbf{s}\}} \left\{\prod_r \delta_{s_r, t_r}\right\} \exp\{-H^{(0)}(\mathbf{s})\} \tag{6.177}$$

where $\mathbf{t}$ is the new spin configuration and $\mathbf{s}$ is the old spin array. We observe that the partition function is invariant under a Kadanoff transformation

$$Z = \sum_{\{\mathbf{t}\}} \exp\{-H^{(1)}(\mathbf{t})\} = \sum_{\{\mathbf{s}\}} \exp\{-H^{(0)}(\mathbf{s})\} \tag{6.178}$$

The second and following steps in the Kadanoff construction are similar to the first one. In general,

$$\exp\{-H^{(n)}(\mathbf{t})\} = \sum_{\{\mathbf{s}\}} \left\{\prod_r \delta_{s_r, t_r}\right\} \exp\{-H^{(n-1)}(\mathbf{s})\} \tag{6.179}$$

In the Kadanoff procedure[78] the summation in the first step can be performed exactly since the nearest neighbors of the spins being summed over are held fixed. In the second and later steps, a perturbation expansion can be used to calculate the various terms. A key condition is that the interactions must be dominated by a small number of short-range nearest-neighbor couplings.

6.15.2 The Gidas Algorithm

In applying real-space renormalization methods to digital image reconstruction, we use a two-stage procedure similar to that followed for the two-dimensional Ising model. First, we generate a cascade of renormalized hamiltonians by means of a coarsening procedure. Second, we find the optimal configuration at each stage in the cascade, starting with the lowest-resolution image and ending with the highest-resolution image. In more detail, we begin by generating a sequence $n = 1, 2, \ldots, N$ of progressively coarser lattices, starting with the original lattice Λ^0 and terminating with the coarsest lattice Λ^N:

$$\Lambda^{(0)} \to \Lambda^{(1)} \to \cdots \to \Lambda^{(N)}$$

These lattices are in registration with one another, and each is characterized by a cell size a factor of l larger then the preceding lattice. The cell-pixels produced by coarsening procedures such as decimation or block construction constitute the elements of the images asociated with the lattices. Thus at each stage in the renormalization procedure we generate a coarsened image $\mathbf{f}^{(n)}$ from the preceding less-coarse image $\mathbf{f}^{(n-1)}$:

$$\mathbf{f}^{(0)} \to \mathbf{f}^{(1)} \to \cdots \to \mathbf{f}^{(N)}$$

At the same time we compute iteratively a sequence of renormalized hamiltonians

$$H^{(0)} \to H^{(1)} \to \cdots \to H^{(N)}$$

In this sequence $H^{(0)} = (1/T_0)H$ assumes the role of the posterior hamiltonian. The construction of the renormalized is done by choosing a conditional probabilities $p^{(n)}(\mathbf{f}^{(n)}|\mathbf{f}^{(n-1)})$ and then computing the quantities

$$\exp\{-H^{(n)}_{T_0,\ldots,T_n}(\mathbf{f}^{(n)})\} = \sum_{\{\mathbf{f}^{(n-1)}\}} p^{(n)}(\mathbf{f}^{(n)}|\mathbf{f}^{(n-1)}) \exp\left\{-\left(\frac{1}{T_n}\right) H^{(n-1)}_{T_0,\ldots,T_{n-1}}(\mathbf{f}^{(n-1)})\right\}$$

(6.180)

The quantity $p^{(n)}(\mathbf{f}^{(n)}|\mathbf{f}^{(n-1)})$ is the probability that the configuation of the system at the nth stage in the cascade is $\mathbf{f}^{(n)}$ given that the configuration of the lattice system at the $(n-1)$th stage is $\mathbf{f}^{(n-1)}$. The quantities $T_n, n = 0, \ldots, N$, are scale parameters linked to the hamiltonians $H^{(n)}$. These scale parameters have been written in a form suggestive of a temperature.

The images $\mathbf{f}^{(n)}, n = 0, \ldots, N$ form a cascade that is processed in a hierarchical manner starting with the coarsest image $\mathbf{f}^{(N)}$ and proceeding to the next coarsest image $\mathbf{f}^{(N-1)}$, and so on, until we reach the final $\mathbf{f}^{(0)}$ image. The processing of information at the $(n-1)$th level is constrained by the information obtained in the preceding nth stage. In this second, optimization stage, we begin by finding the global minimum of $H^{(N)}$. Then we find the global minimum of $H^{(N-1)}$ by searching the subspace of configurations constrained by the global minimum $\mathbf{f}^{(N)}$ of $H^{(N)}$. This is accomplished using the constraint equations

$$p^{(n)}(\mathbf{f}^{(n)}|\mathbf{f}^{(n-1)}) = \max_{\{\mathbf{f}^{(n)}\}} p^{(n)}(\mathbf{f}^{(n)}|\mathbf{f}^{(n-1)})$$

(6.181)

and the control parameters $T_0, \ldots, T_N$. The products of this processing stage are a set of optimal hamiltonians and a set of associated configurations (images). The outcome at each level of resolution is related to the result at the next level through the renormalization group transformations, and the sequence leads to a maximum a posteriori estimate of the undegraded image. In the processing stage the expression

$$\pi^{(n)}_{T_0,\ldots,T_n}(\mathbf{f}^{(n)}) = \frac{\exp\{-H^{(n)}_{T_0,\ldots,T_n}(\mathbf{f}^{(n)})\}}{Z(T_0,\ldots,T_n)}$$

(6.182)

assumes the role of the posterior distribution with partition function

$$\begin{aligned} Z(T_0,\ldots,T_n) &= \sum_{\{\mathbf{f}^{(n-1)}\}} \exp\left\{-\left(\frac{1}{T_n}\right) H^{(n-1)}_{T_0,\ldots,T_{n-1}}(\mathbf{f}^{(n-1)})\right\} \\ &= \sum_{\{\mathbf{f}^{(n-1)}\}} \exp\{-H^{(n)}_{T_0,\ldots,T_n}(\mathbf{f}^{(n)})\} \end{aligned}$$

(6.183)

In the coarsening stage we may choose a set of conditional probabilities to be the analog of those used in the Ising lattice. That is, we may select

$$p^{(n)}(\mathbf{f}^{(n)}|\mathbf{f}^{(n-1)}) = \prod_{i \in \Lambda^{(n)}} \delta_{f_i^{(n)} f_i^{(n-1)}} \tag{6.184}$$

More generally, we may write

$$p^{(n)}(\mathbf{f}^{(n)}|\mathbf{f}^{(n-1)}) = \prod_{i \in \Lambda^{(n)}} q^{(n)}(f_i^{(n)}|f_i^{(n-1)}) \tag{6.185}$$

where the $q^{(n)}$ are a set of local conditional probabilities. For coarsening by decimation we may use the representation in terms of kronecker deltas. In the case of block construction a useful form for the local conditional probabilities is

$$q^{(n)}(f_i^{(n)}|f_i^{(n-1)}) = \frac{r(f_i^{(n)}, f_i^{(n-1)})}{\sum_{\{f_i^{(n)}\}} r(f_i^{(n)}, f_i^{(n-1)})} \tag{6.186}$$

with

$$r(f_i^{(n)}, f_i^{(n-1)}) = \exp\left\{\rho f_i^{(n)} \sum_{j=1}^{4} f_j^{(n-1)}\right\} \tag{6.187}$$

In the above expression the f_j's are the four elements of the square block in the $(n-1)$th lattice replaced in the nth lattice by the single element labeled by the index i and centered on the block. The quantity ρ is an adjustable parameter.

The renormalization group simulated annealing algorithm is of the general form:

1. Choose (a) a coarsening procedure, (b) a set of local conditional probabilities $q^{(n)}(f_i^{(n)}|f_i^{(n-1)})$, and (c) set of positive constants $T_0, \ldots, T_N$.
2. Compute the renormalized hamiltonians $H^{(n)}$ for $n = 1, \ldots, N$.
3. Set $n = N$, and find the global minimum $\mathbf{f}^{(n)}$ of $H^{(n)}$.
4. Decrease n, replacing n with $n - 1$, and find the global minimum $\mathbf{f}^{(n-1)}$ of $H^{(n-1)}$ among all possible configurations satisfying the constraint equation by annealing the Gibbs distribution

$$\frac{\exp\{-[1/T(t)]H_{T_0,\ldots,T_n}^{(n)}(\mathbf{f}^{(n)})\}}{Z(T_0, \ldots, T_n)}$$

   following a selected annealing schedule for $T(t)$.
5. Repeat step 4 until $n = 1$.

The annealing procedure, step 4, can be executed in a number of different ways. In some instances the minimization can be done exactly and deterministically. In other cases approximation methods such as iterated conditional modes may be used. The constants $T_0, \ldots, T_N$ are related to the depths of the local minima of H which in turn depend on the noise characteristics. The optimal values for these constants must be determined empirically for a given restoration. The most difficult part of the process is the construction of an appropriate RG transformation producing the set of coarsened hamiltonians. As is the case for two-dimensional Ising spin systems, exact solutions are not possible. We find that noise-related inhomogeneities and long-range correlations may be treated using perturbative approximations and cumulant expansions. Several examples of the application of the above-mentioned techniques to image reconstruction were given by Gidas.

6.16 REFERENCES

1. Lévy, P. (1956). A special problem of brownian motion and a general theory of Gaussian random functions. In *Proceedings Third Berkeley Symposium on Mathematical Statistics and Probability*, vol. 2. Berkeley: University of California Press, pp. 133–175.
2. Dobruschin, P. L. (1968). The description of a random field by means of its conditional probabilities and conditions on its regularity. Theory of Probability and its Applications (English trans. of Teoriya Veroyatnostei i ee Primeneniya), **13**, 197–224.
3. Wong, E. (1968). Two-dimensional random fields and representation of images. SIAM J. Appl. Math., **16**, 756–770.
4. Woods, J. W. (1972). Two-dimensional discrete Markovian fields. IEEE Trans. Inform. Theory, **IT-18**, 232–240.
5. Whittle, P. (1954). On stationary processes on a plane. Biometrika, **41**, 434–449.
6. Brook, D. (1964). On the distinction between the conditional probability and the joint probability approaches in the specification of nearest-neighbour systems. Biometrika, **51**, 481–483.
7. Besag, J. (1972). Nearest-neighbor systems and the auto-logistic model for binary data. J. Roy. Statist. Soc., **B34**, 75–83.
8. Hammersley, J. M., and Clifford, P. (1971). Markov fields on finite graphs and lattices. Unpublished manuscript.
9. Besag, J. (1974). Spatial interaction and the statistical analysis of lattice systems (with discussion). J. Roy. Statist. Soc., **B36**, 192–236.
10. Grimmett, G. R. (1973). A theorem about random fields. Bull. London Math. Soc., **5**, 81–84.
11. Moussouris, J. (1974). Gibbs and Markov random systems with constraints. J. Stat. Phys., **10**, 11–33.
12. Spitzer, F. (1971). Markov random fields and Gibbs distributions. Amer. Math. Monthly, **78**, 142–154.

13. Abend, K., Harley, T. J., and Kanal, L. N. (1965). Classification of binary random patterns. IEEE Trans. Inform. Theory, **IT-11**, 538–544.
14. Pickard, D. K. (1977). A curious binary lattice process. J. Appl. Prob., **14**, 717–731.
15. Derin, H., Elliott, H., Cristi, R., and Geman, D. (1984). Bayes smoothing algorithms for segmentation of binary images modeled by Markov random fields. IEEE Trans. Pattern Anal. Machine Intell., **PAMI-6**, 707–719.
16. Moran, P. A. P. (1973a). A Gaussian Markovian process on a square lattice. J. Appl. Prop., **10**, 54–62.
17. Moran, P. A. P. (1973b). Necessary conditions for Markovian processes on a lattice. J. Appl. Prob., **10**, 605–612.
18. Chellappa, R. (1985). Two-dimensional discrete Gaussian Markov random field models for image processing. In L. N. Kanal and A. Rosenfeld (eds.), *Progress in Pattern Recognition 2*. New York: Elsevier, North-Holland, pp. 79–112.
19. Sharma, G., and Chellappa, R. (1986). Two-dimensional spectral estimation using noncausal autoregressive models. IEEE Trans. Inform. Theory, **IT-32**, 268–275.
20. Chellappa, R., and Kashyap, R. L. (1982). Digital image restoration using spatial interaction models. IEEE Trans. Acoustics, Speech and Signal Processing, **ASSP-30**, 461–472.
21. Kashyap, R., and Chellappa, R. (1983). Estimation and choice of neighbors in spatial-interaction models of images. IEEE Trans. Inform. Theory, **IT-29**, 60–72.
22. Kashyap, R. L. (1984). Characterization and estimation of two-dimensional ARMA models. IEEE Trans. Inform. Theory, **IT-30**, 736–745.
23. Geman, S., and Geman, D. (1984). Stochastic relaxation, Gibbs distributions and the Bayesian restoration of images. IEEE Trans. Pattern Anal. Machine Intell, **PAMI-6**, 721–741.
24. Marroquin, J., Mitter, S., and Poggio, T. (1986). Probabilistic solution of ill-posed problems in computational vision. J. Am. Stat. Assoc., **82**, 76–89.
25. Besag, J. (1986). On the statistical analysis of dirty pictures (with discussion). J. Roy. Statist. Soc. **B48**, 259–302.
26. Jeng, F.-C., and Woods, J. W. (1991). Compound Gauss-Markov random fields for image estimation. IEEE Trans. Signal Processing, **39**, 683–697.
27. Bilbro, G. L., Snyder, W. E., Garnier, S. J., and Gault, J. W. (1992). Mean-field annealing: A formalism for constructing GNC-like algorithms. IEEE Trans. Neural Networks, **3**, 131–138.
28. Geiger, D., and Girosi, F. (1991). Parallel and deterministic algorithms from MRF's: Surface reconstruction, IEEE Trans. Pattern Anal. Machine Intell., **13**, 401–412.
29. Blake, A., and Zisserman, A. (1987). *Visual Reconstruction*. Cambridge: MIT Press.
30. Gidas, B. (1989). A renormalization group approach to image processing problems. IEEE Trans. Pattern Anal. Machine Intell., **11**, 164–180.
31. Martignon, L., Hasseln, H. von, Grün, S., Aertsen, A., and Palm, G. (1995). Detecting higher-order interactions among spiking events in a group of neurons. Biol. Cybern., **73**, 69–81.
32. Kindermann, R., and Snell, J. L. (1980). *Markov Random Fields and Their Applications*. Providence: American Mathematical Society.

33. Sherrington, C. (1941). *Man on His Nature.* Cambridge: Cambridge University Press.
34. Hebb, D. O. (1949). *The Organization of Behavior.* New York: Wiley.
35. Gerstein, G. L., Bedenbaugh, P., and Aertsen, A. M. H. J. (1989). Neuronal assemblies. IEEE Trans. Biomedical Eng., **36**, 4–14.
36. Gerstein, G. L., Perkel, D. H., and Dayhoff, J. E. (1985). Cooperative firing activity in simultaneously recorded populations of neurons: detection and measurement. J. Neurosci., **5**, 881–889.
37. Miller, J. W., and Goodman, R. (1993). Probability estimation from a database using a Gibbs energy model. In S. J. Hanson, J. D. Cowan, and C. L. Giles (eds.), *Advances in Neural Information Processing Systems*, vol. 5. San Mateo: Morgan Kaufmann, pp. 531–538.
38. Derin, H., and Elliott, H. (1987). Modeling and segmentation of noisy and textured images using Gibbs random fields. IEEE Trans. Pattern Anal. Machine Intell., **PAMI-9**, 39–55.
39. Hassner, M., and Sklansky, J. (1980). The use of Markov random fields as models of texture. Comput. Graph. and Image Processing, **12**, 357–370.
40. Cross, G. R., and Jain, A. K. (1983). Markov random field texture models. IEEE Trans. Pattern Anal. Machine Intell., **PAMI-5**, 25–39.
41. Hunt, B. R. (1973). The application of constrained least squares estimation to image restoration by digital computer. IEEE Trans. Comput., **C22**, 805–812.
42. Andrews, H. C., and Hunt, B. R. (1977). *Digital Image Restoration.* Englewood Cliffs: Prentice Hall, ch. 7.
43. Gonzales, R. C., and Wintz, P. (2nd. ed.,1987). *Digital Image Processing.* Reading: Addison-Wesley, ch. 5.
44. Poggio, T., Torre, V., and Koch, C. (1985). Computational vision and regularization theory. Nature, **317**, 314–319.
45. Geman, D, Geman, S., Graffigne, C., and Dong, P. (1990). Boundary detection by constrained optimization. IEEE Trans. Pattern Anal. Machine Intell., **12**, 609–628.
46. Gidas, B. (1991). Metropolis-type Monte Carlo simulation algorithms and simulated annealing. Division of Applied Mathematics, Brown University. Preprint.
47. Gidas, B. (1985). Nonstationary Markov chains and convergence of the annealing algorithm. J. Stat. Phys., **39**, 73–131.
48. Marroquin, J. L. (1985). Probabilistic solution of inverse problems. Ph.D. dissertation. Department of Electrical Engineering and Computer Science, Massachusetts Institute of Technology.
49. Dubes, R. C., and Jain, A. K. (1989). Randon field models in image analysis. J. Appl. Stat., **16**, 131–164.
50. Konrad, J., and Dubois, E. (1991). Comparison of stochastic and deterministic solution methods in Bayesian estimation of 2D motion. Image and Vision Comput., **9**, 215–228.
51. Rignot, E., and Chellappa, R. (1991). Sgementation of synthetic-aperture-radar complex data. J. Opt. Soc. Am., **A8**, 1499–1509.
52. Rignot, E., and Chellappa, R. (1992). Segmentation of polarimetric synthetic aperture radar data. IEEE Trans. Image Proc., **1**, 281–300.

53. Rignot, E., and Chellappa, R. (1993). Maximum a posteriori classification of multifrequency, multilook, synthetic aperture radar intensity data. J. Opt. Soc. Am., **A10**, 573–582.
54. Geman, D., and Reynolds, G. (1992). Constrained restoration and the recovery of discontinuities. IEEE Trans. Pattern Anal. Machine Intell., **14**, 367–383.
55. Gibson, J. J. (1950a). *The Perception of the Visual World.* Boston: Houghton Mifflin.
56. Gibson, J. J. (1950b). The perception of visual surfaces. Am. J. Psych., **63**, 367–384.
57. Barrow, H. G., and Tenenbaum, J. M. (1981). Computational vision. Proc. IEEE, **69**, 572–595.
58. Marr, D. (1982). *Vision.* San Francisco: W. H. Freeman.
59. Ullman, S. (1979). *The Interpretation of Visual Motion.* Cambridge: MIT Press.
60. Poggio, T., Gamble, E. B., and Little, J. J. (1988). Parallel integration of vision modules. Science, **242**, 436–440.
61. Gamble, E. B., Geiger, D., Poggio, T., and Weinshall, D. (1989). Integration of visual modules and labeling of surface discontinuities. IEEE Trans. Systems, Man, Cybern., **19**, 1576–1581.
62. Beckerman, M., and Sweeney, F. J. (1994). Segmentation and cooperative fusion of laser radar image data. In *Proceedings SPIE Conference on Sensor Fusion and Aerospace Applications II*, **2233**, 88–98.
63. Derin, H., Kelly, P. A., Vzina, G., and Labitt, S. G. (1990). Modeling and segmentation of speckled images using complex data. IEEE Trans. Geosci. Remote Sensing, **28**, 76–87.
64. Murray, D. W., and Buxton, B. F. (1987). Scene segmentation from visual motion using global optimization. IEEE Trans. Pattern Anal. Machine Intell., **PAMI-9**, 220–228.
65. Konrad, J., and Dubois, E. (1992). Bayesian estimation of motion vector fields. IEEE Trans. Pattern Anal. Machine Intell., **14**, 910–927.
66. Wright, W. A. (1989). A Markov random field approach to data fusion and colour segmentation. Image Vision Comput., **7**, 144–150.
67. Nadabar, S. G., and Jain, A. K. (1992). Edge detection and labeling by fusion of intensity and range images. In *Procceedings SPIE Conference on Applications of Artificial Inetlligence X: Machine Vision and Robotics*, **1708**, 108–119.
68. Modestino, J. W., and Zhang, J. (1992). A Markov random field model-based approach to image intepretation. IEEE Trans. Pattern Anal. Machine Intell., **14**, 606–615.
69. Ramachandran, V. S. (1990). Visual perception in people and machines. In A. Blake and T. Troscianko (eds.), *AI and the Eye.* New York: Wiley, pp. 21–77.
70. Jeng, F.-C., and Woods, J. W. (1990). Simulated annealing in compound Gaussian random fields. IEEE Trans. Inform. Theory, **36**, 94–107.
71. Barnard, S. T. (1989). Stochastic stereo matching over scale. Int. J. Comput. Vision, **3**, 17–32.
72. Julesz, B. (1971). *The Foundation of Cyclopean Perception.* Chicago: University of Chicago Press.
73. Bilbro, G. L., and Snyder, W. E. (1990). Applying mean field annealing to image noise removal. J. Neural. Network Comput., Fall, 5–17.

74. Blake, A. (1989). Comparison of the efficiency of deterministic and stochastic algorithms for visual reconstruction. IEEE Trans. Pattern Anal. Machine Intell., **11**, 2–12.
75. Geiger, D., and Girosi, F. (1990). Coupled Markov random fields and mean field theory. In D. S. Touretzky (ed.), *Advances in Neural Information Processing Systems*, vol. 2. San Mateo: Morgan Kaufmann, pp. 660–667.
76. Niemeyer, Th., and van Leeuwen, J. M. J. (1974). Wilson theory for 2-dimensional Ising spin systems. Physica, **71**, 17–40.
77. Kadanoff, L. P., and Houghton, A. (1975). Numerical evaluations of the critical properties of the two-dimensional Ising model. Phys. Rev. **B11**, 377–386.
78. Wilson, K. G. (1975). The renormalization group: Critical phenomena and the Kondo problem. Rev. Mod. Phys., **47**, 773–784.
79. Derin, H., and Kelly, P. A. (1989). Discrete-index Markov-type random processes. Proc. IEEE, **77**, 1485–1510.

7

THE APPROACH TO EQUILIBRIUM

In the last few chapters, we have explored processes in which a system, initially prepared in a nonequilibrium state, evolved by small changes through stochastic relaxation into equilibrium with its surroundings. Sequences of changes generated by a sampling algorithm that simulates an equilibrium dynamics with a gradually decreasing temperature allowed the system to relax into one of a number of favored, stable low-energy states. In this chapter we shift our emphasis from the properties of the stable states to the processes and principles that guide the approach to equilibrium.

7.1 NONEQUILIBRIUM DYNAMICS

7.1.1 Relaxation and Equilibrium Fluctuations

The dynamic variables, or parameters, that describe the behavior of a system at equilibrium are not constant in time but instead fluctuate about their steady state values. The spontaneous fluctuations of a variable $A(t)$ about its equilibrium value, namely $\delta A(t) = A(t) - \langle A \rangle$, are depicted in Fig. 7.1 as function of time t. One of the striking characteristics of a system not too far from equilibrium is that if left undisturbed, it will relax into an equilibrium state in a manner that is indistinguishable from that of the decay of temporal correlations among the dynamic variables in a system at equilibrium. This principle was first articulated by Onsager, who postulated that the relaxation of a macroscopic nonequilibrium perturbation is governed by the same rules

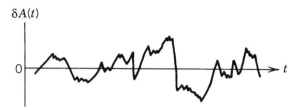

Figure 7.1. Spontaneous fluctuations of a parameter A in a system at equilibrium. (From Chandler[1]. Reprinted with permission of Prof. David Chandler.)

as the regression of spontaneous fluctuations, meaning the decay of correlations, in a system at equilibrium.

This connection between the relaxation of a nonequilibrium system and equilibrium fluctuations, formalized by Onsager as the regression hypothesis, can be subsumed within a general framework relating the microscopic fluctuations in a system at equilibrium to the macroscopic parameters governing the evolution of that system toward equilibrium. The theory relating fluctuations and disssipation as developed by Callen and Welton,[2] Callen and Greene,[3] Kubo and co-workers,[4] and others,[4] relates the linear response of a system to an external disturbance to the fluctuation properties of the system in thermal equilibrium. This general theory has a number of antecedents. These include the Einstein relation[5] and Nyquist's theorem[6] relating fluctuations in voltage to resistance in a linear electrical circuit. The fluctuation-dissipation theorems, together with the reciprocity theorems of Onsager,[7] serve as a foundation for nonequilibrium dynamics. One of our primary goals in this chapter will be to see how the relationships formalized as the regression hypothesis and the fluctuation-dissipation theorems emerge in an analysis of brownian motion.

7.1.2 Brownian Motion

Brownian motion is named after the botanist Robert Brown who in 1826, using a light microscope, observed that small specs of dust or pollen suspended in a fluid seemed to move about constantly in a random zigzag manner. The explanation for this natural phenomenon was given by Einstein in 1905. In his analysis Einstein showed that brownian motion arises from interactions of the dust particle with a heat bath, more specifically, from random collisions of the dust particles with the molecules of the fluid. Einstein produced a description of brownian motion in terms of the diffusion equation, and derived a relation connecting the viscous friction of the medium to the diffusion constant. The theory of brownian motion developed by Einstein,[5] Smoluchowski,[8] and Langevin,[9] together with Perrin's brownian motion experiments used to deduce values for Boltzmann's constant and Avogardo's number, and Millikan's oil drop experiment, helped gain a rapid acceptance of the atomic theory of matter. This happened not long after Boltzmann, in 1906, reportedly in dispair over a perceived slow acceptance of the atomic-based, kinetic theory by the scientific establishment, commited suicide.

7.1.3 Markov Processes

Before discussing brownian motion and its mathematical models, the Langevin and Fokker-Planck equations, let us briefly discuss the notion of a Markov process. Our starting point is the introduction of time-dependent random variables $X(t)$, which may assume values $x_0, x_1, x_2, \ldots$ at times $t_0, t_1, t_2, \ldots$, and where $t_0 \leq t_1 \leq t_2 \leq \ldots$. Systems characterized by these variables will evolve probabilistically in time by means of a stochastic process. Associated with this process are joint and conditional probabilities related to one another through expressions of the general form

$$p(\ldots; x_0, t_0; x_1, t_1; x_2, t_2; x_3, t_3; \ldots)$$
$$= p(x_2, t_2; x_3, t_3; \ldots | x_1, t_1; x_0, t_0; \ldots) p(x_1, t_1; x_0, t_0; \ldots) \quad (7.1)$$

We now make the Markov assumption that the conditional probability at any time is determined by the state of the system at its immediate preceding time:

$$p(x_2, t_2; x_3, t_3; \ldots | x_1, t_1; x_0, t_0; \ldots) = p(x_2, t_2; x_3, t_3; \ldots | x_1, t_1) \quad (7.2)$$

Using these relations, we find that the joint probabilities associated with a Markov process can be decomposed into a product of two-time conditional probabilities and a single one-time probability:

$$p(x_0, t_0; x_1, t_1; \ldots; x_{n-1}, t_{n-1}; x_n, t_n)$$
$$= p(x_n, t_n | x_{n-1}, t_{n-1}) \ldots p(x_1, t_1 | x_0, t_0) p(x_0, t_0) \quad (7.3)$$

and in particular, we find that

$$p(x_1, t_1; \ldots; x_{n-1}, t_{n-1}; x_n, t_n | x_0, t_0)$$
$$= p(x_n, t_n | x_{n-1}, t_{n-1}) \ldots p(x_1, t_1 | x_0, t_0) \quad (7.4)$$

We now recall that we may eliminate a variable from a joint probability by either summing or integrating over all possible values of that variable. If we apply this property in integral form to a Markov process involving three time steps, integrating over all possible states at the intermediate time step, we obtain the Chapman-Kolmogorov equation:

$$p(x_2, t_2 | x_0, t_0) = \int p(x_2, t_2 | x_1, t_1) p(x_1, t_1 | x_0, t_0) dx_1 \quad (7.5)$$

The Markov assumption and the reasoning leading to expressions such as the Chapman-Kolmogorov equation were employed by Einstein in his derivation of the diffusion equation for brownian motion.

7.2 OUTLINE OF THE CHAPTER

In studying brownian motion, we are focusing our attention on a class of systems that undergoes departures from equilibrium that are linearly related to the perturbations producing the departures. These systems are never far from equilibrium and return to adjusted equilibrium states subsequent to the disturbances. We will begin our exploration of the approach to equilibrium in such systems with a derivation in Section 7.3 of the Fokker-Planck equation.

Einstein's analysis for nonviscous media was generalized to viscous fluids by Langevin. In his work[9] Langevin introduced a new element, a stochastic force, into the classical equations of motion. We introduce the Langevin equation for brownian motion and derive its formal solution for the temporal evolution of the velocities in Section 7.4. We then calculate the mean square displacement of the brownian particle. An analysis of the result shows that there are two distinct time scales, a short-time inertial regime and a long-time diffusive regime. We next examine the properties of the random force introduced by Langevin and establish a connection between its properties and those of the system at equilibrium.

The Langevin equation is the prototypic example of a stochastic differential equation. The rapidly fluctuating random term in a stochastic differential equation is usually referred to in the literature as a *Wiener process* or brownian motion. We will characterize Wiener processes as they appear in stochastic differential equations in Section 7.5. In Chapter 4 we discussed how Langevin diffusions may be used as a continuum method for global optimization that is complementary to the discrete simulated annealing algorithm. We now provide an interpretative framework that helps us understand the operation of the Langevin diffusion method for global optimization.

The remaining sections of the chapter are devoted to a further exploration of the physical principles governing the approach to equilibrium as illustrated by brownian motion. We will relate the diffusion coefficient describing the macroscopic transport properties of the system to the friction constant and thence to the microscopic fluctuating random force, and we will examine the emergence of Onsager's regression of spontaneous fluctuations in the velocity correlation function.

7.3 THE FOKKER-PLANCK EQUATION

7.3.1. Brownian Motion

In his development of the theory of brownian movement, Einstein considered the probability law that must be obeyed for collisions that take place in a time interval τ, which is small compared to observation times but large enough that movements in two successive time intervals can be considered to be statistically independent events. The probability $p(x, t + \tau)$ that the particle is at position x, at time $t + \tau$ is regarded as depending only on the position at the immediate preceding time t and not on its position at any other earlier time. This is our Markovian assumption, and associated with it are continuous-time analogues to the rule for n-step discrete transition probabilities presented in Chapter 2, such as the Chapman-Kolmogorov equation. Making these assumptions, Einstein proceeded to show that the probabilities obey the diffusion equation

$$\frac{\partial p(x, t)}{\partial t} = D \frac{\partial^2 p(x, t)}{\partial x^2} \tag{7.6}$$

where D is the diffusion coefficient. The diffusion equation is a special case of a class of equations, termed *Fokker-Planck equations*, that describes systems evolving along continuous paths in sample space. We will now apply the above-mentioned argument of Einstein, slightly generalized, to an examination of the probability that a particle has a velocity v at time t. Our basic assumption is that in a small time interval, the velocity of a particle can only change by a small amount. This mimics the brownian motion of a particle, in which the dust specs undergo small zigzag movements.

7.3.2 Fokker-Planck Equation for Brownian Motion

We begin by defining our notation. Let $p(v, t)dv$ denote the probability that a particle has a velocity between v and $v + dv$ at time t. We model the motion of a particle as a Markovian process in which the probability $p(v, t)$ depends on the velocity v_0 of the particle at an earlier time t_0. This is done by rewriting the probability $p(v, t)$ as a conditional probability $p(v, t | v_0, t_0)$. We are interested in the general situation where a particle starts at time t_a with velocity v_a, at time t', passes through some intermediate velocity v', and ends up at time t_b with velocity v_b. The probability for this two-step sequence of events is given by the product of the two conditional probabilities $p(v', t' | v_a, t_a)$ and $p(v', t' | v_b, t_b)$. The probability of starting at time t_a with velocity v_a and ending up at time t_b with velocity v_b is obtained by integrating over all values of the intermediate velocity v':

$$p(v_b, t_b | v_a, t_a)dv_b = \int_{-\infty}^{\infty} p(v_b, t_b | v', t')p(v', t' | v_a, t_a)dv_b dv' \tag{7.7}$$

We now asssume that the stochastic properties described by the conditional probabilities are independent of the time chosen for their evaluation; that is, we assume that the conditional probabilities are functions of the time differences. If we introduce the positive-definite time differences,

$$\tau_a = t' - t_a, \quad \tau_b = t_b - t' \tag{7.8}$$

258   THE APPROACH TO EQUILIBRIUM

then under this assumption we are able to replace Eq. (7.7) with the expression

$$p(v_b, \tau_a + \tau_b | v_a) = \int_{-\infty}^{\infty} p(v_b, \tau_b | v') p(v', \tau_a | v_a) dv' \quad (7.9)$$

Having used the Markovian assumption to construct a Chapman-Kolmogorov-like equation called the Smoluchowski equation, we now now make use of our second main assumption that brownian motion occurs through a sequence of small time steps. We now expand the left-hand side in a Taylor's series about τ_b small:

$$\frac{\partial p(v_b, \tau_a | v_a)}{\partial \tau_a} \tau_b = -p(v_v, \tau_a | v_a) + \int_{-\infty}^{\infty} p(v_b, \tau_b | v') p(v', \tau_a | v_a) dv' \quad (7.10)$$

We next define the difference $\Delta = v' - v_b$ and rewrite the right-hand side in the form

$$\frac{\partial p(v_b, \tau_a | v_a)}{\partial \tau_a} \tau_b = -p(v_b, \tau_a | v_a) + \int_{-\infty}^{\infty} p(v_b, \tau_b | v_b - \Delta) p(v_b - \Delta, \tau_a | v_a) d\Delta$$

$$(7.11)$$

We now invoke our assumption that in a small time interval the velocity can only change by a small amount. This means that the probabilities are only appreciable when Δ is small, and we may expand the integrand in a Taylor's series in Δ about the value

$$p(v_b + \Delta, \tau_b | v_b) p(v_b, \tau_a | v_a)$$

retaining only the lowest-order terms. The resulting expression is

$$p(v_b, \tau_b | v_b - \Delta) p(v_b - \Delta, \tau_a | v_a) = \sum_{n=0}^{\infty} \frac{(-\Delta)^n}{n!} \frac{\partial}{\partial v_b^n} [p(v_b + \Delta, \tau_b | v_b) p(v_b, \tau_a | v_a)]$$

$$(7.12)$$

Inserting this in our previous expression yields the formula

$$\frac{\partial p(v_b, \tau_a | v_a)}{\partial \tau_a} \tau_b = -p(v_b, \tau_a | v_a)$$

$$+ \sum_{n=0}^{\infty} \frac{(-1)^n}{n!} \frac{\partial}{\partial v_b^n} \left[p(v_b, \tau_a | v_a) \int_{-\infty}^{\infty} \Delta^n p(v_b + \Delta, \tau_b | v_b) d\Delta \right]$$

$$(7.13)$$

The first term ($n = 0$) in the expansion is

$$p(v_b, \tau_a | v_a) \int_{-\infty}^{\infty} p(v_b + \Delta, \tau_b | v_b) d\Delta = p(v_b, \tau_a | v_a) \quad (7.14)$$

since the probabilities are taken as normalized to unity. The next two terms in the expansion give, upon dropping the subscripts,

$$A(v, \tau) \equiv \frac{1}{\tau} \int_{-\infty}^{\infty} \Delta p(v + \Delta, \tau | v) d\Delta = \frac{1}{\tau} \langle \Delta v(\tau) \rangle \quad (7.15)$$

and

$$B(v, \tau) \equiv \frac{1}{\tau} \int_{-\infty}^{\infty} \Delta^2 p(v + \Delta, \tau | v) d\Delta = \frac{1}{\tau} \langle [\Delta v(\tau)]^2 \rangle \quad (7.16)$$

and the expansion assumes the form

$$\frac{\partial p(v, \tau | v_0)}{\partial \tau} = -\frac{\partial}{\partial v}[A(v, \tau) p(v, \tau | v_0)] + \frac{1}{2} \frac{\partial}{\partial v^2}[B(v, \tau) p(v, \tau | v_0)] \quad (7.17)$$

This is the one-dimensional form of the Fokker-Planck equation for the probabilities $p(v, t | v_0)$. This equation involves the first two moments of the probability distribution evaluated over an infinitesimal time interval. The quantity $A(v, t)$ is commonly referred to as the *drift coefficient* and $B(v, t)$ as the *diffusion coefficient*. When generalized to more than one dimension, $A(v, t)$ becomes the drift vector and $B(v, t)$ the diffusion matrix. We will evaluate the mean and mean square velocity increments shortly, using the Langevin equation. When we use the results of the evaluation to solve the Fokker-Planck equation, we will obtain a Gaussian distribution of velocities that approaches a Maxwellian probability distribution over time, and thus the system equilibrates to an ensemble of states described by a Gibbs distribution.

7.4 THE LANGEVIN EQUATION

We now consider our second way of studying brownian motion. In Langevin's treatment of brownian motion, we start with an equation of motion for the brownian particle in contact with the heat bath. We decompose the forces exerted on the particle and giving rise to the irrregular motion into two terms. One of these describes a slowly varying force operating on a long time scale, and the second a rapidly varying fluctuating force $F(t)$ acting on a short time scale. We model the first term as a velocity-dependent frictional force of the usual form $-\alpha dx/dt$. The friction coefficient α for a spherical particle is related to the viscosity η and particle radius r_0 by the standard hydrodynamic expression $\alpha = 6\pi\eta r_0$. The resulting Langevin equation for the velocity v is

$$m\frac{dv}{dt} = -\alpha v + F(t) \tag{7.18}$$

The frictional force appears in the above equation with a minus sign, so it operates in a manner that restores the velocities to their equilibrium values whenever the brownian particles are perturbed. The fluctuating force $F(t)$ is completely random; its mean value vanishes independent of position and velocity, and the forces at different times are uncorrelated. We will discuss this term in more detail shortly.

7.4.1 Formal Solution

The Langevin equation is of the general form

$$y' + p(t)y = q(t) \tag{7.19}$$

where the prime (') denotes the derivative with respect to t. To solve this equation, we first note that the homogeneous equation, obtained by setting $q(t) = 0$, is separable

$$\frac{dy}{y} = -p(t)dt \tag{7.20}$$

with solution

$$y = y_0 \exp\left\{-\int p(t)dt\right\} \tag{7.21}$$

where y_0 is a constant of integration. We are now ready to solve the inhomogeneous equation. We define

$$A = \int p(t)dt \tag{7.22}$$

so that $y\exp(A) = y_0$, and then differentiate the quantity $y \cdot \exp(A)$:

$$\frac{d(ye^A)}{dt} = y'e^A + ye^A\frac{dA}{dt} = (y' + py)e^A \tag{7.23}$$

If we now invoke our original equation by replacing $y' + py$ with q, we have

$$\frac{d(ye^A)}{dt} = qe^A \tag{7.24}$$

which can be immediately integrated to give the general solution

$$y = e^{-A} \int qe^A dt + Ce^{-A} \quad (7.25)$$

We now return to the Langevin equation and make the substitutions $A = -(\alpha/m)t$ and $q(t) = F(t)$. Our solution to Eq. (7.18) is given by the expression

$$v(t) = v_0 e^{-\alpha t/m} + \int_0^t e^{-\alpha(t-t')/m} F(t') dt' \quad (7.26)$$

where we have absorbed a factor of $1/m$ into the definition of $F(t)$ which now represents an acceleration rather than a force.

7.4.2 The Mean Square Displacement

If we calculate the mean square displacement of a particle operated on by the two forces, we find that over time the particle executes a diffusive random walk. To establish this result, we rewrite the Langevin equation in terms of the positions

$$m\frac{d^2 x}{dt^2} = -\alpha \frac{dx}{dt} + F(t) \quad (7.27)$$

Multiplying both sides of Eq. (7.30) by x yields

$$\frac{m}{2}\frac{d(x^2)}{dt^2} - m\left(\frac{dx}{dt}\right)^2 = -\frac{\alpha}{2}\frac{d(x^2)}{dt} + xF(t) \quad (7.28)$$

We now consider the mean behavior of a large number of particles. Since the mean value of the fluctuating force vanishes, we have

$$\langle xF(t) \rangle = \langle x \rangle \langle F(t) \rangle = 0 \quad (7.29)$$

If we assume, as Einstein did, that the brownian particles are in thermal equilibrium with their surroundings, namely with a heat bath, then they must have mean energy given by the equipartition theorem of thermodynamics. For motion in one dimension, the equipartition theorem states that

$$\frac{m}{2}\left\langle \left(\frac{dx}{dt}\right)^2 \right\rangle = \frac{k_B T}{2} \quad (7.30)$$

THE APPROACH TO EQUILIBRIUM

Upon making these substitutions, we find that

$$\frac{m}{2}\frac{d\langle x^2\rangle}{dt^2} + \frac{\alpha}{2}\frac{d\langle x^2\rangle}{dt} = k_B T \tag{7.31}$$

This is a simple, linear, first-order, inhomogeneous differential equation of the form given by Eq. (7.19) for the variable

$$y = \frac{1}{2}\frac{d\langle x^2\rangle}{dt} \tag{7.32}$$

since in terms of this variable Eq. (7.31) becomes

$$\frac{dy}{dt} + \frac{\alpha}{m} y = \frac{k_B T}{m} \tag{7.33}$$

If we now specialize our solution, Eq. (7.26), to the mean square displacement by setting

$$e^A = \exp\frac{\alpha t}{m}, \quad q = \frac{k_B T}{m} \tag{7.34}$$

then we have

$$y = \frac{k_B T}{\alpha} + C \exp\left\{\frac{-\alpha t}{m}\right\} \tag{7.35}$$

To evaluate the constant of integration, we require that $y = 0$ at $t = 0$. This condition gives $C = -k_B T/\alpha$, and upon substituting back our original variables, we obtain the expression

$$\frac{1}{2}\frac{d\langle x^2\rangle}{dt} = \frac{k_B T}{\alpha}\left[1 - \exp\left\{\frac{-\alpha t}{m}\right\}\right] \tag{7.36}$$

Integration of Eq. (7.36) yields our final result

$$\langle x^2\rangle = \frac{2k_B T}{\alpha}\left[t - \frac{m}{\alpha}\left(1 - \exp\left\{\frac{-\alpha t}{m}\right\}\right)\right] \tag{7.37}$$

where the constant of integration has been evaluated by setting the mean square position to zero at time $t = 0$. If we consider Eq. (7.37) in the limiting cases of short and long time intervals relative to the characteristic time m/α, we find that

$$\langle x^2 \rangle = \begin{cases} \left(\dfrac{k_B T}{m}\right) t^2, & t \ll \dfrac{m}{\alpha} \\ \left(\dfrac{2k_B T}{\alpha}\right) t, & t \gg \dfrac{m}{\alpha} \end{cases} \qquad (7.38)$$

Over a short time interval the brownian particles move inertially with constant thermal velocity $(k_B T/m)^{1/2}$. The picture changes when viewed over long time intervals. For large times the particles move diffusively, and since they undergo a sequence of random collisions, they do not translate as far in a given time interval as they would if moving freely. Values for the mean square displacement were measured by Perrin in his brownian motion experiments. Perrin utilized these experimental results together with values for the radius r_0 of the brownian particles, and the viscosity η of the medium (recall that $\alpha = 6\pi\eta r_0$) to deduce values for Boltzmann's constant k_B and, given an estimate of the gas constant R, Avogadro's number. Perrin's estimates were based on Eq. (7.38) in the diffusive ($t \gg m/\alpha$) limit.

7.4.3 Properties of the Random Force

There are a number of observations we can make concerning the rapidly fluctuating random force. First, the random force term appearing in the Langevin equation is needed to maintain the irregular motion of the Brownian particle. Let us consider an equation of motion which in the absence of a random term only contains a force derived from a potential $V(v)$:

$$m\frac{dv}{dt} = F(v) = -\frac{d}{dv} V(v)$$

A particle moving under the influence of this type of force will settle into a position corresponding to a local minimum of the potential associated with the force. This is contrary to the correct situation described by a Maxwellian (Gibbs) distribution of velocities of the form $\exp(-V(v)/k_B T)$, where T is the temperature of the heat bath. The addition of the random force in the Langevin equation prevents the particle from settling into a minimum position.

Second, we have already noted that the random force is uncorrelated with the position and velocity of a brownian particle, and its values at different times are uncorrelated; that is,

$$\langle F(t)F(t') \rangle = \kappa \delta(t - t') \qquad (7.39)$$

where κ is a strength parameter. This last statement needs some qualification, specifically the time interval $t - t'$ should be large compared to the time between individual collisions of the brownian particles with the molecules of the fluid.

Third, we have just noted the role of the random force in maintaining the canonical equilibrium of the system. We can explore the relationship between the random force and the evolution of the system toward equilibrium by making use of our solution to the Langevin equation. If we require that the probability distribution for the velocities $v - v_0 \exp(-\alpha t/m)$ approaches a Maxwellian distribution, then according to Eq. (7.26) the probability distribution for the quantities

$$\lim_{t \to \infty} \int_0^t e^{-\alpha(t-t')/m} F(t') dt'$$

must be Maxwellian as well. Let us consider a small time interval τ during which all physical variables except the fluctuating forces are approximately constant. The net acceleration experienced by a brownian particle during this time interval is given by

$$A(\tau) = \int_t^{t+\tau} F(t') dt' \tag{7.40}$$

Then it can be shown that the distribution of velocities $v - v_0 \exp(-\alpha t/m)$ will be Maxwellian provided that the accelerations $A(\tau)$ are Gaussian distributed. In more detail, if the accelerations are Gaussian distributed with zero mean and variance $2q\tau$ according to

$$p(A(\tau)) = \frac{1}{(4\pi q \tau)^{1/2}} \exp\left\{-\frac{(A(\tau))^2}{4q\tau}\right\} \tag{7.41}$$

where $q = \alpha k_B T/m^2$, then the velocities are governed by the probability distribution

$$p(v, t | v_0) = \left[\frac{m}{2\pi k_B T (1 - e^{-2\alpha t/m})}\right]^{1/2} \exp\left\{-\frac{m(v - v_0 e^{-\alpha t/m})^2}{2 k_B T (1 - e^{-2\alpha t/m})}\right\} \tag{7.42}$$

and therefore

$$p(v, t | v_0) \to \left(\frac{m}{2\pi k_B T}\right)^{1/2} \exp\left(-\frac{mv^2}{2 k_B T}\right), \quad t \to \infty \tag{7.43}$$

which is the Maxwell velocity distribution. The derivation of this result has been given by Chandrasekhar[10] in his classic review of brownian motion. We conclude by noting that if we take τ to be arbitrarily small, we may also consider the forces $F(t)$ to be Gaussian distributed with zero mean and with correlations given by Eq. (7.39) with $\kappa = 2q$.

7.5 STOCHASTIC DIFFERENTIAL EQUATIONS

As mentioned in the introductory remarks, the Langevin equation is the prototypical example of a stochastic differential equation. Let us consider a Langevin equation for a single variable $y(t)$, two specified functions $a(y, t)$ and $b(y, t)$, and a rapidly fluctuating term $\xi(t)$

$$\frac{dy(t)}{dt} = a(y, t) + b(y, t)\xi(t) \tag{7.44}$$

As before we assume that the rapidly fluctuating quantities $\xi(t)$ have zero mean and have values at different times that are uncorrelated

$$\langle \xi(t)\xi(t') \rangle = \delta(t - t') \tag{7.45}$$

7.5.1 The Wiener Process

In order to explore further the properties of the Langevin equation, we introduce a set of variables $W(t)$ defined as the time integrals of the $\xi(t)$:

$$W(t) = \int_0^t \xi(t')dt' \tag{7.46}$$

The sequence of random variables $W(t_0) = w_0$, $W(t_1) = w_1, \ldots$ defined in this manner is a Markov process. That is, the values of $W(t + \tau)$ are determined by the values of $W(t)$ and not by the values of $W(t')$ at any earlier time $t' < t$. As a consequence the conditional probabilities $p(w, t \mid w_0, t_0)$ associated with these random variables satisfy the Fokker-Planck equation. We can determine the drift and diffusion coefficients for this Markov process by calculating the mean and mean square increments of W according to Eqs. (7.15) and (7.16). We find that

$$\langle \Delta W(t) \rangle = \left\langle \int_t^{t+\tau} \xi(t')dt' \right\rangle = 0 \tag{7.47}$$

and

$$\begin{aligned} \langle (\Delta W(t))^2 \rangle &= \int_t^{t+\tau} dt' \int_t^{t+\tau} dt'' \langle W(t')W(t'') \rangle \\ &= \int_t^{t+\tau} dt' \int_t^{t+\tau} dt'' \delta(t' - t'') = \tau \end{aligned} \tag{7.48}$$

Therefore the drift coefficient $A(w, t)$ is zero, the diffusion coefficient $B(w, t)$ is unity, and the conditional probabilities $p(w, t \mid w_0, t_0)$ are a solution to the diffusion equation considered by Einstein, namely

$$\frac{\partial p(w, t \mid w_0, t_0)}{\partial t} = \frac{1}{2} \frac{\partial^2 p(w, t \mid w_0, t_0)}{\partial w^2} \qquad (7.49)$$

Upon imposing the initial condition that $p(w, t_0 \mid w_0, t_0) = \delta(w - w_0)$, we find that the variables are Gaussian distributed with mean w_0 and variance $t - t_0$. Markov processes of this form are known as Weiner processes. Our stochastic equation can be rewritten in terms of the Weiner process as

$$dy(t) = a(y, t)dt + b(y, t)dW \qquad (7.50)$$

with

$$dW = W(t + \tau) - W(t) = \xi(t)dt \qquad (7.51)$$

7.5.2 Global Optimization

In Chapter 4 we discussed the global optimization technique known as Langevin diffusions. Stochastic differential equations of the form of Eq. (7.50) form the basis for this technique. To use Eq. (7.50) for global optimization of a function $U(y)$, we replace the friction coefficient $a(y, t)$ by the gradient of the function to be optimized. This gradient term controls the drift of the system toward equilibrium. We then replace the coefficient $b(y, t)$ of the Weiner process with the quantity $[2k_B T(t)/\mu]^{1/2}$. That is, we introduce a time-dependent temperature $T(t)$ that controls the magnitude of the fluctuations, or stochastic noise. If we set the Boltzmann constant k_B and the reduced mass μ equal to unity so that $b(y, t)$ is simply $[2T(t)]^{1/2}$, our stochastic differential equation for global optimization becomes

$$dy(t) = -\nabla U(y)dt + \sqrt{2T(t)}\, dW \qquad (7.52)$$

We now derive the associated Fokker-Planck equation by evaluating the mean and mean square incremental changes in $y(t)$, assuming that T is constant. The evaluations are straightforward. We assume that the gradient of U does not change appreciably over a small time interval. The incremental change in the Wiener process vanishes according to Eq. (7.48), and thus we have a drift coefficient $A(y, t) = -\nabla V(y)$. To evaluate the mean square incremental change, we observe that the random and drift terms are uncorrelated, and the correlation function for ∇U vanishes. Upon using Eq. (7.48), we obtain a diffusion coefficient $B(y, t) = 2T$. The Fokker-Planck equation for the conditional probabilities $p(y, t \mid y_0)$ is

$$\frac{\partial p(y, t \mid y_0)}{\partial t} = \frac{\partial}{\partial y}\left(\frac{\partial U}{\partial y} p(y, t \mid y_0)\right) + T \frac{\partial^2 p(y, t \mid y_0)}{\partial y^2} \qquad (7.53)$$

for the one-dimensional case, and more generally

$$\frac{\partial p}{\partial t} = \nabla \cdot (p\nabla U) + T\Delta p \tag{7.54}$$

where Δ is the Laplacian. As discussed by Aluffi-Pentini et al.,[11] Gidas,[12] and Chiang et al.,[13] the solution to this equation for long times, subject to the initial condition

$$p(y, t \mid y_0) \to \delta(y - y_0), \qquad t \to 0 \tag{7.55}$$

and standard conditions (defined in Chapter 4) on the boundedness of U and ∇U, is the Gibbs distribution

$$p(y, t \mid y_0) \to \frac{1}{Z_T} \exp\left(-\frac{U(y)}{T}\right), \qquad t \to \infty \tag{7.56}$$

where Z_T is the partition function. Thus the continuous-state chains generated by Langevin diffusions converge to Gibbs distributions as do the Markov chains generated by Metropolis-type algorithms. We may now reintroduce the time-dependent temperature and an accompanying temperature (annealing) schedule. The result is a Langevin diffusion algorithm for global optimization in which large and rapid random fluctuations dominate the short-time behavior and permit the escape from local minima or metastable states, and a slow drift controls the long-time behavior and pushes the system into a near-optimal low-energy state.

7.6 FLUCTUATIONS AND DISSIPATION

The fluctuation-dissipation theorem takes several forms. One of the simplest of these is the inverse relationship between the diffusion coefficient and the friction constant. Our goal in this section is to establish this relationship. In the process we will relate the diffusion coefficient to the velocity correlation function that exhibits a characteristic exponential decay with increasing time. These general results will be studied further in the following sections where we derive an expression for the decay of velocity correlations that represents Onsager's regression of spontaneous fluctuations in the context of brownian motion.

Let us begin by defining the diffusion constant D in terms of the mean square displacement of a particle in a large time interval t. At equilibrium this displacement is, again assuming one-dimensional motion,

$$D = \frac{1}{2t} \langle [x(t) - x(0)]^2 \rangle \tag{7.57}$$

Using the relationship

$$x(t) - x(0) = \int_0^t v(t')dt \tag{7.58}$$

Equation (7.57) can be rewritten in terms of the velocity correlation function as

$$D = \frac{1}{2t}\int_0^t dt_1 \int_0^t dt_2 \langle v(t_1)v(t_2)\rangle \tag{7.59}$$

To evaluate this expression, we observe that the velocity correlation function should only depend on the time difference $s = t_2 - t_1$:

$$\langle v(t_1)v(t_2)\rangle = \langle v(0)v(s)\rangle \tag{7.60}$$

We then have

$$D = \frac{1}{2t}\int_0^t dt_1 \int_{-t_1}^{t-t_1} ds \langle v(0)v(s)\rangle \tag{7.61}$$

The evaluation of the integral can be simplified by noting that the velocity correlation function does not depend on t_1. The integration over t_1 becomes straightforward once we switch the order of integration. The result is the double integral

$$D = \frac{1}{2t}\int_0^t ds \int_0^{t-s} dt_1 \langle v(0)v(s)\rangle + \frac{1}{2t}\int_{-t}^0 ds \int_{-s}^t dt_1 \langle v(0)v(s)\rangle \tag{7.62}$$

The next step is to note that since the velocity correlation function is time translation invariant, it obeys the symmetry condition

$$\langle v(0)v(s)\rangle = \langle v(0)v(-s)\rangle \tag{7.63}$$

Carrying out the integration over t_1 gives

$$\begin{aligned}D &= \frac{1}{2t}\int_0^t ds \langle v(0)v(s)\rangle (t-s) + \frac{1}{2t}\int_{-t}^0 ds \langle v(0)v(s)\rangle (t+s) \\ &= \frac{1}{t}\int_0^t ds \langle v(0)v(s)\rangle (t-s)\end{aligned} \tag{7.64}$$

where we have changed the integration variable from s to $-s$ and then used the symmetry property, Eq. (7.63), in the last step. One way of evaluating the integrand is to use the Langevin equation to generate a differential equation

FLUCTUATIONS AND DISSIPATION

for the velocity correlation function. Our procedure in this case is to start with Eq. (7.18) and integrate over a small time interval τ to obtain

$$mv(s+\tau) - mv(s) = -\alpha v(s)\tau + \int_s^{s+\tau} F(t)dt \qquad (7.65)$$

If we multiply each term in Eq. (7.65) by $v(0)$ and form their expectations, we find that

$$\langle v(0)v(s+\tau)\rangle - \langle v(0)v(s)\rangle = -\left(\frac{\alpha}{m}\right)\langle v(0)v(s)\rangle\tau + \frac{1}{m}\left\langle v(0)\int_s^{s+\tau} F(t)dt\right\rangle \qquad (7.66)$$

But

$$\left\langle v(0)\int_s^{s+\tau} F(t)dt\right\rangle = \langle v(0)\rangle\left\langle \int_s^{s+\tau} F(t)dt\right\rangle = 0 \qquad (7.67)$$

and thus the velocity correlation function obeys the differential equation

$$\frac{d\langle v(0)v(s)\rangle}{ds} = -\left(\frac{\alpha}{m}\right)\langle v(0)v(s)\rangle \qquad (7.68)$$

which has the solution

$$\langle v(0)v(s)\rangle = \langle v^2(0)\rangle e^{-(\alpha/m)|s|} = \frac{k_BT}{m}e^{-(\alpha/m)|s|} \qquad (7.69)$$

where we have used the equipartition theorem in the last equality. Inserting this solution back into Eq. (7.64) produces the term

$$\frac{1}{t}\int_0^t ds\, e^{-(\alpha/m)|s|}(t-s) = \frac{1}{t}\int_0^\infty ds\, e^{-(\alpha/m)|s|}t = \frac{m}{\alpha} \qquad (7.70)$$

where we have considered the limit as t becomes large in evaluating the integral. Thus we have as our final expression for the diffusion coefficient

$$D = \frac{k_BT}{\alpha} \qquad (7.71)$$

which relates the mobility or inverse of the friction constant to the diffusion constant and thus to the fluctuations in the velocity of the brownian motion.

7.7 THE REGRESSION OF SPONTANEOUS FLUCTUATIONS

7.7.1 Stationary Processes

If a system settles down in time to a steady state, in which its stochastic properties are independent of the time that they are measured, then the process is said to be *stationary*. More formally, a process is stationary if the joint probabilities are only functions of time differences. We have seen that the joint probabilities can be decomposed into a product of two-time conditional probabilities and a one-time probability. A process is stationary if the one-time probabilities are time independent,

$$p(x_0, t_0) \to p(x_0) \tag{7.72}$$

and if the two-time conditional probabilities are dependent on the time differences

$$p(x_j, t_j | x_i, t_i) \to p(x_j, t_j - t_i | x_i, 0) \tag{7.73}$$

and similarly for the joint probabilities. The basic idea is that in the stationary state regime the system has settled down to its steady state behavior where all properties are independent of the time of their measurement. In a stationary process, time-dependent averages are functions of the time differences. We assumed this type of temporal behavior in our derivation of the Fokker-Planck equation in our passage from Eq. (7.7) to Eq. (7.9), and again in our treatment of the velocity correlation function when we went from Eq. (7.59) to Eq. (7.61). In the stationary state regime, time correlation functions of physical quantities

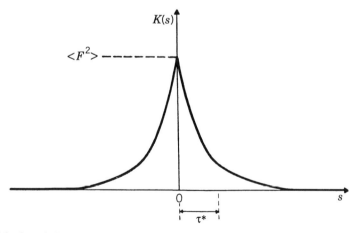

Figure 7.2. Correlation function $K(s) = F(t)F(t + s)$ of a random function $F(t)$. The quantity τ^* represents the correlation time. (From Reif[14]. Reprinted with permission of McGraw-Hill.)

are functions of time differences, and the correlations decay exponentially as the time differences grow larger. This point is illustrated in Fig. 7.2 where we have plotted the correlation function $K(s) = F(t)F(t + s)$ of a random function $F(t)$. We observe that the correlation function is symmetric about the origin, consistent with Eq. (7.63), and we note the exponential decay characterized by a correlation time τ^*. We will probe this aspect further in the next subsection where we calculate the velocity correlation function at large times.

7.7.2 The Velocity Correlation Function

A somewhat more revealing derivation of the velocity correlation function can be given for the case where we have two times t and t', both large so that the first term in Eq. (7.26) vanishes but still with a small time difference $t' - t$. Our starting point here is the equation

$$v(t) = \int_0^t e^{-\alpha(t-\tau)/m} F(\tau) d\tau \tag{7.74}$$

We now use this form to calculate the velocity correlation function:

$$\langle v(t)v(t') \rangle = \int_0^t e^{-\alpha(t-\tau)/m} d\tau \int_0^{t'} e^{-\alpha(t'-\tau')/m} \langle F(\tau)F(\tau') \rangle d\tau' \tag{7.75}$$

We can make use of the fact that the random forces are uncorrelated. Inserting Eq. (7.40) into Eq. (7.75) gives

$$\langle v(t)v(t') \rangle = \kappa \int_0^t d\tau e^{-\alpha(t-\tau)/m} \int_0^{t'} e^{-\alpha(t'-\tau')/m} \delta(\tau - \tau') d\tau'$$

$$= \kappa e^{-\alpha(t+t')/m} \int_0^t e^{2\alpha\tau'/m} d\tau' = \frac{m\kappa}{2\alpha} [e^{-\alpha(t-t')/m} - e^{-\alpha(t+t')/m}] \tag{7.76}$$

and therefore

$$\langle v(t)v(t') \rangle = \frac{m\kappa}{2\alpha} e^{-\alpha|t-t'|/m} \tag{7.77}$$

In this expression we observe that the time that it takes for the velocity correlation to decline by a factor of e is $\tau_e = 1/\alpha$. As illustrated in Fig. 7.2, the velocities bcome rapidly decorrelated as the time interval increases,. We now see from Eq. (7.77) that the greater the friction (dissipation), the shorter is the period of correlation.

7.8 THE VELOCITY INCREMENTS

We now turn to the construction of expressions for the mean and mean square velocity increments; that is, to the small changes in velocity brought on by collisions of the brownian particles with the molecules of the fluid in which they are emersed. The evaluations of these quantities are of interest for several reasons. First, we will establish that the particle velocities perturbed by the collisions will return to their equilibrium values with a time constant m/α. Second, we will be able to use the results of the evaluation to solve the Fokker-Planck equation. Third, in the course of the calculations we will find that the friction coefficient is related to the strength of the fluctuations, thereby establishing that the frictional and fluctuating force terms in the Langevin equation reflect two aspects of the same collision process.

7.8.1 The Mean Incremental Change in Velocity

To achive the aforementioned objectives we start with the equation containing a single force term $G(t)$:

$$m\frac{dv}{dt} = G(t) \tag{7.78}$$

Our goal is to construct an expression for the mean velocity increment during a finite time interval τ that is microscopic yet large compared to the time between individual collisions of the brownian particles with the molecules of the fluid. We use the above equation to relate the mean velocity change to the integral over the time interval of the mean force:

$$m\langle \Delta v \rangle = m\langle v(t+\tau) - v(t) \rangle = \int_t^{t+\tau} \langle G(t') \rangle dt' \tag{7.79}$$

The probability that a small system consisting of our brownian particle and the molecules of the fluid in the particle's immediate vicinity is found at a time t' in a particular state is proportional to the corresponding density of states of the heat bath. The probability that the system is in that same state at some later time is related to the probability at the earlier time through the ratios of the densities of states, $\exp(\beta \Delta E)$ where as usual $\beta^{-1} = k_B T$. We assume that the brownian particle moving with velocity v is in equilibrium with the heat bath at time t and again when moving with velocity $v + \Delta v$ at the later time t'. To obtain an expression for the mean value of the force G at time t', we consider the change in the density of states of the heat bath due to the velocity increment. If we denote the density of states of the heat bath at time t by $\Omega(E)$,

THE VELOCITY INCREMENTS   273

and that at time t' by $\Omega(E + \Delta E)$, then we may approximate the latter by the quantity $\Omega(E)(1 + \beta \Delta E)$. In our model of brownian motion, we regard the mean value of G at time t to be zero, but at time t' it will be nonzero. We then have

$$\langle G(t') \rangle = \sum_\tau \Omega(E + \Delta E) G(t')$$

$$= \sum_\tau \Omega(E)(1 + \beta \Delta E) G(t') = \beta \langle G(t') \Delta E \rangle_0 \tag{7.80}$$

where the subscript 0 denotes that we are computing the mean value of $G(t')\Delta E$ in terms of the equilibrium probabilities at time t. The change in energy ΔE during the time interval $t' - t$ is given by the integral

$$\Delta E = -\int_t^{t'} v(t'') G(t'') dt'' = -v(t) \int_t^{t'} G(t'') dt'' \tag{7.81}$$

representing the negative of the work done by the force G on our brownian particle. In the above expression we have assumed that $v(t'')$ is slowly varying over the time interval from t to t'' and can be removed from the integral. Combining this result with our expression for the mean value of $G(t')$ yields the relation

$$\langle G(t') \rangle = -\beta \bar{v}(t) \int_t^{t'} \langle G(t') G(t'') \rangle_0 dt'' \tag{7.82}$$

where we have separately replaced $v(t)$ by its average value $\bar{v}$ at time t, assuming that the velocity is slowly varying compared to the force. If we now put this result for the mean value of the force into Eq. (7.76), we obtain the expression

$$m \langle v(t + \tau) - v(t) \rangle = -\beta \bar{v}(t) \int_t^{t+\tau} dt' \int_t^{t'} dt'' \langle G(t') G(t'') \rangle_0$$

$$= -\beta \bar{v}(t) \int_t^{t+\tau} dt' \int_{t-t'}^0 ds \langle G(t') G(t' + s) \rangle_0 \tag{7.83}$$

In the second step of Eq. (7.83), we have introduced the variable $s = t'' - t'$ which represents a physically meaningful time difference. The significance of the equilibrium evaluation of the correlation function of the force is that this quantity depends on time difference s alone and is independent of the time t'. If we set the time t' equal to zero in the correlation function, then it is clear

274 THE APPROACH TO EQUILIBRIUM

that the integration over t' is trivial once we exchange the order of integration. We therefore proceed as follows:

$$\int_t^{t+\tau} dt' \int_{t-t'}^0 ds \langle G(0)G(s) \rangle_0 = \int_{-\tau}^0 ds \int_{t-s}^{t+\tau} dt' \langle G(0)G(s) \rangle_0$$

$$= \int_{-\tau}^0 ds \langle G(0)G(s) \rangle_0 (s+\tau) = \tau \int_{-\tau}^0 ds \langle G(0)G(s) \rangle_0$$

$$= \frac{\tau}{2} \int_{-\infty}^{\infty} ds \langle G(0)G(s) \rangle_0 \tag{7.84}$$

In the second step we have exchanged the order of integration as we did previously in Section 7.6. We then carried out the integration over t'. The result of this integration is to bring in a factor $(s+\tau)$ into the remaining integral. The first term may be neglected compared to the second over the range of integration, and we obtain our final expression by extending the range of integration and then using the fact that the force correlation function is symmetric under the transformation of s to $-s$, as was the velocity correlation function. Combining our results gives our desired expression for the velocity increment

$$\langle \Delta v(\tau) \rangle = -\frac{\alpha}{m} \bar{v}(t) \tau \tag{7.85}$$

with

$$\alpha = \frac{1}{2k_B T} \int_{-\infty}^{\infty} ds \langle G(0)G(s) \rangle_0 \tag{7.86}$$

If we equate the velocity increment divided by the finite time interval τ with the time derivative of the average velocity, we obtain the equation of motion

$$m \frac{d\bar{v}}{dt} = -\alpha \bar{v} \tag{7.87}$$

and we now identify the quantity α as our friction coefficient, which we now see is directly related to the fluctuations in the force. This equation can be integrated to give

$$\bar{v} = \bar{v}_0 e^{-(m/\alpha)t} \tag{7.88}$$

and therefore, whenever the Brownian particles are perturbed, their average velocities return to their equilibrium values $\bar{v}_0$ with a time constant m/α.

7.8.2 The Mean Square Incremental Change in Velocity

The procedure for generating an expression for the mean square change in velocity is similar to that used for the mean velocity increment. Our starting point is Eq. (7.76), and we find that

$$
\begin{aligned}
m^2 \langle (\Delta v)^2 \rangle &= \int_t^{t+\tau} dt_1 \int_t^{t+\tau} dt_2 \langle G(t_1)G(t_2) \rangle_0 \\
&= \int_t^{t+\tau} dt_1 \int_{t-t_1}^{t+\tau-t_1} ds \langle G(t_1)G(t_1+s) \rangle_0 \\
&= \int_0^{\tau} ds \int_t^{t+\tau-s} dt_1 \langle G(0)G(s) \rangle_0 \quad (7.89) \\
&\quad + \int_{-\tau}^{0} ds \int_{t-s}^{t+\tau} dt_1 \langle G(0)G(s) \rangle_0 \\
&= \tau \int_{-\infty}^{\infty} ds \langle G(0)G(s) \rangle_0
\end{aligned}
$$

where $s = t_2 - t_1$. In the above sequence we have exploited the translational invariance of the correlation function with respect to t_1, exchanged the order of integration, and by integrating over t_1 brought in a factor of $\tau - s$ in the first integral and a factor of $\tau + s$ in the second. The ranges of integration are similar to those encountered in evaluating the velocity correlation function. We obtain the last equality by neglecting the s terms compared to the τ terms, and extending the ranges of integration utilizing the symmetry of the correlation function. Our final form for the mean square incremental change in velocity can be written in terms of the friction coefficient as

$$
\langle (\Delta v)^2 \rangle = \frac{\tau}{m^2} 2 k_B T \alpha \quad (7.90)
$$

We may now replace the drift and diffusion coefficients appearing in the Fokker-Planck equation with the results of the evaluations of the mean and mean square incremental changes in velocity. If we solve the resulting equation subject to the initial condition

$$
p(v, t | v_0) \to \delta(v - v_0), \quad t \to 0 \quad (7.91)
$$

we recover Eq. (7.42), and therefore in the limit of large times $p(v, t | v_0)$ becomes Maxwellian. We will encounter stochastic differential equations and their associated Fokker-Planck equations again in Chapter 9 when we discuss the dynamics of globally coupled phase oscillators in the presence of noise.

276 THE APPROACH TO EQUILIBRIUM

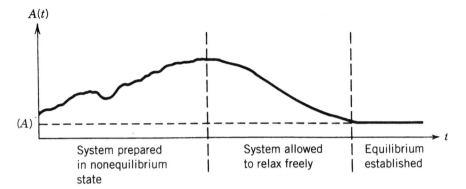

Figure 7.3. Restoration of equilibrium in a system after it receives a nonequilibrium perturbation. (From Chandler[1]. Reprinted with permission of Prof. David Chandler.)

One of the important results of our evaluation of the first two moments of the incremental changes in velocity is Eq. (7.86) which directly relates the friction coefficient to the correlation function of the rapidly fluctuating random force in the equilibrium regime. The friction constant α controls the rate of approach to equilibrium. The relationship between the friction constant and the fluctuating force given by Eq. (7.86) tells us that the greater the fluctuations, the faster is the approach to equilibrium. The fluctuations are generated by the coupling interactions between the brownian particles and the molecules of the heat bath. Thus equilibration will occur more rapidly if the particles are strongly coupled to the heat bath than if they are weakly coupled.

To conclude, the spontaneous fluctuations of a system in equilibrium with its surroundings are indistinguishable from deviations from equilibrium produced by small perturbations of the system. In other words, the regression (decay) of spontaneous fluctuations in a system at equilibrium occurs through the same mechanisms that promote the relaxation of macroscopic nonequilibrium perturbations of that system. In brownian motion the system undergoes departures from equilibrium that are linearly related to the perturbations producing the departures. The system is never far from equilibrium ard returns to an adjusted equilibrium state subsequent to the disturbance. This property is depicted in Fig. 7.3 where we observe the preparation of a system in a nonequilibrium state followed by its relaxation back into equilibrium over time.

7.9 FURTHER READING

One topic mentioned in the introduction but not pursued further in this chapter is Onsager's reciprocity relations. These relations represent a generalization of the fluctuation-dissipation theorem for brownian motion to a principle that applies to a broad range of systems. A discussion of the Onsager reciprocity relations and also the Nyquist relations can be found in the text by Reif.[14] The fluctuation-dissipation theorem and its connection with linear response theory have been elucidated by Chandler,[1] and a detailed exposition of stochastic differential equations has been presented by Gardner.[15]

7.10 REFERENCES

1. Chandler, D. (1987). An Introduction to Modern Statistical Mechanics. New York: Oxford University Press.
2. Callen, H. B., and Welton, T. A. (1951). Irreversibility and generalized noise. Phys. Rev. **83**, 34–40.
3. Callen, H. B., and Greene, R. F. (1952). On a theorem of irreversible thermodynamics. I: Phys. Rev., **86**, 702–710; II: **88**, 1387–1391.
4. Kubo, R. (1966). The fluctuation-dissipation theorem. Rep. Prog. Phys., **29**, 255–284.
5. Einstein, A. (1905). Über die von der molekular-kinetischen Theorie der Wärme geforderte Bewegung von in ruhenden Flüssigkeiten suspenierten Teilchen. Ann. Phys. (Leipzig). ser 4, **17**, 549–560 (English trans. in A. Einstein, Investigations on the Theory of Browniam Movement, A. D. Cowper (trans.), R. Fürth (ed.), New York: Dover, 1956, reprint of the 1926 edition).
6. Nyquist, H. (1928). Thermal agitation of electric charge in conductors. Phys. Rev., **32**, 110–113.
7. Onsager, L (1931, 1932). Reciprocal relations in irreversible processes. I: Phys. Rev., **37**, 405–426; II: **38**, 2265–2279.
8. Smoluchowski, M. von (1906). Zur Kinetischen Theorie der Brownschen Molekularbewegung und der Suspensionen. Ann. Phys. (Leipzig). **21**, 756–780.
9. Langevin, P. (1908). Sur la théorie du mouvement brownien. Comptes Rendus, (Paris) **146**, 530–533.
10. Chandresekhar, S. (1943). Stochastic problems in physics and astronomy. Rev. Mod. Phys., **15**, 1–89.
11. Aluffi-Pentini, F., Parisi, V., and Zirilli, F. (1985). Global optimization and stochastic differential equations. J. Optim. Theory Appl., **47**, 1–16.
12. Gidas, B. (1986). The Langevin equation as a global minimization algorithm. In E. Bienenstock et al. (eds.), Disordered Systems and Biological Organization. Berlin: Springer-Verlag, pp. 321–326.
13. Chiang, T.-S., Hwang, C.-Y., and Sheu, S.-J. (1987). Diffusions for global optimization in Rn. SIAM J. Control. Optim., **25**, 737–753.
14. Reif, F. (1965). Fundamentals of Statistical and Thermal Physics. New York: McGraw-Hill.
15. Gardiner, C. W. (1983). Handbook of Stochastic Methods. Berlin: Springer-Verlag.

8

SYNAPTIC PLASTICITY

8.1 SYNAPTIC PLASTICITY IN THE VISUAL CORTEX

8.1.1 Forms of Synaptic Plasticity

The term "synaptic plasticity" encompasses a variety of processes. It embraces anatomical changes such as the sprouting of new axonal and dendritic structures, the pruning, or retraction and disappearance, of old arbors, and changes in the membrane properties of existing growth. Anatomical modifications such as these are thought to be necessary for sustaining long-term changes in synaptic efficiency.

The appellation synaptic plasticity includes naturally occurring cell death. This is one of the most dramatic aspects of vertebrate neural development. Naturally occurring cell death is a regressive process, distinct from alterations and removal of transient cell structures, or programmed cell death. Naturally occurring cell death is the name given[1] the massive reductions in cell number taking place late in development following the initial propagation of axons to postsynaptic targets and in some cases subsequent to initiation of afferent synaptic input. Losses during late prenatal and early postnatal development range from about 15% in some neural populations to as much as 85%, and average about 50%.[2,3]

Naturally occurring cell death in vertebrates may serve two important functions. First, it provides a way for numerically matching pre- and postsynaptic target structures. In many parts of the nervous system, neurons are initially overproduced. This overproduction serves a useful purpose — a large number of presynaptic cells guarantees adequate synaptic input to the target structure. This initial overproduction is remedied by numerically matching pre- and postsynaptic neural structures. The mechanism underlying this matching may involve some form of competition by afferents for some target-size-related factor or resource. One possible mechanism for naturally occurring cell death is that of a competition for a trophic factor whose supply is proportional to the target size.[2,3]

The second function served by naturally occurring cell death is to provide a way for selectively removing cells whose axons project to inappropriate targets. For example, in studying the retinocollicular projection in the albino rat, we find[4] that targeting errors are present in about 14% of the initial connections. These topographically incorrect projections are selectively removed during the period of naturally occurring cell death. In these studies we observe a second important mechanism underlying cell death. By using TTX to block retinal ganglion action potentials, we are able to infer that ganglion afferents are directly involved in some local, activity-dependent way in the elimination of incorrectly projecting cells, perhaps through a Hebbian mechanism.

The term synaptic plasticity embraces use-dependent changes in synaptic efficiency, produced by manipulations of visual input during the critical period of postnatal development in the cat and monkey and generated by spontaneous prenatal electrical activity in the mammalian retina. It includes LTP and LTD induced by tetanic stimulation in hippocampal rat brain slices, and it entails adult plasticity[5-8] in which cortical motor and sensory maps are reorganized either following injury or through use-dependent factors. The last form of plasticity, only recently discovered, demonstrates that the mature cortex is far from static. Finally, the term synaptic plasticity includes the dramatic use-dependent modifications of synaptic connectivity observed in *Xenopus* and discussed by us in Chapter 5, and a variety of changes in synaptic efficiency found in *Aplysia* and other creatures more modest than humans.

8.1.2 Use-Dependent Changes in Synaptic Efficiency

The efficiency of a synapse can be assessed by determining the postsynaptic response to a constant afferent volley. The appellation "Hebbian" invoked in the previous subsection refers to changes in synaptic efficiency consistent with the Hebb-Stent rule presented in Section 1.8. It is usually taken to imply a highly local (quasi-local) process of synaptic modification driven by correlated pre- and postsynaptic activity. Thus a Hebbian synapse is one in which changes in synaptic state are brought about by cooperative processes initiated by spatiotemporally correlated afferent input. We briefly sketched some key elements of the neurophysiological substrate involved in the initiation of synaptic modification, such as NMDA receptor activation, in Chapter 1. We will discuss these components in relation to the theory of Bienenstock, Cooper, and Munro (BCM) later in this chapter. At this point we simply note that NMDA receptor activity, as revealed through blockade studies, is an important experimental indicator that a Hebbian mechanism is operative. The segregation of eye-specific afferent inputs in three-eyed frogs,[9] and the refinement of topographical map in two-eyed frogs,[10] are striking examples of experience-dependent, NMDA receptor-mediated, synaptic modification. We find similar NMDA receptor blockade evidence during the critical period of postnatal visual development in newborn kittens.[11,12]

Cells in area 17 selectively respond to certain features in the external visual environment. They are position, ocular, and orientation selective. As was the case for the development of the topographically mapped eye-specific lamination in the LGN, and ocular dominance stripes in three-eyed tadpoles, selectivity arises from use-dependent synaptic modification. In these instances of developmental, electrical-activity-driven modification, initial sets of diffuse connections (and strengths) are refined into precise sets of mature connections that endow the cortical cells with the aforementioned properties.

The critical period for visual experience in kittens begins at about the third week, is peaked at the fourth and fifth weeks, and extends to the third month after birth. Monkeys are more mature visually at birth, and as a consequence the critical period starts earlier for them than for the cat. For the monkey the critical period is the first six weeks of postnatal life. The critical period has been a favored subject for study for several decades. As a result we have a considerable body of classical rearing data including normal rearing, monocular deprivation, reverse suture, strabismus, binocular deprivation, and normal rearing following monocular deprivation. In this chapter we will examine models of, and underlying mechanisms for, the Hebbian, experience-dependent changes in synaptic efficiency during the critical period in kittens.

8.2 MODELS OF SYNAPTIC MODIFICATION

8.2.1 Feature Selectivity

In the context of feature extraction, we may define selectivity as the discriminatory ability of a cell with respect to its environment. To express this property mathematically, we represent the environmental input as an input activity vector **d**. This vector, assumed to be a random variable characterized by a probability or sample distribution, is the quantity manipulated in the classical rearing experiments. The input activity vector may be taken as random noise, correlated noise, or patterned input. A typical example of a feature is a bar in the receptive field of a neuron, oriented at a some particular angle. Patterned input could be a random sequence of presentations of this bar, each one at a different orientation as sampled from a uniform distribution over the angular range from 0 to 180 deg. In general, the vector $\mathbf{d} = \mathbf{d}(t)$ is a stationary stochastic process, with a time-invariant distribution.

We may formally represent the selectivity of the synaptic weight vector **m** to the environmental input **d** through a relation of the form[13]

$$Sel_d(\mathbf{m}) = 1 - \frac{\langle \mathbf{m} \cdot \mathbf{d} \rangle}{\max(\mathbf{m} \cdot \mathbf{d})} \qquad (8.1)$$

where "max" denotes the peak response to a particular input, and the averaging (expectation), denoted by angle brackets, is taken over all presentations of the input. We see in this expression that a sharply peaked response to a particular **d**, plus a generally low background response to other input presentations, will yield a selectivity near unity. If the response curve for the various presentations is broadly tuned, the selectivity will be less than unity. If the response, or output activity, is the same for all presentations, then the mean and maximum response will coincide. In these situations the selectivity is zero, and the system is said to be *nonselective*.

As noted by Barlow,[14] the operation that is of prime importance for all higher neural functions is the detection of those associations that he calls "suspicious coincidences." These associations are unexpected combinations of occurrences of events that would not happen by chance very often and are therefore statistically significant. The type of visual feature discussed in the previous paragraphs may be identified as candidate suspicious coincidences. Features such as oriented edges are examples of specific patterns or sequences of events that occur more often than expected if entirely due to random events. These features are selected using information locally available at the cell from the convergent afferents and from connections with neighboring cells.

The earliest modeling efforts by Anderson,[15,16] Kohonen,[17] and Cooper[18] served to introduce the notion of a linear associator in which the integrative activity of a neuron is linearly dependent on the dot product of **m** and **d**. The first study devoted to the question of how feature selectivity is generated was carried out by Malsburg.[19] In his work, excitatory and inhibitory cells were arranged in two-dimensional grid and endowed with lateral, center-surround connections. Synaptic strengths were modified according to a Hebbian rule, appended by a constraint on the total synaptic strength introduced to prevent unbounded increases in the weights. The resulting modifiable cortical cells were found to develop orientation selectivity and related properties resembling those seen experimentally. The ideas presented in these early papers were further developed Pérez, Glass, and Shlaer[20] and by Nass and Cooper.[21]

8.2.2 Stability

The theory of synaptic plasticity of Bienenstock, Cooper, and Munro,[13] which we will examine in detail in this chapter, was proposed to explain the development of orientation selectivity and binocular interaction observed experimentally in kitten visual cortex. In its original formulation BCM theory was a single-cell theory. That is, the modifications were considered at synapses of a single neuron receiving afferent input, relayed through the LGN, from the two eyes. The theory was extended to a network of excitatory geniculocortical connections, and excitatory and inhibitory corticocortico connections, by Cooper and Scofield,[22] compared in great detail to experimental data by Clothiaux, Bear, and Cooper,[23] and presented in an objective function formulation by Intrator and Cooper.[24]

The dynamic equation for synaptic modification in BCM theory, first presented by Cooper, Liberman, and Oja,[25] takes the form of a product of the input activity and a nonlinear function of the summed postsynaptic response. This modification function may be positive, leading to strengthening of the connections, or negative producing a weakening, depending on whether or not the summed postsynaptic response exceeds a modification threshold. The threshold is dynamic and moves as a faster-than-linear function of the total postsynaptic activity. The nonlinear equations for synaptic modification and threshold adaptation are the key elements of BCM theory. These features enable the cells to adapt to the level of recent activity and allow the synaptic strengths to stabilize.

8.2.3 Information Processing in the Visual System

In this chapter we will examine BCM theory from three viewpoints. First, we will examine the single-cell and network formulations of the theory, and discuss the predictions of the theory in relation to the classical rearing data. We will then outline the objective function formulation of the theory. Second, we will establish a connection between the theory and the neurophysiological and biochemical substrate. Third, we will examine the information-processing functions carried out by BCM and other model neurons. One of our goals in modeling synaptic plasticity is to provide some insight into the significance of the feature selective stable states. In the geniculocortical mapping we do not encounter a replica of the retinal image, but instead we find that the image has been reduced in some manner. We would like to understand the information-processing principles underlying the transformation of the retinal image under a Hebbian rule, with the hope that organizational principles may emerge that provide guidance into the processes that may be at work elsewhere in the central nervous system. In our earlier chapters we observed that highly organized stable states are optimal with respect to some quantity. A number of candidate optimality, or organizational, principles for the operations carried out under the Hebbian rules by the genicolocortical circuitry have been identified.

One principle arises in synaptic systems possessing constraints on the square of the summed synaptic strength. As pointed out initially by Oja,[26] and expanded upon by Sanger,[27] these systems function as principal component analyzers. That is, we have a principle of variance maximization, or as Linsker[28] has noted, we have a constrained optimization problem whose solution is a system designed for maximal information preservation. In the dynamic BCM theory the cells become maximally selective to significant input presentation, and the system is optimal with respect to classification. The task performed by a BCM neuron involves moments of the projected distribution higher than second.

A number of models of synaptic modification have been developed during the past few years. These models differ from one another in their choice of learning rule, their network connectivity, assumptions about the input environment, and in their dynamic regimes. Noteworthy among the recent models are those of Linsker,[29-31] Miller, Keller, and Stryker,[32] and Miller.[33] As a result of these differences, each model provides a unique set of insights. We will examine some of the findings provided by synaptic modification models in an analysis of a simplified correlational model due to Shouval and Cooper.[34]

8.3 OUTLINE OF THE CHAPTER

In this chapter, as well as the next one, we are taking some first steps toward understanding the dynamic behavior of self-organizing biological systems. As in the previous chapters, we are focusing on those special ordered stable states to which a system may self-organize itself into in response to internal and external signals. Therefore we will begin in Sections 8.4 and 8.5 with a brief overview of the phase space behavior of dynamic systems. The deterministic treatment of phase space, culminating with Liouville's theorem, will serve as a complement to our stochastic treatment presented in the last chapter. Our goal will be to introduce phase space portraits, which in BCM theory are characterized by the existence of several fixed points. These are the non-equilibrium dynamic counterparts to the equilibrium states discussed in previous chapters.

In Sections 8.6 to 8.8 we will examine how feature selective states develop from the Hebbian, BCM modification rule, a fixed network architecture, and patterned visual input. The key element in this theory, the dynamic modification threshold, will be examined in Section 8.6. The kinetics and the equilibrium states reached by the system, starting from appropriate input stimuli describing classical rearing conditions, will be discussed in Section 8.7, and a mean-field formulation will follow in Section 8.8.

We will then begin to examine the neurophysiological basis for the theory, starting with the molecular mechanisms that may possibly underly synaptic modification. In Section 8.9 we will return to the properties of NMDA receptors, and hippocampal LTP and LTP mentioned in Chapter 1. Kirkwood et al.[35] have found that LTP can be evoked in the superficial layers of the neocortex under the same experimental protocols as LTP produced in hippocampal CA1 preparations, and with the same input specificity and NMDA receptor dependence. It follows from this finding that BCM theory may be applicable to the associative and cooperative forms of hippocampal LTP and LTD produced in area CA1. Thus, although we are focusing our attention on the visual system, the principles we derive may be far more general. We continue our exploration of the synaptic substrate in Section 8.10, where we discuss some recent findings involving the enzyme CaM kinase and link together synaptic plasticity, learning, and memory. The data, involving genetic manipulations in the mouse and in *Drosophila*, provide additional insight into the biochemical mechanisms underlying BCM theory.

284 SYNAPTIC PLASTICITY

The last three sections are devoted to inquiries into the information-processing activities carried out by model neurons in the visual system. We begin this third part of the chapter in Section 8.11 with a discussion of principal component neurons. We then examine phenomenological spin models and a particular class of synaptic modification models. We find in Section 8.12 that under some simplifying assumptions synaptic modification models formulated in terms of activity correlations become equivalent to spin models. We conclude in Section 8.13 with an objective function formulation of BCM theory which enables us to examine the statistical procedures carried out by BCM neurons.

8.4 DYNAMICS

8.4.1 Phase Space

A particularly useful way of characterizing the dynamics of a system is through its phase space behavior. Phase space is an abstract space whose coordinates X_i are the variables of interest. In this space the dynamics of a system are described by a set of first-order, time evolution equations of the general form

$$\frac{dX_i}{dt} = F_i(\{X_i\}) \tag{8.2}$$

We will discuss a number of model neural systems whose dynamics are described by coupled nonlinear differential equations of the form given above. In the Wilson-Cowan and König-Schillen models (Chapter 9), the relevant phase space is the $E - I$ plane, where E and I represent excitatory and inhibitory activities, respectively. In the BCM theory of synaptic plasticity, the phase space of interest is the $d\mathbf{m}/dt - \mathbf{m}$ plane, where $\mathbf{m}$ is the synaptic strength.

To clarify the meaning and significance of our time evolution equation, let us consider the following familiar physical example: Newton's equation of motion for an object of constant mass m located at position $\mathbf{x}$ at time t, and moving with velocity $\mathbf{v} = d\mathbf{x}/dt$, is

$$m\frac{d^2\mathbf{x}}{dt^2} = \mathbf{F}\left(\mathbf{x}, \frac{d\mathbf{x}}{dt}\right) \tag{8.3}$$

where $\mathbf{F}$, the force, is regarded for illustrative purposes as a function of position and velocity. This second-order differential equation can be cast into the form of Eq. (8.2) by introducing a new variable $\mathbf{y} = d\mathbf{x}/dt$. With the introduction of this additional variable, we have the two equations of motion

$$\frac{d\mathbf{x}}{dt} = \mathbf{y} \tag{8.4a}$$

and

$$\frac{d\mathbf{y}}{dt} = \left(\frac{1}{m}\right) \mathbf{F}(\mathbf{x}, \mathbf{y}) \tag{8.4b}$$

The phase space variables are the pair of quantitites $\mathbf{y} = d\mathbf{x}/dt$ and $\mathbf{x}$.

8.4.2 Trajectories in Phase Space

A state of a dynamic system is represented by a point in phase space. The sequence of such points, describing the evolution of the system, forms a unique trajectory in phase space. The simplest situation that can occur is when

$$\frac{dX_i}{dt} = 0 \tag{8.5}$$

for $i = 1, \ldots, N$. In these cases the dynamics are represented by singular points in phase space, called *fixed points*. As already noted, phase space fixed points are the counterpart to the equilibrium states of thermodynamic systems. It should be observed that thermodynamic equilibrium states are not static in the mechanical sense, but rather thermodynamic system continully undergo changes in state. The equilibrium character of the system is maintained by the detailed balance condition obeyed by the transitions between states. Fixed points also represent static states of mechanical equilibrium and the steady states reached by dissipative systems. Another type of phase space trajectory is the *limit cycle*. These are curves that close back upon themselves and are representative of periodic motion such as that of a nonlinear oscillator. We will encounter limit cycle behavior in Chapter 9 when we study rhythmic and oscillatory systems. These two types of dynamics, fixed points and limit cycles,

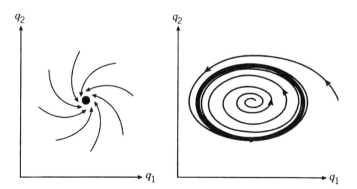

Figure 8.1. Phase space trajectories: (a) Trajectories terminating in a fixed point; (b) trajectories approaching a stable limit cycle from outside and inside the limit cycle.

are illustrated schematically in Fig. 8.1 for an arbitrary pair of phase space variables q_1 and q_2.

8.4.3 Conservative and Dissipative Systems

Phase space trajectories tend to form highly informative stereotypic patterns in phase space. Two properties of phase space promote the clarity of the descriptions. First, as already noted, trajectories in phase space are unique. That is, trajectories remain distinct from one another and do not have intersections. Second, collections of neighboring points, forming volumetric elements, obey certain invariance conditions, and volume elements for conservative systems evolve in phase space differently from those for dissipative system. The first property, uniqueness, follows from a theorem of Cauchy's on systems of differential equations. The second aspect, invariance, arises from Liouville's theorem applied to phase space.

8.5 LIOUVILLE'S THEOREM

8.5.1 Hamilton's Equations

Liouville's theorem tells us that, for hamiltonian, energy-conserving systems, the volume occupied by a collection of phase space points remains constant in time. Before proceeding with a derivation of Liouville's theorem, let us characterize a conservative system through Hamilton's equations. To accomplish this task, suppose we have the energy conserving hamiltonian

$$H = \frac{1}{2} m \left(\frac{dx}{dt} \right)^2 + V(x) \tag{8.6}$$

In writing Eq. (8.6), we have assumed that there is a single pair of phase space variables X_1 and X_2, with

$$\begin{aligned} X_1 &= x \equiv q \\ X_2 &= m \frac{dx}{dt} \equiv p \end{aligned} \tag{8.7}$$

We have further stipulated that the potential is a function of the coordinate q alone. This restriction rules out dissipative forces such as the velocity-dependent friction that was central to brownian motion. Upon rewriting the hamiltonian in terms of p and q, we find that

$$\begin{aligned} \frac{\partial H}{\partial p} &= \frac{p}{m} = \frac{dq}{dt} \\ \frac{\partial H}{\partial q} &= \frac{dV}{dq} = -F = -\frac{dp}{dt} \end{aligned} \tag{8.8}$$

This result may be generalized to any number of phase space coordinates $X_1, \ldots, X_{2n}$, yielding Hamilton's equations

$$\frac{\partial H}{\partial p_i} = \frac{dq_i}{dt}$$

$$\frac{\partial H}{\partial q_i} = -\frac{dp_i}{dt}$$

(8.9)

where $i = 1, \ldots, n$.

8.5.2 Liouville's Theorem of Volume Conservation

Let us now consider a finite volume element $dV = dX_1 dX_2 \ldots dX_N$, located in an $N = 2n$ dimensional phase space between X_1 and $X_1 + dX_1$, X_2 and $X_2 + dX_2, \ldots,$ and let ρ denote the density of points enclosed within this volume at time t. This volume element is depicted in two dimensions in Fig. 8.2. We see in this figure that the number of points in the volume at that time is given by the product of ρ and dV:

$$\rho(X_1, X_2, \ldots, X_N, t) dX_1 dX_2 \ldots dX_N$$

During a unit time interval the number of points flowing into the phase space volume through face "$X_1 = $ constant" is

$$\rho \frac{dX_1}{dt} dX_2 \ldots dX_N = \rho F_1 dX_2 \ldots dX_N$$

Similarly we see that the number of points flowing out through face "$X_1 +$

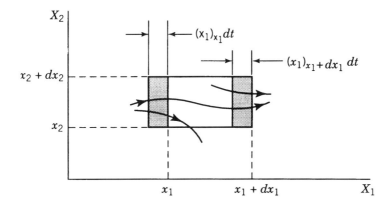

Figure 8.2. Flow of points in a volume element in phase space, entering from the left and leaving to the right.

$dX_1 = $ constant" is

$$\rho(X_1 + dX_1, X_2, \ldots, X_N, t) F_1(X_1 + dX_1, X_2, \ldots, X_N) dX_2 \ldots dX_N$$

We can expand this last expression in a Taylor's series about the point X_1, keeping only the leading terms, to give

$$\rho(X_1, \ldots) F_1(X_1, \ldots) dX_2 \ldots dX_N$$
$$+ \frac{\partial \rho(X_1, \ldots)}{\partial X_1} F_1(X_1 \ldots) dV + \rho(X_1, \ldots) \frac{\partial F_1(X_1, \ldots)}{\partial X_1} dV$$

If we now subtract the number of points leaving the volume element from the number of points entering, we obtain the contribution

$$\frac{\partial \rho(X_1, \ldots, X_N, t)}{\partial t} dV$$

due to flow through the X_1 faces. Similar contribution are found for the other faces of the volume element, and adding these contributions together yields the expression

$$\frac{\partial \rho}{\partial t} = - \sum_{i=1}^{N} \left(\frac{\partial \rho}{\partial X_i} F_i + \rho \frac{\partial F_i}{\partial X_i} \right) \qquad (8.10)$$

We note the the total derivative of ρ with respect to time is

$$\frac{d\rho}{dt} = \frac{\partial \rho}{\partial t} + \sum_{i=1}^{N} \frac{\partial \rho}{\partial X_i} \frac{dX_i}{dt} \qquad (8.11)$$

Upon combining these last two equations, we find that

$$\frac{d\rho}{dt} = \frac{\partial \rho}{\partial t} + \sum_{i=1}^{N} \frac{\partial \rho}{\partial X_i} F_i = - \sum_{i=1}^{N} \rho \frac{\partial F_i}{\partial X_i} \qquad (8.12)$$

For a conservative system we can evaluate the last equality using Hamilton's equations. We see that in this case

$$\sum_{i=1}^{N} \left(\frac{\partial F_i}{\partial X_i} \right) = \sum_{i=1}^{n} \left(\frac{\partial}{\partial q_i} \frac{dq_i}{dt} + \frac{\partial}{\partial p_i} \frac{dp_i}{dt} \right) = \sum_{i=1}^{n} \left(\frac{\partial}{\partial q_i} \frac{\partial H}{\partial p_i} - \frac{\partial}{\partial p_i} \frac{dH}{\partial q_i} \right) = 0 \qquad (8.13)$$

We find therefore that the density of phase space points is constant in time for a conservative system:

$$\frac{d\rho}{dt} = \frac{\partial \rho}{\partial t} + \sum_{i=1}^{N} \frac{\partial \rho}{\partial X_i} F_i = 0 \qquad (8.14)$$

In contrast, for a dissipative system there is an overall contraction of the volume in phase space as the system evolves toward its attractor.

8.5.3 Initial Conditions

Initial conditions, or where a system starts from in phase space, can be crucial in determining the eventual steady state condition reached by that system. To illustrate this point, let us consider a simple pendulum described by an equation of motion of the simplified form

$$\frac{d^2 x}{dt^2} + \sin x = 0$$

where x denotes the angular deviation of the pendulum from its stable equilibrium position. The corresponding phase space equations are

$$\frac{dy}{dt} = -\sin x$$

$$\frac{dx}{dt} = y$$

Upon using the conservation of energy expression to sketch the phase space trajectories for the pendulum, we obtain the picture displayed in Fig. 8.3. We see in this plot that the phase space is partitioned into several distinct regions. The center is a unique point of stable equilibrium corresponding to the pendulum sitting at rest at the bottom of its arc. The nested ellipses (circles) drawn around this point denote simple harmonic motion. The fixed points at the apparent intersections of the hyperbolic trajectories represent points of unstable equilibrium, and further out beyond the separatrix are regions of free rotation. Since trajectories in phase space cannot cross one another, once the simple physical pendulum is initially situated in a particular region of phase space, it is destined to remain there.

In Chapter 5 we examined experimental data and energy minimization, simulated annealing calculations that demonstrated how an adaptive system, the retinotectal projection, can self-organize into different stable states depending on the mix of participating interactions. To those considerations we must now add that the initial conditions are often crucial in determining which steady states are dynamically accessible. In the next two chapters we will

290 SYNAPTIC PLASTICITY

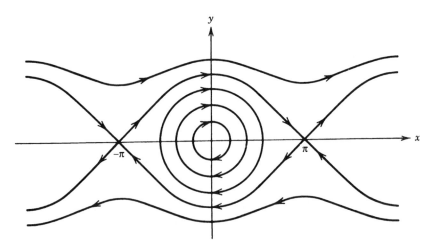

Figure 8.3. Phase space portrait for a simple pendulum.

observe several instances of this key aspect and will also see that dynamic modulation by neural systems allows for switching from one region to another.

8.6 BCM THEORY: SINGLE-CELL FORMULATION

The starting point in any developmental model of visual selectivity is the pattern of intensities falling on the retina produced by light reflecting from physical surfaces (the optic array), by spots of light produced in a controlled experiment, or by some other physical means. We assume that there is a one-to-one correspondence between these retinal intensity patterns and the sensory input through the LGN to the cortical neurons in the afferent layer 4 of area 17. We next specify the dynamic variables. These are taken to be the instantaneous pre- and postsynaptic firing rates; that is, the dynamic variables are the mean firing frequencies and not either the individual spikes or the instantaneous membrane potentials.

8.6.1 The Linear Integrator

Let $m_j(t)$ denote the efficiency at time t of the synapse of the axon of the jth presynaptic neuron with the postsynaptic cell under consideration. We view this synaptic junction as an ideal one, representing the net effect of either a single presynaptic neuron or of a functional microcircuit. We will discuss this point further when we introduce the mean-field approximation. For the present we only note that the junction strength and sign may vary during development. The contribution to the postsynaptic activity from the jth afferent is denoted by the variable $c_j(t)$. This quantity is related to the input activity, or mean firing rate, $d_j(t)$ and the synaptic efficiency through the simple relationship

$$c_j(t) = m_j(t)d_j(t) \tag{8.15}$$

The total activity $c(t)$ due to the $j = 1, 2, \ldots, N$ afferents is then obtained by summing over all contributions of the form given by Eq. (8.15), that is,

$$c(t) = \sum_j c_j(t) = \sum_j m_j(t)d_j(t) \tag{8.16}$$

We can regard the quantities $d_j(t)$ and $m_j(t)$ as the components of the vectors $\mathbf{d}(t)$ and $\mathbf{m}(t)$. In other words, we define an input activity vector

$$\mathbf{d}(t) = (d_1(t), d_2(t), \ldots, d_N(t))$$

and a synaptic strength vector $\mathbf{m}(t) = (m_1(t), m_2(t), \ldots, m_N(t))$. This allows us to cast the above expression, Eq. (8.16), into the form of an inner product

$$c(t) = \mathbf{m}(t) \cdot \mathbf{d}(t) \tag{8.17}$$

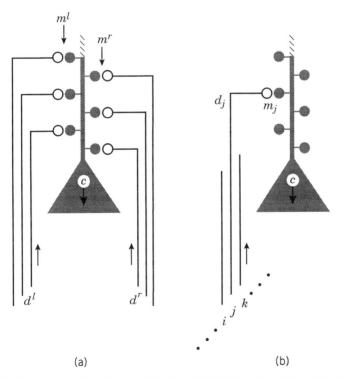

Figure 8.4. Synapses undergoing modification: (*a*) Initial system; (*b*) system after selectivity has developed. Symbols are explained in the text. (From Bear, Cooper, and Ebner[36]. Reprinted with permission of the American Association for the Advancement of Science.)

and the vector $\mathbf{m}(t)$ characterizes the state of the system. In writing the above expression, we assume that the neuron is operating above threshold and below saturation, in a linear regime where the output activity of the cell is directly proportional to the input activity and synaptic strength. If we now take the contributions from both left and right eyes into account, the postsynaptic activity becomes

$$c(t) = \mathbf{m}^l(t) \cdot \mathbf{d}^l(t) + \mathbf{m}^r(t) \cdot \mathbf{d}^r(t) \tag{8.18}$$

The summation of the inputs to give the output c is portrayed in Fig. 8.4. The goal of the BCM formalism is to explain how synapses modify in time as a function of input activity and postsynaptic depolarization.

8.6.2 Rule for Synaptic Modification

The rule for synaptic modification in BCM theory is of the form of a product of the presynaptic activity $\mathbf{d}$ and a function ϕ, which depends on the summed postsynaptic activity c and a modification threshold Θ_M. The rule can be written in the form

$$\frac{d\mathbf{m}(t)}{dt} = \eta \phi(c(t), \Theta_M(t)) \mathbf{d}(t) \tag{8.19}$$

or, for each afferent fiber,

$$\frac{dm_j(t)}{dt} = \eta \phi(c(t), \Theta_M(t)) d_j(t) \tag{8.20}$$

where η is a positive constant that sets the overall magnitude of the possible synaptic modification for a single iteration. The function ϕ is zero when the postsynaptic activity vanishes, that is, when $c(t) = 0$. It is also zero when the summed activity exactly equals the threshold value Θ_M. The modification threshold Θ_M separates the function into two regimes depending on whether the postsynaptic activity is greater than or less than the modification threshold:

$$\phi(c(t), \Theta_M) < 0, \qquad c(t) < \Theta_M \tag{8.21a}$$

and

$$\phi(c(t), \Theta_M) > 0, \qquad c(t) > \Theta_M \tag{8.21b}$$

The dependence of ϕ on $c(t)$ is illustrated in Fig. 8.5. We observe that at low levels of activity just above zero the slope of ϕ is negative, while in the vicinity of the modification threshold the slope of ϕ is positive. This activity depend-

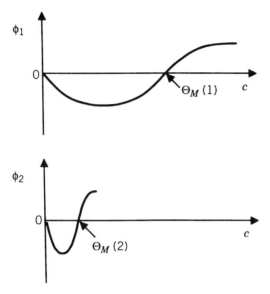

Figure 8.5. The modification function for two different values of the modification threshold. (From Intrator and Cooper[24]. Reprinted with permission of Elsevier Sciences, UK.)

ence ensures that when $d_j > 0$ and c is large enough to exceed the modification threshold, the synaptic strength m_j will increase. Alternatively, when $d_j > 0$ and the summed postsynaptic activity is less than the threshold level, m_j will decrease. In this low activity regime the changes in synaptic strength are negative; weakly supported signals produce a weakening of the synapse junction.

The second important aspect of the model is the dynamic stabilization. To prevent unbounded increases in synaptic strength, and also unlimited decreases, the threshold for strengthening versus weakening moves, or slides, as a function of the time-average activity in the postsynaptic cell. The time-averaging takes place during a time interval preceding the current time t. The time-averaging occurs over a time interval much longer than the membrane time constant τ, describing the passive exponential rise and decay of the membrane potential, and therefore the time average of $c(t)$ evolves more slowly than $c(t)$.

The dependence of Θ_M on the time-averaged activity is necessarily nonlinear. More specifically, the modification threshold changes with the time-averaged activity $\bar{c}(t)$ according to a power law with an exponent greater than one. By this means the modification threshold provides stability, preventing unbounded increases or decreases in strength. One simple choice for the nonlinear function is to use the square of the time-averaged activity

$$\Theta_M = (\bar{c}(t))^2 \to (\mathbf{m}(t) \cdot \bar{\mathbf{d}}(t))^2 \qquad (8.22)$$

As indicated above, this quantity can be approximated by forming the spatial average $\bar{\mathbf{d}}(t)$ over all input patterns. In making this substitution, we assume that $\mathbf{d}(t)$ is stationary, and the time spent by $\mathbf{d}(t)$ at any point in phase space is proportional to the weight of the distribution of $\mathbf{d}$ at that point. We further assume that the synaptic strengths vary slowly compared to time scale for averaging.

8.6.3 Spatial and Temporal Competition

Let us compare the BCM modification equation to that of a model in which the summed strengths are constrained to remain constant. In Malsburg's model[19] we have a rule for synaptic change of the form

$$m_j(t + 1) = a(t + 1)[m_j(t) + bcd_j(t)] \tag{8.23}$$

In this expression we see that the change in strength of the jth synapse is proportional to the product of the summed postsynaptic activity and the input activity. The factor $a(t + 1)$ is the renormalization constant, and b is the constant of proportionality for the change in synaptic strength. The summed postsynaptic activity c is understood to be thresholded, and in this simplified version of Malsburg's model, we have neglected the fixed intracortical connections. If we let s denote the the summed strength so that

$$\sum_j m_j(t + 1) = \sum_j m_j(t) = s \tag{8.24}$$

then we have

$$a(t + 1) = \frac{s}{s + bcd} \tag{8.25}$$

where d denotes the summed input activity. Inserting this relationship into Malsburg's synaptic modification rule yields the expression

$$m_j(t + 1) - m_j(t) = K(t)\left[\left(\frac{d_j}{d}\right) - \left(\frac{m_j}{s}\right)\right] \tag{8.26}$$

with

$$K(t) = \frac{sbcd}{s + bcd} \tag{8.27}$$

We observe that the direction of the change in synaptic strength does not depend on the summed postsynaptic activity $c(t)$ but instead depends on input activity d_j. This type of competition is spatial in character and differs from the temporal competition between patterns promoted by the dependence on the whole cell quantity $c(t)$ in the BCM modification rule.

8.6.4 Stability of the Fixed Points

To understand the behavior of the system as it evolves under the dynamic rule for synaptic modification in BCM theory, let us consider a case where there are two equally likely input patterns, designated by the vectors $\mathbf{d}^1$ and $\mathbf{d}^2$ having

$$p(\mathbf{d} = \mathbf{d}^1) = p(\mathbf{d} = \mathbf{d}^2) = \frac{1}{2}$$

From our definition of selectivity, Eq. (8.1), we see that the maximum selectivity is 1/2. States of this type are realized when $\mathbf{m}$ becomes orthogonal to one of the input pattern vectors, so that its dot product with that vector vanishes while having a nonzero inner product with the other input vector. At the other end of the range of possible values for the selectivity, we find that the minimum selectivity is zero, and states with this property are attained when $\mathbf{m} \cdot \mathbf{d}^1 = \mathbf{m} \cdot \mathbf{d}^2$.

Assuming that $\cos(\mathbf{d}^1 \cdot \mathbf{d}^2) \geqslant 0$, we have the important result that there are exactly four fixed points, $\mathbf{m}^0$, $\mathbf{m}^1$, $\mathbf{m}^2$, and $\mathbf{m}^{1,2}$ such that $sel_d(\mathbf{m}^1) = sel_d(\mathbf{m}^2) = 1/2$ and $sel_d(\mathbf{m}^0) = sel_d(\mathbf{m}^{1,2}) = 0$. Furthermore $\mathbf{m}^1$ and $\mathbf{m}^2$ are stable, $\mathbf{m}^0$ and $\mathbf{m}^{1,2}$ are unstable, and the system evolves with probability one to either $\mathbf{m}^1$ or $\mathbf{m}^2$. The phase space portrait for this two-dimensional system is shown in Fig. 8.6. In this portrait we observe the convergence of the dynamic trajectories starting from a variety of initial conditions to one or the other of these fixed points. Suppose that the two input patterns correspond to presentations of a horizontal and a vertical bar. Then the significance of the above-mentioned theorem is that, although both bar orientations may occur with equal frequencies, one or the other of the patterns (orientations) will be selected as a consequence of noise, fluctuations, or other vagaries in the presentations. This situation is analogous to spontaneous symmetry breaking in Ising spin systems, with the horizontal and vertical orientations of the bar in the role of the positive and negative alignments of the spins.

The temporal competition between input patterns in BCM theory operates in the following manner: The nonlinear threshold dependence on the time-averaged output activity $\bar{c}$ can be written

$$\Theta(\bar{c}) = \left(\frac{\bar{c}}{c_0}\right)^p \bar{c} \qquad (8.28)$$

where c_0 is a normalization constant and p is a number greater than unity. The system, initially untuned, will start with a value of $\bar{c}$ that is much less than c_0. As a result the system is automatically in the positive regime where the synaptic strengths are increased, namely $\phi(c, \bar{c}) > 0$ for all input presentations. As the synaptic strengths grow, the time-averaged output activity increases, and therefore the threshold moves upward. If there is a large input d_j, to the jth

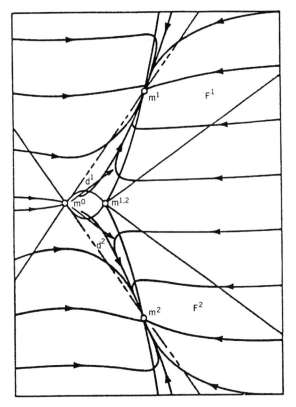

Figure 8.6. Phase space portrait in an environment consisting of two inputs. (From Bienenstock, Cooper, and Munro[13]. Reprinted with permission of the Society for Neuroscience.)

synapse but the pattern is unfavored, then Eq. (8.27) tells us that the change in synaptic weight for the corresponding synapse is negative: $dm_j/dt = \eta\phi d_j < 0$. The nonlinearity guarantees that some input patterns will produce an output activity that falls below the threshold activity, leading to a weakening of future responses to that pattern, while others will produce an output that exceeds the threshold for strengthening of the synaptic response. The response to unfavored patterns will decay to zero where it stabilizes, and the response to favored patterns increases until it too stabilizes. In the foregoing discussion the time-averaged output activity may be defined as

$$\bar{c}(t) = \frac{1}{\tau} \int_{-\infty}^{t} c(t') \exp\left\{-\frac{(t-t')}{\tau}\right\} dt' \qquad (8.29)$$

measured with respect to the spontaneous firing rate.

8.7 CORTICAL RESPONSE PROPERTIES

8.7.1 Circular Environment

If the visual fields subtended by cells in the striate cortex are small, then patterned, or contoured, input will resemble noise-corrupted edges oriented in various directions. We may then use as our patterned input a generalization of the two-pattern set, $\{\mathbf{d}^1, \mathbf{d}^2\}$, used in our discussion of Section 8.5, to a K input patterns set, $\{\mathbf{d}^1, \mathbf{d}^2, \ldots, \mathbf{d}^K\}$. The environment is then modeled as a circular matrix of inner products of the vectors $\mathbf{d}^1, \ldots, \mathbf{d}^K$. In this representation the input vectors are treated as random variables, uniformly distributed over a set of one-parameter orientation coding, circularly symmertic family of points $\mathbf{d}^i$ with $i = 1, \ldots, K$.

The previous stability results can be generalized to K discrete input vectors. Assuming that the probabilities

$$p(\mathbf{d} = \mathbf{d}^1) = \cdots = p(\mathbf{d} = \mathbf{d}^K) = \frac{1}{K} \tag{8.30}$$

we now have 2^K fixed points with selectivities $0, 1/K, 2/K, \ldots, (K-1)/K$, and there are K fixed points of maximum selectivity, $(K-1)/K$, with respect to $\mathbf{d}$. If the vectors $\mathbf{d}^1, \mathbf{d}^2, \ldots, \mathbf{d}^K$ are orthogonal, or close to orthogonal, or perhaps even far from orthogonal, the K fixed points of maximum selectivity, $\mathbf{m}^1, \mathbf{m}^2, \ldots, \mathbf{m}^K$, will be stable, and the system will converge to one of these fixed points regardless of its starting point in phase space. Displayed in Fig. 8.7 is an example of the evolution of the synaptic system in a circular environment to a selective stable state.

8.7.2 Classical Rearing

The critical period for postnatal development in the kitten is peaked at the second month. Classical rearing refers to a series of manipulations to the kittens visual environment during this period of time during when (modifiable) synaptic geniculocortico and lateral corticocortico connections rapidly adjust their transmission efficiencies in response to changes in the visual stimulus. The classical rearing conditions include normal rearing (NR), monocular deprivation (MD), reverse suture (RS), strabismus (ST), or artificial squint, binocular deprivation (BD), and the restoration of normal vision following the period of deprived visual input (RE).

Under normal rearing, neurons are binocular and develop a selectivity to bars or contours of a particular orientation. If one eye is sutured closed (MD) after a period of normal rearing, then there is a rapid (in as little as four to eight hours) loss of responsiveness to stimulation of the visually deprived eye. Selectivity to oriented input from the other eye is retained. The ocular dominance shifts can be produced using a diffusing lens in place of suturing

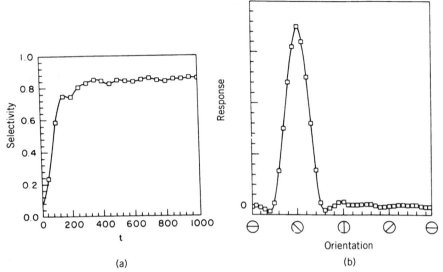

Figure 8.7. Synaptic evolution: (a) Development of selectivity; (b) tuning curve at $t = 1000$. (From Bienenstock, Cooper, and Munro[13]. Reprinted with permission of the Society for Neuroscience.)

and is therefore due to a loss of patterned vision and not to a decline in overall activity. If vision in the previously deprived eye is restored and the open eye is sutured closed (RS), then there will be a shift in responsiveness to that of the newly opened eye. If vision from the two eyes is misaligned (ST), then there will be a loss of binocular vision. Cells in the striate cortex will respond to one eye or the other, but not both.

8.7.3 Orientation Selectivity and Binocular Interactions

The total imput activity to the cortical cells includes contributions from signals produced by external stimuli, from spontaneous activity, and from non-LGN noise. The input activity appearing in the previous expressions is the mean firing rate relative to the average spontaneous activity present in the absence of retinal stimulation. Assuming that the spontaneous activity is a time-, afferent-, and eye-independent constant d_{sp}, we have for the left and right eyes

$$d_j^l(t) = d_{a,j}^l(t) - d_{sp} \tag{8.31a}$$

and

$$d_j^r(t) = d_{a,j}^r(t) - d_{sp} \tag{8.31b}$$

where the subscript a denotes the observed activity. In vector notation,

$$\mathbf{d}^l(t) = \mathbf{d}_a^l(t) - \mathbf{d}_{sp} \tag{8.32a}$$

and

$$\mathbf{d}^r(t) = \mathbf{d}_a^r(t) - \mathbf{d}_{sp} \tag{8.32b}$$

where the spontaneous activity vector is simply $\mathbf{d}_{sp} = (d_{sp}, d_{sp}, \ldots, d_{sp})$.

We will consider the two types of afferent stimulation-random noise and patterned activity. Let us indicate random noise by n_j and patterned activity term arising from external input as d_j^ω. In monocular and binocular deprivation, the observed activity fluctuates about the spontaneous activity level; that is, the activity in each afferent fluctuates about zero. We describe this situation by the relation

$$d_j^l(t) = n_j^l(t), \quad d_j^r(t) = n_j^r(t) \tag{8.33}$$

For patterned activity the corresponding expressions for the left eye are

$$d_j^l(t) = d_{a,j}^{\omega,l}(t) - d_{sp}(t) + n_j^l(t) = d_j^{\omega,l}(t) + n_j^l(t) \tag{8.34}$$

with the definition

$$d_j^{\omega,l}(t) \equiv d_{a,j}^{\omega,l}(t) - d_{sp}(t) \tag{8.35}$$

and similarly for the right eye. Finally, we may write expressions for the response of our cortical neurons to spontaneous and patterned activity as

$$c_{sp}(t) = \mathbf{m}^l(t) \cdot \mathbf{d}_{sp}(t) + \mathbf{m}^r(t) \cdot \mathbf{d}_{sp}(t) \tag{8.36}$$

and

$$c(t) = \mathbf{m}^l(t) \cdot \mathbf{d}^l(t) + \mathbf{m}^r(t) \cdot \mathbf{d}^r(t) + c_n(t) \tag{8.37}$$

where we have appended a term $c_n(t)$ for non-LGN noise to the formula for patterned activity.

During normal rearing (NR) each eye receives the same sequence of randomly selected, patterned input appended by uncorrelated noise. The cortical neurons become binocularly driven and selective. The synaptic strengths, initially small, increase, and the cells become selective to the same orientation for each eye. The input patterns are assumed to be equally likely. If there is some initial orientation preference, the cells develop selectivity to that orientation. If, subsequent to synaptic stabilization, patterned input to one eye is removed (MD), the synaptic strengths related to the deprived eye decline. The cortical neurons become selective and, generally, driven by the open eye

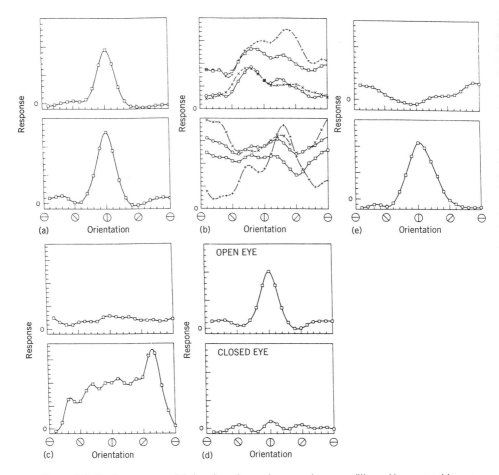

Figure 8.8. Tuning curves obtained under various rearing conditions. Upper and lower panels are results for the two eyes: (a) Normal rearing; (b) dark rearing showing the absence of selectivity. Cells are binocular: (c) binocular deprivation. Cells are sometimes monocular: (d) monocular deprivation. Cells are monocular and selective: (e) uncorrelated rearing. Both monocular and binocular final selective states are encountered. (From Bienenstock, Cooper, and Munro[13]. Reprinted with permission of the Society for Neuroscience.)

only. Ocularity and orientation are coupled. The more selective the cell becomes, the more it is driven exclusively by the open eye, and nonselective neurons remain binocular. This coupling is analyzed in detail by Bienenstock, Cooper, and Munro[13] and by Clothiaux, Bear, and Cooper.[23]

If closed and open eye MD inputs are switched, as in reverse suture (RS), and the modification threshold does not drop too quickly, the newly closed eye synaptic strengths weaken prior to strengthening of the synapses for the recovering open eye. If the patterned input to the two eyes is uncorrelated, as in strabismus (ST), the previously binocular cells become monocular. The eye with the weaker synaptic strengths at the time strabismus is initiated becomes decoupled and is no longer able to drive the cortical neurons. Shown in Fig. 8.8 are a representative set of tuning curves corresponding to a number of different classical rearing conditions. Both kinetics and equilibria of the system, as predicted by the theory, are in good agreement with the experimental data of Mioche and Singer[37] and earlier experimental findings. The selection of model parameters, and the sensitivity of the results to these choices, is discussed in detail by Clothiaux, Bear, and Cooper.[23]

8.7.4 Receptive Field Properties

In Chapter 5 we discussed the role of correlated, spontaneous activity in the patterning of the neural connections of lower vertebrates and in the prenatal development of eye-specific lamination in the mammalian lateral geniculate nucleus. We now note that some orientation selectivity is present at birth as a consequence of spontaneous symmetry breaking. The symmetry breaking may be driven by locally correlated and also random uncorrelated spontaneous activity, and by geometric asymmeties. This aspect was initially investigated by Linsker in his modeling study[30-32] and then explored further by Kammen and Yuille,[38] Yuille, Kammen, and Cohen,[39] and MacKay and Miller.[40]

In Linsker's analysis, a feedforward network of self-organizing cells develops a succession of more sophisticated feature analyzing properties layer by layer under the influence of spontaneous activity. The receptive fields evolve from a symmetric center-surround form in the outermost layer to a form possessing a measure of orientation selectivity in the deeper layers. In the energy function studies by Kammen and Yuille[38] and by Kammen, Yuille, and Cohen,[39] the two-dimensional receptive fields, representing a cell's spatial response function, develop orientation-selective properties as a consequence of random (noise) fluctuations, namely of spontaneous symmetry breaking, and also as a result of small asymmetries in the spatial geometry of the system that appear in an appropriate correlation function. MacKay and Miller[40] presented a detailed study of the dynamics in Linsker's model, showing how different receptive field properties emerge in different dynamic parameter regimes.

Several studies have emphasized that the importance of natural patterned input in developing receptive field properties. Liu and Shouval[41] and Shouval and Liu[42] studied receptive field properties of principal component neurons in natural environments, while Law and Cooper[43] studied receptive field properties of BCM neurons. In the remainder of the present section, we will briefly look at receptive field formation using BCM theory in order to elucidate further the roles of spontaneous and patterned activity in developmental processes.

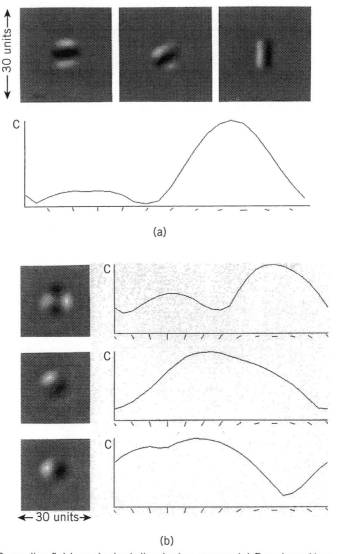

Figure 8.9. Receptive fields and orientation tuning curves: (*a*) Developed in a realistic neural environment; (*b*) developed in correlated noise for three cells under different initial conditions. (From Law and Cooper[43]. Reprinted with permission of the National Academy of Sciences.)

In the study by Law and Cooper[43] receptive field formation was studied within the framework of BCM theory by including retinal processing of natural images. A center-surround Gaussian form was incorporated to model the retinal receptive fields, stereotypic retinal input patterns were replaced by patches from natural scenes, and the sliding modification threshold was defined as the second moment of the cell activity in a manner consistent with Eq. (8.29);

that is, as[24]

$$\Theta(t) = \frac{1}{\tau} \int_{-\infty}^{t} c^2 \exp\left\{-\frac{(t-t')}{\tau}\right\} dt' \qquad (8.38)$$

In Eq. (8.38) the scale constant τ controls the overall rate of movement of the threshold.

The results of calculations with this model are as follows: Previous results obtained using stereotypic visual stimuli are largely unchanged when going to more realistic input patterns. At the end of normal rearing, BCM neurons are binocular with response properties resembling simple cells in the striate cortex. As shown in Fig. 8.9a, the receptive fields of the cells have alternating excitatory and inhibitory bands that produce orientation-selective responses to bars of light. Consistent with experimental results, and correlation of activity calculations, correlated spontaneous activity generates some orientation selectivity. As can be seen in Fig. 8.9b, the receptive fields at this earlier stage of development contain excitatory and inhibitory regions, but the orientation selectivity is not as well developed as that arising later when there is patterned input.

8.8 MEAN-FIELD NETWORK

8.8.1 Mean-Field Approximation

The BCM theory presented in Section 8.6 was formulated for a single cortical target cell. In this section we present the extension by Cooper and Scofield[22] of the BCM formalism to a network of excitatory and inhibitory neurons. In the network model, LGN neurons project to cortical excitatory and inhibitory cells, which in turn interact with one another through cortical-cortical synapses. The cortical-cortical interactions in this approach are mediated by a mean-field in a manner similar to ferromagnetic spins interacting through a Weiss molecular field.

The LGN-cortico-cortico network of interest is depicted schematically in Fig. 8.10. As before, the input activity from the LGN is represented by the vectors **d**. In place of a single **d** vector, with components d_1 to d_n, we now have an array of vectors $\mathbf{d}_i$, one for each cortical cell. We consider the simple situation where the input from the two eyes are constant in time, and the same for all cortical cells so that $\mathbf{d}_i = \mathbf{d}$ for all i. That is, we assume that each cell sees the same portion of the visual field. As depicted in the figure, the LGN-cortical synaptic weights **m** become arrays of weights with each row representing a weight vector $\mathbf{m}_i$ for the ith cortical cell. The cortico-cortico synaptic weights are arranged as a matrix L with matrix elements L_{ij}, denoting the connection strengths between axons of cell j and their dendritic targets on

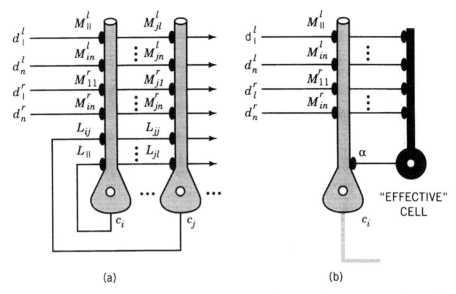

Figure 8.10. (a) Network showing inputs from left and right eyes, together with geniculocortico and cortico-cortico synapses. (b) Mean field approximation. All other cortical cells are replaced by an effective neuron coupled with an average strength α, denoted in the text as "**a**" and M_{ij} are the synaptic strengths $\mathbf{m}_j$. (From Bear and Cooper[93]. Reprinted with permission of Lawrence Erlbaum Associates.)

cell i. The activity, or firing rate, of cell i is the sum of contributions from the geniculocortico and cortico-cortico pathways:

$$c_i = \mathbf{m}_i \cdot \mathbf{d} + \sum_j L_{ij} c_j \tag{8.39}$$

where the dot product of $\mathbf{m}_i$ with $\mathbf{d}$ is intended to represent contributions from both eyes.

We now introduce the mean activity $\langle c \rangle$ as the spatially averaged firing rate of all cortical neurons:

$$\langle c \rangle = \frac{1}{N} \sum_i c_i \tag{8.40}$$

The mean-field approximation consists in replacing c_j in the cortico-cortico sum by the spatially averaged firing rate to yield

$$c_i = \mathbf{m}_i \cdot \mathbf{d} + \langle c \rangle \sum_j L_{ij} \tag{8.41}$$

To be consistent, the mean firing rate must satisfy the relation

$$\langle c \rangle = \langle \mathbf{m} \rangle \cdot \mathbf{d} + \langle c \rangle L_0 = (1 - L_0)^{-1} \langle \mathbf{m} \rangle \cdot \mathbf{d} \tag{8.42}$$

where

$$\langle \mathbf{m} \rangle = \frac{1}{N} \sum_i \mathbf{m}_i \tag{8.43}$$

and

$$L_0 = \frac{1}{N} \sum_{ij} L_{ij} \tag{8.44}$$

The firing rate for the ith cortical cells then becomes

$$c_i = \left(\mathbf{m}_i + (1 - L_0)^{-1} \langle \mathbf{m} \rangle \sum_j L_{ij} \right) \cdot \mathbf{d} \tag{8.45}$$

The indexes used to label our cortical cells are intended to reflect the relative positions of one cell to another in the network. Therefore the difference, $i - j$, is a measure of neighborness, and is not a function of the absolute positions of the cortical neurons. If we now assume that the cortico-cortico synaptic strengths are a function of $i - j$ alone, then L_{ij} becomes a circulant matrix in which each row is a circular shift of the preceding row, and the first row is a circular shift of the last one. In this situation we have

$$\sum_i L_{ij} = \sum_j L_{ij} = L_0 \tag{8.46}$$

and

$$c_i = (\mathbf{m}_i + L_0 (1 - L_0)^{-1} \langle \mathbf{m} \rangle) \cdot \mathbf{d} \tag{8.47}$$

Thus, in the mean-field approximation, the effect of the cortical network on the output activity is to shift the synaptic weight vector $\mathbf{m}$ by the mean-field $\mathbf{a}$:

$$c_i(\mathbf{a}) = (\mathbf{m}_i - \mathbf{a}) \cdot \mathbf{d} \tag{8.48}$$

where

$$\mathbf{a} = -L_0 (1 - L_0)^{-1} \langle \mathbf{m} \rangle \tag{8.49}$$

In this model the network is assumed to be mildly inhibitory: $L_0 < 0$ with $|L_0| < 1$ so that $\langle c \rangle$ remains positive. We have therefore a model involving excitatory geniculocortical synapses and weakly inhibitory cortico-cortico synapses. Experimental observations support the notion that the geniculocortico synapses, located on dendrite spines, modify rapidly, while the cortico-cortico synapses modify more slowly.

8.8.2 The Cortical Network

In the mean-field network described in the last subsection, all synapses modify at the same rate. The model can be generalized to include synapses that modify rapidly and others that modify slowly or not at all. In this approach we have modifiable (**m**) and nonmodifiable (**z**) synapses

$$c_i = \mathbf{m}_i \cdot \mathbf{d} + \sum_j L_{ij} c_j = \mathbf{m}_i \cdot \mathbf{d} + \langle c \rangle L_0 \tag{8.50}$$

and

$$c_k = \mathbf{z}_k \cdot \mathbf{d} + \sum_j L_{kj} c_j = \mathbf{z}_k \cdot \mathbf{d} + \langle c \rangle L_0 \tag{8.51}$$

The synaptic evolution equations are

$$\frac{d\mathbf{m}_i(t)}{dt} = \eta \phi(c(t), \Theta_M^i(t)) \mathbf{d}(t) \tag{8.52}$$

$$\frac{d\mathbf{z}_k}{dt} = 0 \tag{8.53}$$

and

$$\frac{dL_{ij}}{dt} = 0 \tag{8.54}$$

Solving this set of coupled, nonlinear equations in the mean-field approach gives the following result with respect to the position and stability of the fixed points: There is a mapping

$$m_i(\mathbf{a}) \leftrightarrow m_i(\mathbf{a}) - \mathbf{a} \tag{8.55}$$

with the property that for every $m_i(\mathbf{a})$ there is a mapped point satisfying the original zero-mean-field (single-cell) synaptic evolution equation. The modification threshold is unaltered in this mapping. For every fixed point of the single-cell system, there is a correponding fixed point of the mean-field network with the same selectivity and stability properties. As before, only selective fixed points are stable.

The mean-field results given by Eq. (8.48) or equivalently, by Eq. (8.55) have a simple interpretation. The synaptic strengths are shifted by an amount $-\mathbf{a}$ that represents the average inhibitory influence of the intracortical network. In the presence of mean-field inhibition patterned stimulation that would otherwise lead to a potentiation of the synaptic strength may instead result in a depression. Conversely, excitatory inputs that were previously ineffective may be able to exert an influence on the synaptic strengths once intracortical inhibition is removed or blocked. This mean-field model supports the notions of unmasking (Section 5.9.1) and plastic gates (Section 8.10.4).

8.9 NEUROPHYSIOLOGICAL BASIS

In chapter 5 we studied how axons establish their initial synaptic contacts and then refine those contacts. The emphasis in the ensuing discussion of the electrophysiological and biochemical mechanisms that drive the synaptic processes was on the axonal growth cone. We will now switch our focus to examine postsynaptic mechanisms, beginning with the observation that dendritic spines are the main postsynaptic targets of excitatory synaptic input and, as such, are the locus of the modifiable synapses of BCM and other theories of synaptic plasticity. We will then discuss the excitatory amino acid receptors found in dendritic spines, and anatomical and morphological changes that involve cell adhesion signaling. In the next section we will continue our exploration of the molecular substrate by looking at one key element, the enzyme CaMKII, in the chain of biochemical events that alters the efficiency of synaptic transmission. We will relate the activity of the enzyme to the crossover point in BCM theory and to the predicted prior-activity-dependent movement of the threshold.

8.9.1 Dendritic Spines

Dendritic spines are small structures found on pyramidal and stellate cells in the neocortex, hippocampus, and other cortical regions.[44] The number of spines on a single neuron can be quite large, ranging from about 15,000 spines on a pyramidal cell in the visual cortex to as many as 175,000 on a cerebellar Purkinje cell. Densities of spines along a dendrite are typically on the order of 2 spines per micron (μm) of length in layer V pyramidal cells in the visual cortex, 3 spines per micron on CA1 pyramidal cells in the hippocampus, and 13 spines per micron on rat Purkinje cells. Although variable in size and shape, a spine may be thought of as having a head and a neck, which may be either long or short, that provides an open channel for ions to diffuse to the dendrite shaft. As was the case for axonal growth cones, dendritic spines have a dense network of actin filaments, contain actin-activity-mediating proteins such as calmodulin, and can change their morphology in response to signals.

There is both theoretical and experimental evidence to support the idea that dendritic spines can serve as neuronal compartments for synaptic modification by providing a biochemically focused environment in which only coactive synapses will become potentiated. In experimental investigations by Müller and Conner[45] and by Guthrie, Segal, and Kater,[46] microfluoremetry was used to directly study the evolution of calcium following presynaptic stimulation. Evidence was presented in those studies that buffering and calcium pumps can restrict diffusion through spine necks. In theoretical studies by Zador, Koch, and Brown[47] and by Gold and Bear,[48] a heuristic reaction-diffusion model adapted from cable theory was used to study calcium and Ca^{2+}-calmodulin dynamics. Results from these studies, showing that concentration gradients can be maintained, support the notion of biochemical compartmentalization by dendritic spines.

8.9.2 NMDA and AMPA Receptors

Both α-amino-3-hydroxy-5-methyl-4-isoxazolepropionate (AMPA) and NMDA receptors are colocated on dendritic spines. AMPA receptors allow monovalent cations Na^+ and K^+, and to a much lesser extent divalent cations such as Ca^{2+}, to move in and out of the dendritic spines. Their kinetics are fast, on the order of milliseconds (ms), and are activated by ligand binding.

NMDA receptors have a slow kinetics that operate on a time scale on the order of 100 ms. This type of glutamergic excitatory amino acid receptor is dual ligand and voltage gated; its ionophore is more permeable to Ca^{2+} ions than to either K^+ or Na^+ ions, and its activation allows for a substantial Ca^{2+} influx. This receptor also has allosteric properties — glycine binding is necessary for its activation. However, glycine is thought to be readily available in the synaptic cleft, and it is the dual gating and preferential Ca^{2+} entry that determine its role in synaptic plasticity. In more detail, NMDA receptors are activated when presynaptic activity (ligand gating) is correlated within a time window of about 100 ms with sustained postsynaptic depolarization (voltage gating). The postsynaptic depolarization can be most effectively produced when AMPA receptors, localized on postsynaptic targets such as dendritic spines, receive convergent synchronous input from many fibers. When the synaptic membrane is sufficiently depolarized by this means to relieve a Mg^{2+} block, the NMDA receptor is activated, and a graded entry of Ca^{2+} is initiated.

The voltage-gating requirement endows NMDA receptor with Hebbian properties, and the activation of NMDA receptors is a first step in the sequence of molecular events leading to changes in synaptic efficiency. Examples where NMDA activity has been linked to synaptic plasticity include the requirement for NMDA activation in the induction of associative LTP in CA1, in the experience-dependent plasticity and stabilization in the developing visual pathways of amphibians, fish, and mammals, and in the activity-dependent maturation of neurons.

In the next few paragraphs we will examine some of the steps that follow NMDA-receptor-mediated calcium entry into the postsynaptic cell. Synaptic modification processes may well include morphological and anatomical changes, as was mentioned in the introduction to this chapter. In Chapter 5 we examined the role of cell surface molecules such as NCAM in developmental processes. We end our preliminary discussion with the observation that the cell surface molecule signaling system is not rendered inert once synaptic contacts are established but instead remains available to promote and guide changes in adult synaptic morphology. For example, there is evidence in support of a role of NCAM and L1 in simple adult forms of synaptic plasticity such as LTP.[49-53] In the studies by Bailey et al.[49] and Mayford et al.,[50] changes in the long-term facilitation of synaptic efficiency were observed in the marine molusc *Aplysia*. This modification process is accompanied by the formation of new synaptic contacts. The growth was associated with the downregulation in the presynaptic cell of the cell adhesion molecule apCAM. In another investigation, of rat hippocampal LTP by Lüthl et al.,[51] NCAM and L1 were found to modulate the development and stabilization of the changes in synaptic efficiency, and in still another study by Cremer et al.,[52] inactivation of NCAM was found to lead to deficits in spatial learning. Lastly, a calcium-dependent mechanism for rapidly degrading the cytoplasmic domain of NCAM was found by Sheppard et al.[53] This mechanism could enable the rapid uncoupling of pre- and postsynaptic membranes.

8.9.3 The LTP/LTD Crossover and the BCM Modification Threshold

To recap briefly, we have seen that the threshold level for postsynaptic activation required for synaptic modification in the CNS is related to a voltage-dependent Ca^{2+} entry into cortical dendrites which acts as a second messenger to trigger the molecular changes needed for a modification of synaptic efficiency. In the BCM model, whenever a cortical neuron is depolarized below a modification threshold the synaptic strength is increased, and whenever the depolarization is inadequate, the strength is decreased. The modification threshold in this model floats as a function of the average postsynaptic activity. The dynamic constraint on total (average) postsynaptic activity, embodied in the modification threshold, prevents saturation by introducing a temporal competition between input patterns.

Electrophysiological evidence in support of the BCM modification rule has been obtained by Dudek and Bear[54] in a study of hippocampal LTP and LTD. The hippocampus, and in particular CA1 and dentate gyrus, are favored choices for synaptic plasticity study due to the ease and long-lasting character of the changes that can be induced. The tetanic-stimulus protocols used to initiate hippocampal LTP simulate the effects of many convergent inputs. In the experiments by Dudek and Bear, the Shaffer collateral projection to CA1 was stimulated electrically at frequencies ranging from 0.5 to 50 Hz. At low

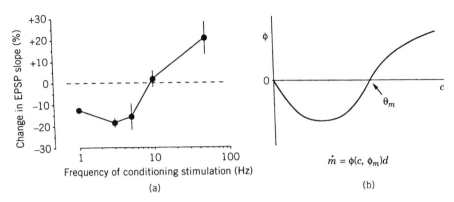

Figure 8.11. Comparison of electrophysiological data with the BCM modification threshold. (From Dudek and Bear[54]. Reprinted with permission of Prof. Mark A. Bear.)

conditioning frequencies a long-term depression was produced. At higher stimulation frequencies little response was evoked, while at the highest frequencies long-term potentiation was observed. As can be seen in Fig. 8.11, a plot of change in the excitatory postsynaptic potential (EPSP) slope versus frequency of conditioning stimulation is in dramatic agreement with BCM theory.

In the last section we noted that NMDA receptor activation allows for the graded entry of calcium into the postsynaptic cell. A considerable body of data has been acquired on the biochemical processes triggered by the calcium entry. One of the important results of these studies is the unification of LTP and LTD within a common biochemical framework. It seems that hippocampal LTP is induced by activation of the calcium-calmodulin-dependent protein kinase II (CaM kinase or CaMKII) as first reported by Malinow et al.[55] and Malenka et al.,[56] and hippocampal LTD is induced by activation of serine-threonine protein phosphatases, as shown by Mulkey et al.[57] The balance between the phosphorylation by CaM kinase and dephosphorylation by serine-threonine protein phosphatases is determined by the amount of calcium entering through the NMDA receptors. At low concentrations of calcium entry, the effect is one of dephosphorylation leading to LTD. At high calcium entry concentrations, the consequence is phosphorylation leading to LTP. Thus the balance between these two processes regulates synaptic changes in a manner predicted by BCM theory. The biochemical and electrophysiological findings have been summarized by Bliss and Collingridge[58] and by Bear and Malenka.[59]

8.10 FROM GENES TO BEHAVIOR

With the results of Dudek and Bear,[54] we are able to relate the balance between phosphorylation and dephosphorylation to the threshold behavior of the BCM ϕ function. We will now explore further some of the steps that follow calcium entry into the postsynaptic cell, focusing on calcium-calmodulin-dependent protein kinase II (CaM kinase or CaMKII) mediated processes. We will discuss the mechanisms for changing synaptic strength by looking at some reinforcing, dramatic experimental results of Griffith[60] and coworkers studying behavioral and synaptic defects in *Drosophila*. We will briefly discuss the work of Silva et al.,[61,62] Mayford et al.,[63] and Bach et al.[64] using genetically altered mice, and Malinow et al.[55,65] studying LTP in hippocampal CA1. One goal of these studies is the linking of CaM kinase activity, and lack thereof, with learning deficits. Another is the identification of possible targets of the kinase activities. In the studies by Mayford et al.[63] and Bach et al.,[64] we will find evidence that points toward a CaM kinase mechansm for the sliding BCM threshold.

8.10.1 CaM Kinase

Multifunctional calcium-calmodulin-dependent protein kinase II (CaM kinase) is the most abundant protein kinase in the brain and is highly concentrated in neocortical areas and the hippocampus. This enzyme phosphorylates a variety of target proteins and mediates many processes triggered by synaptic calcium entry.[66] It influences neurotransmitter release, membrane excitability, and synaptic strength. This enzyme becomes active when calcium-calmodulin (CaM) complexes, formed in the dendrite spines, subsequently bind to one or more of the 10 (to 12) identical subunits of the kinase. As depicted by Michelson and Schulman,[67] these subunits are arranged topologically in a ring. Upon binding to a CaM complex, a CaM kinase subunit undergoes transitions from one to another of its five activation states. Transitions between the various states of the ring form a nonhomogeneous Markov chain, and there is considerable resemblance to an MRF process in that we again have a spatial neighborhood system. The state of a given subunit at a given time depend on the concentration of the CaM complex, the state of the subunit at an earlier time step, and the state of its right-hand nearest neighbor.[67]

The primary proteins phosphorylated by CaM kinase are its own subunits. One of its most interesting properties[68,69] of CaM kinase is that it can become independent of its nomal regulators after brief activation and therefore act in a graded fashion as a molecular switch; that is, as a multistate molecular switch. This property allows CaMKII to transition to a calcium-independent state, remaining active after the decay of the initiating transient calcium signal and serving as a molecular memory. This property, coupled to a variety of results which we will shortly discuss, supports the notion that synaptic plasticity, learning, and memory share a common substrate. Before discussing this further, we briefly note one more property of CaM kinase, namely, calmodulin trapping.[70] This capability presents itself after autophosphorylation and the decay of the transient calcium signal. Its significance is that it provides the cellular substratum with a molecular potentiation of the calcium transients and may enable detection of the frequency of the calcium signals. We will discuss oscillatory properties of calcium signaling in the next chapter.

Now let us discuss CaM kinase in the hippocampus. In experiments in which protein kinase blockers were administered, Malinow, Schulman, and Tsien[55] showed that postsynaptic CaM kinase was a necessary component of the biochemical substrate underlying hippocampal LTP. In a more recent work in which CaM kinase was introduced into hippocampal cells using a recombinant vaccinia virus, Pettit, Perlman, and Malinow[65] demonstrated that postsynaptic CaM kinase activity was both necessary and sufficient for generating LTP in CA1 hippocampal brain slices. Supporting evidence for the CaM kinase links has been obtained in genetic knockout experiments in mice by Silva et al.[61,62] In these studies genetically disrupted CaM kinase α-subunits were shown to produce deficiencies in the induction of LTP in the hippocampus[61] and deficits in spatial learning.[62]

We may recall that AMPA receptors mediate fast EPSP's by enabling the inflow of sodium ions and the outflow of potassium ions. A possible locus for the phosphorlyating action of CaM kinase is the AMPA subunit, GLuR1. The idea[55,65] is that CaM kinase may possibly activate GluR1 receptors at synapses with no active GLuR1 receptors or increase the number of already active GLuR1 receptors. (AMPA subunits, designated as GluR1 to GLuR6, exist in two forms, known as flip and flop.) Thus we have a chain of steps that starts with convergent, AMPA-receptor-mediated synaptic activity that leads to NMDA receptor activation followed by calcium entry, calcium-calmodulin binding, and CaM kinase activation, resulting in phosphorylation at the sites of AMPA subunits, thereby changing channel and membrane properties and potentiating synaptic transmission.

8.10.2 Griffith *Drosophila* Data

In a series of experiments carried out by Griffith et al.,[60] we gain further insights into synaptic and behavioral plasticity. Courting in *Drosophila melanogaster* has been studied in the past to explore how genes affect behaviors. A general conclusion is that *Drosophila* behaviors are regulated by a large number of interacting, multipurpose genes. In the investigations by Griffith et al., genetic manipulations were carried out in order to (1) probe the dependence of courting behavior (learning) on CaM kinase and (2) to identify the protein targets of CaM kinase activity.

Male flies learn not to waste time on females who have already mated and are therefore unresponsive. The males are discouraged by a pheromone that leads to a calcium buildup activating CaM kinase. Griffith et al. showed that removal of CaM kinase makes the males forgetful. This was done by engineering a strain of fly whose CaM kinase activity levels could be controlled through heat shock. The amount of forgetfulness was proportional to the extent of the CaM kinase suppression. When CaM kinase was strongly reduced, mutant males continued to pursue mated females. That is, they did not learn at all.

The second goal, identifying the molecular targets, had as its starting point previous observations that manipulations of potassium channels in *Drosophila* and *Aplysia* produce alterations in learning and memory. *Eag* is a *Drosophila* gene whose protein product may be a potassium channel regulatory subunit. Griffith et al. found that *eag* deficient mutants have learning failures similar to those of the heat shock sensitive flies. Results of three additional sets of experiments strengthen the connection between CaM kinase and the *eag* protein product. Synaptic electrophysiology reveals that there are significant defects in responses to stimulation at the larval neuromuscular junction, double mutant investigations demonstrate that CaM kinase and *eag* effects are not additive, and *in vitro* studies show that there are strong biochemical interactions. These four sets of results provide us with impressive evidence in support of the supposition that the downstream target of CaM kinase phosphorylation is an *eag* protein product subunit that putatively regulates membrane potassium channel outflow.

To summarize, a tentative hypothesis arising from the data discussed in the last two sections is one of strongly cooperative coupling of NMDA and AMPA receptors. First, it appears that activity-dependent modulation of synaptic strength in CA1 is controlled, in part, by the balance between activities of protein kinases and serine-threonine protein phosphatases in the postsynaptic cell. Second, a possible substrate for this competition between phosphorylation and dephosphorylation is the AMPA receptor subunit GluR1. Thus the activity-dependent changes, initiated when sufficient cooperativity among the AMPA receptors occurs to remove the Mg^{2+} block leading to an entry of Ca^{2+} through NMDA receptors, set off a sequence of events that produces alterations in the structural properties of the AMPA receptors.

8.10.3 Movement of the BCM Threshold

The ability of CaMKII to remain active in a Ca^{2+}-independent state for long times through autophosphorylation has led to the suggestion by Miller and Kennedy[68,69] that it may serve as a memory molecule. A model of how this may possibly operate has been put forth by Lisman.[71] In recent studies of CaMKII in LTP and LTD in hippocampal synapses by Mayford et al.[63] and by Bach et al.,[64] evidence for CaMKII regulation of a sliding LTP/LTD threshold was discovered. There are two components to the their findings. The first is the establishment that prior activity produces shifts in the crossover point from LTD to LTP in the manner predicted by BCM theory. The second is the demonstration that CaMKII activity regulates the movement of the threshold.

Evidence for a dependence on prior activity has been found in earlier studies. For example, in the work by Huang et al.,[72] prior tetanic stimulation of CA1 rat hippocampal cells inhibited later induction of LTP. In the study by

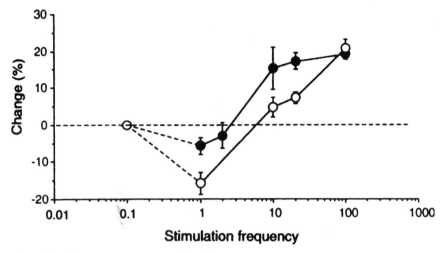

Figure 8.12. Frequency-response function from visual cortex of light-deprived (filled circles) and normal-reared (open circles) rats. Plotted are the percentage changes in the field potentials relative to baseline as a function of stimulation frequency. (From Kirkwood, Rioult, and Bear[73]. Reprinted with permission of Nature. © 1996 Macmillan Magazines Limited.)

Mayford et al., similar results were obtained. Low-frequency stimulation of the hippocampal slices elicited a potentiation of synaptic efficiency, but when preceded by a high-frequency stimulation, produced a depression of synaptic efficiency.

To establish the connection with CaMKII, the frequency-response properties of transgenic mice that express autonomous CaMKII were compared by Mayford et al. to those of wild-type mice. Although high-frequency stimulation produced the same response in the two groups of mice, clear differences were observed when low-frequency stimulation was applied to the two groups of mice. Low-frequency stimulation that produced a modest potentiation in wild-type mice generated a strong depression in transgenic mice. Further analysis of data for immature and adult animals showed that the amount of Ca^{2+}-independent CaMKII was correlated with the magnitude of the depression produced by low-frequency stimulation. In a further study of CaMKII influences, Bach et al. found that the mutant mice had spatial hippocampal-dependent learning deficits. To conclude, the data support the notion that the level of autonomous CaMKII activity, reflecting prior synaptic activity, regulates the crossover point from LTD to LTP.

A final piece of data in support of the notion that prior activity regulates changes in synaptic efficiency is provided by experimental studies by Kirkwood et al.[73] These authors compared visual cortex slices from rats reared in normal visual environments to those raised in the dark. They found that LTD was nearly absent in dark reared rats. The effects of dark rearing could be reversed by placing dark reared rats in a normal visual environment for a period of time as short as two days. To obtain a more quantitative assessment of the effects of visual experience on LTP and LTD, Kirkwood[73] measured frequency response functions in visual cortex slices. Their results, plotted in Fig. 8.12, provide evidence that the modification threshold slides as a function of prior activity in a manner predicted by BCM theory.

8.10.4 Plastic Gates

A possible explanation, advanced by Kirkwood et al.[74] and Kirkwood and Bear,[75] for the reduction in synaptic plasticity at the end of the critical period is that it is due to late-maturing layer IV inhibitory circuitry. The basic idea is that maturing layer IV inhibitory circuitry makes the type of activity-dependent plasticity seen early in the critical period more difficult to produce. Membrane- or circuit-based mechanisms may be responsible for this reduction in plasticity. In the former instance, inhibitory synapses onto layer III cells reduce synaptic plasticity by shunting excitatory postsynaptic currents. In the latter case, the patterns and amounts of afferent activity reaching the layer III cells are modulated by the inhibition. This possibility was discussed in the context of a mean-field network in Section 8.8.

8.11 PRINCIPAL COMPONENT NEURONS

In the next three sections we will explore the information-processing functions of model neurons in the visual system, the third of our goals. We begin with a discussion of principal component neurons, then examine a special class of synaptic modification models and their relation to phenomenological spin models, and end with the objective function formulation of BCM theory.

8.11.1 Introductory Remarks

We have already noted that images of natural scenes are not random patterns but rather are highly organized spatial structures that have some common statistical properties. The alterations in receptive fields produced by manipulating visual experience provides support for the notion that the development of the visual system is influenced by the statistical properties of the images. It follows that having some knowledge of the statistical properties of natural scenes will be of benefit to us in understanding the behavior of cells in the visual system. Several studies of image statistics have been carried out, motivated by this line of reasoning. In the studies by Field,[76] Ruderman and Bialek,[77] and Ruderman,[78] we find that there is no preferred angular scale in natural images. This scale invariance resembles Kadanoff scaling discussed in Chapter 3. It can be demonstrated by carrying out a decimation procedure with the grey-valued pixels of the image assuming the role of the spins. When this is done,[77,78] we find that the probability distributions of image contrasts and image gradients are unchanged (invariant) as we vary the amount of decimation (scale).

Another way of representing the scale invariance present in natural images is through the covariance matrix

$$Q_{ij} = \langle I_i - \langle I_i \rangle \rangle \langle I_j - \langle I_j \rangle \rangle$$

for the input intensities $\{I_i\}$. Field[76] found that the Fourier transform $Q(k)$ of the covariance matrix has the form $Q(k) = c/k^2$, while Ruderman[78] found a similar result that $Q(k) = c/k^{2-\eta}$, where η is small and c is a constant. This result places a constraint on the form of the covariance matrix and serves as the starting point for the principal component analyses of Liu and Shouval[41] and Shouval and Liu.[42] Their investigations were preceded by an earlier study of the principal components of natural images by Hancock, Baddeley, and Smith.[79]

8.11.2 Principal Components

The simplest way to depict the information-processing functions of the visual circuitry is in terms of projections of two-dimensional point-cloud data onto a set of coordinate axes. Shown in Fig. 8.13 is an example of an arbitrary data set and its projections onto two different pairs of coordinate axes. The second,

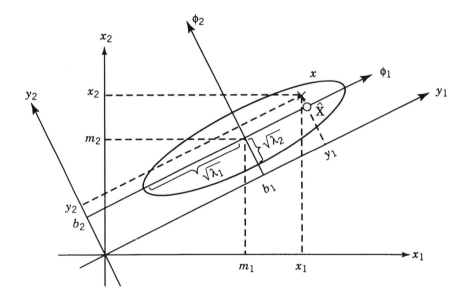

Figure 8.13. Two-dimensional illustration of principal components analysis (From Fukunaga[94]).

rotated coordinate system has the statistical property that the variance of the data projected onto its x-axis is maximal. In formal terms, principal components are linear combinations of random variables that have special properties in terms of variances. The most important of these nomalized linear combinations is the first principal component. This is the combination having the maximum variance, as illustrated above for our two-dimensional system. Thus the basic idea is that we are rotating the coordinate system in order to find projections with desirable statistical properties. The projections are interesting in the sense that they maximally preserve information content while compressing the data.

Principal components are the characteristic vectors of the covariance matrix. In principal component analysis, or the Karhunen-Loéve expansion as it is sometimes called, we calculate the eigenvalues and eigenvectors of the covariance or corelation matrix and project orthogonally onto the space spanned by the eigenvectors belonging to the largest eigenvalues. This procedure compresses the meaningful structure into a few leading components. The mathematics of this procedure is as follows.

8.11.3 Principal Components and Constrained Optimization

Suppose that $\mathbf{X}$ is n-component random vector

$$\mathbf{X} = [x_1 x_2 \ldots x_n]^T \tag{8.56}$$

The correlation matrix $\mathbf{Q}$ associated with this random vector is

$$\mathbf{Q} = \langle \mathbf{X}\mathbf{X}^T \rangle = \begin{bmatrix} \langle x_1 x_1 \rangle & \cdots & \langle x_1 x_n \rangle \\ \vdots & \vdots & \vdots \\ \langle x_n x_1 \rangle & \cdots & \langle x_n x_n \rangle \end{bmatrix} \tag{8.57}$$

If the mean vector associated with $\mathbf{X}$ is zero, then $\mathbf{Q}$ is the covariance matrix; otherwise, it is not. In the following, let us assume that the mean vector is zero. Then let us introduce an n-component fixed vector $\mathbf{a}$ that satisfies the normalization condition

$$\mathbf{a}^T \mathbf{a} = \sum_i a_i^2 = 1 \tag{8.58}$$

where a_i are the components of $\mathbf{a}$; the superscript T indicates that we are taking the transpose, and we are now using matrix notation to denote the inner product. We will use this fixed vector to help us find interesting projections. The quantity that we are interested in is the variance of the inner product of $\mathbf{a}$ with our random vector $\mathbf{X}$. This variance is given by the expression

$$\langle (\mathbf{a}^T \mathbf{X})^2 \rangle = \langle \mathbf{a}^T \mathbf{X} \mathbf{X}^T \mathbf{a} \rangle = \mathbf{a}^T \langle \mathbf{X}\mathbf{X}^T \rangle \mathbf{a} = \mathbf{a}^T \mathbf{Q} \mathbf{a} \tag{8.59}$$

We want to determine the normalized linear combination $\mathbf{a}^T\mathbf{X}$ that has the largest variance. That is, we are using the variance as our measure of interestingness. For this choice it is clear that interesting means projections that have large variances. Therefore we want to find the vector $\mathbf{a}$ that satisfies our normalization condition, Eq, (8.58), and maximizes the variance given as Eq (8.59). As was done in Chapter 2, we use the method of Lagrange multipliers to maximize the variance subject to the constraint. The function to be maximized is

$$\xi = \mathbf{a}^T\mathbf{Q}\mathbf{a} - \lambda\mathbf{a}^T\mathbf{a} \tag{8.60}$$

where λ is a Lagrange multiplier. On carrying out the requisite differentiation, we find that

$$(\mathbf{Q} - \lambda\mathbf{I})\mathbf{a} = 0 \tag{8.61}$$

and $\mathbf{Q} - \lambda\mathbf{I}$ must be singular. Thus the Lagrange multipliers λ must satisfy the condition

$$|\mathbf{Q} - \lambda\mathbf{I}| = 0 \tag{8.62}$$

If we multiply the terms in Eq. (8.61) on the left by $\mathbf{a}^T$, we see that the variance is equal to the eigenvalue λ:

$$\mathbf{a}^T\mathbf{Q}\mathbf{a} = \lambda\mathbf{a}^T\mathbf{a} = \lambda \tag{8.63}$$

Thus, if the vector $\mathbf{a}$ satisfies Eqs. (8.58) and (8.61), then the variance of $\mathbf{a}^T\mathbf{X}$ is λ. There are n roots to Eq. (8.62), and the maximum variance is given by the largest root λ_{max}.

8.11.4 Hebbian Learning and Synaptic Constraints

Let us consider the linear rule, Eq. (8.17), for output activity c given in terms of the input activity vector $\mathbf{d}$ and the synaptic weight vector $\mathbf{m}$. In our matrix notation this rule is

$$c = \sum_i m_i d_i = \mathbf{m}^T\mathbf{d} = \mathbf{d}^T\mathbf{m} \tag{8.64}$$

In its simplest form Hebb's rule for synaptic modification is

$$\frac{dm_i}{dt} = \eta c d_i = \eta \sum_j m_j d_j d_i \tag{8.65}$$

On reaching a fixed point, our Hebbian network would satisfy the relation

$$\left\langle \frac{dm_i}{dt} \right\rangle = 0 \rightarrow \left\langle \sum_j m_j d_j d_i \right\rangle = \mathbf{Qm} = 0 \qquad (8.66)$$

with input correlation matrix

$$\mathbf{Q} = \langle \mathbf{dd}^T \rangle \qquad (8.67)$$

This result would imply that the synaptic weight vector **m** is an eigenvector of the input correlation matrix with eigenvalue equal to zero. However, it is unstable, and the synaptic weights would undergo unbounded growth.

8.11.5 Oja's Solution

The problem of unbounded growth in the synaptic strengths can be avoided by imposing a constraint on the total synaptic strength analogous to the normalization condition on the projection vector **a**. There are several ways of doing this. In Oja's approach we modify the Hebbian rule to read

$$\frac{dm_i}{dt} = \eta c(d_i - cm_i)$$

$$= \eta \sum_j m_j d_j \left(d_i - \sum_k m_k d_k m_i \right) \qquad (8.68)$$

where η is the learning rate. For this modified rule we see that

$$\left\langle \frac{dm}{dt} \right\rangle = 0 \rightarrow \left\langle \sum_j m_j d_j d_i - \sum_{jk} m_j d_j m_k d_k m_i \right\rangle$$

$$= \sum_j Q_{ij} m_j - \left[\sum_{jk} m_j Q_{jk} m_k \right] m_i = 0 \qquad (8.69)$$

or

$$\mathbf{Qm} - [\mathbf{m}^T \mathbf{Qm}]\mathbf{m} = 0 \qquad (8.70)$$

Thus at a fixed point

$$\mathbf{Qm} = \lambda \mathbf{m} \qquad (8.71)$$

with

$$\mathbf{m}^T \mathbf{m} = \sum_i m_i^2 = 1 \qquad (8.72)$$

We therefore arrive at the same eigenvalue and constraint equations as before in our examination of variance maximization. We can summarize the situation briefly by noting that Hebbian modification with a constraint on the sum of the squared synaptic strengths either explicitly or implicitly as in Oja's rule, produces a synaptic vector **m** for which the projection of the input activity $\mathbf{m}^T\mathbf{d}$ has a maximum variance. The synaptic vector lies in the direction of the maximal eigenvector of the correlation matrix. The maximum eigenvalue associated with this eigenvector is equal to the output variance, and the synaptic system may be characterized as performing a principal component analysis of the input data.

Principal component neurons have been the subject of several studies. Linsker[80] has explored the feature analyzing capabilities of principal component neurons, noting that they are optimal with respect to information preservation. In particular, under certain conditions a neuron whose output is the leading principal component of the input vector conveys the maximum Shannon information about the input vector.

8.11.6 Linsker's Model

Another way of limiting unbounded growth in synaptic weights is by clipping. In this approach the sum of the synaptic weights are kept constant, and each synaptic weight is required to lie within a set range. The solutions to this model are not the same as those obtained using Oja's approach, referred to in the literature as the PCA solution, where the sum of the squares of the synaptic strengths are constrained to be constant. In Linsker's model[29-31] the changes in synaptic strength are governed by a rule of the form

$$\frac{dm_i}{dt} = \sum_j Q_{ij} m_j + k_1 + \frac{k_k}{N} \sum_j m_j \tag{8.73}$$

where N is the number of synaptic inputs and k_1 and k_2 are constants. The synaptic weights arep clipped so that they lie in the range from m_- to m_+. We may define two energy functions E_Q and E_k:

$$\begin{aligned} E_Q &= -\left(\frac{1}{2}\right) \langle (c - \langle c \rangle)^2 \rangle = -\left(\frac{1}{2}\right) \sum_{ij} Q_{ij} m_i m_j \\ &= -\left(\frac{1}{2}\right) \mathbf{m}^T \mathbf{Q} \mathbf{m} \end{aligned} \tag{8.74}$$

and

$$E_k = -k_1 \sum_j m_j - \frac{k_2}{2N} \left(\sum_j m_j\right)^2 \tag{8.75}$$

The first term is the variance in the input activity, and the second term may be viewed as a constraint. The sum of these two quantities defines an energy

$$E = E_Q + E_k \qquad (8.76)$$

with the property that its derivative with respect to the synaptic strengths gives the synaptic changes

$$\frac{dm_i}{dt} = -\frac{\partial E}{\partial m_i} \qquad (8.77)$$

for each i. Thus each iteration of the network produces a lowering of the energy function. Stability corresponds to a global near minimum of the energy function, which is equivalent to the maximum in the input variance subject to the constraint.

The dynamics of the model system defined by Eq. (8.73) was analyzed by MacKay and Miller[40] in terms of the eigenvectors of of the covariance matrix Q. Of particular interest was the character of the eigenvectors in Linsker's symmetry-breaking layer $B \to C$. In different regimes for the parameters k_1 and k_2, different receptive field structures dominate. Assuming Gaussian covariances, the principal eigenvector dominates the dynamics for $k_1 = 0 = k_2$. This eigenvector is one for which all synapses have the same sign. The next two eigenvectors in order of importance are a bi-lobed-oriented (symmetry-breaking) eigenvector and a circularly symmetric center-surround eigenvector. As k_1 and k_2 are varied, particular eigenvectors other than the principal one gain in relative importance.

8.12 SYNAPTIC AND PHENOMENOLOGICAL SPIN MODELS

Cells in the primate visual cortex self-organize into ocular dominance columns and iso-orientation patches. The precise patterns of connectivity have been studied using voltage-sensitive dyes and optical imaging techiques by several groups. Results reported include those by Blasdel and Salama,[81] Bonhoeffer and Grinvald,[82] and Blasdel.[83] The patterns observed experimentally are highly ordered. For example, Bonheoffer and Grinvald find well-formed pinwheel iso-orientation patterns. The objective in a number of theoretical studies of synaptic modification has been to explain the emergence of these highly ordered repeating structures.

8.12.1 Phenomenological Spin Models

Simple phenomenological spin models[81,82] can reproduce qualitatively many of the features of the data of Blasdel and Salama, Bonhoeffer and Grinvald, and Blasdel. In a study by Cowan and Friedman,[81] a two-dimensional Ising lattice of eye-specificity encoding spins was considered. An Ising hamiltonian was constructed containing coupling strengths J_{ij} of the form of a difference of Gaussians:

$$J_{ij} = a_+ \exp\left(-\frac{|i-j|}{\sigma_+^2}\right) - a_- \exp\left(-\frac{|i-j|}{\sigma_-^2}\right) \tag{8.78}$$

If we take $a_+/a_- = \sigma_+^2/\sigma_-^2$ with $\sigma_+ < \sigma_-$, then this type of coupling generates a short-range attraction plus a long-range repulsion between terminals from the same eye, and vice versa for terminals from different eyes. A representative ocular dominance pattern generated using simulated annealing is displayed in Fig. 8.14a. To model the formation of iso-orientation patches, Cowan and Friedman treated the spins as continuously variable and independent of ocularity. The hamiltonian for iso-orientation then assumes the form

$$E = -\sum_{i \neq j} J_{ij} |s_i| |s_j| \cos(\theta_i - \theta_j) \tag{8.79}$$

where the angle variable θ_i denotes the planar orientation of the ith spin vector. This hamiltonian is similar to that of the planar XY model discussed in Section 3.3. A typical iso-orientation pattern produced using simulated annealing is shown in Fig. 8.14b.

Ocular dominance patches are somewhat irregular in shape and orientation but have a characteristic periodicity. There is a systematic oscillatory variation in the degree of left-eye versus right-eye dominance across the model cortex describable in terms of a characteristic wavelength. In their study of ocular dominace column, Miller et al.[32] noted that the characteristic wavelength is the one for which the Fourier transform of a cortical interaction function is maximal. Conditions for this statement to hold include a limit on the relative size of the arbors, locality in the positioning of the correlation functions, and excitatory intracortical connections. In the Cowan-Friedman model the characteristic wavelength is determined by the variances σ^+ and σ^-.

8.12.2 Synaptic Models in the Common Input Approximation

If simpified, a number of models of synaptic modification operating under a normalized Hebbian learning rules sensitive to second-order statistics become equivalent to the phenomenological spin models of the Cowan-Friedman or Swindale[82] type. This aspect of the theory has been investigated in a recent study by Shouval and Cooper[34] and has been noted previously by Miller et al.[32] We can investigate this point by considering, as before, an LGN-cortico-cortico network with modifiable geniculocortico synapses and fixed cortico-cortico connections. We now relate the input activity d and modifiable synaptic

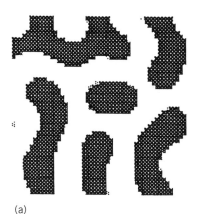 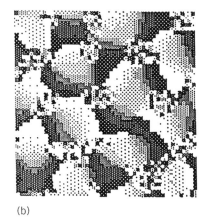

(a)   (b)

Figure 8.14. Ocular dominance columns and iso-orientation patches in the Cowan-Friedman model. (*a*) Pattern of ocular dominance: light and dark shadings encode eye-specificity. (*b*) Pattern of iso-orientation patches: each shading denotes a particular orientation selectivity. (From Cowan and Friedman[81]. Reprinted with permission of Prof. Jack D. Cowan.)

weights m to the output activity c through an expression of the general form

$$c(\mathbf{r}) = \sum_{\mathbf{x},\mu} I(\mathbf{r} - \mathbf{x})A(\mu - \beta(\mathbf{x}))d(\mu)m(\mathbf{x}, \mu) \qquad (8.80)$$

where μ denotes points in the LGN, and $\mathbf{r}$ and $\mathbf{x}$ in the cortex. The quantity $I(\mathbf{r} - \mathbf{x})$, called the *cortical interaction function*, describes the cortico cortico synaptic interactions. More specifically, it describes the net influence on a cortical cell located at position $\mathbf{r}$ of the excitation of a cortical cell located at position $\mathbf{x}$, produced by geniculate input. The quantity $A(\mu - \beta(\mathbf{x}))$ represents a symmetric *arbor function*. It gives the density of synapses as a function of the distance $\mu - \beta(\mathbf{x})$ between $\beta(\mathbf{x})$, the center of the arbor function in the cortex, and the position of the originating cell in the LGN.

The approach taken by Shouval and Cooper in their analysis is to design an energy function such that the fixed points of the network correspond to the minima of the energy function. The energy function selected is of the form

$$\begin{aligned}E &= -\sum_{\mathbf{r}} c^2(\mathbf{r}) - \gamma(\mathbf{m}) \\ &= -\sum_{\mathbf{r},\mathbf{x},\mathbf{x}'} I(\mathbf{r} - \mathbf{x})I(\mathbf{r} - \mathbf{x}') \sum_{\mu,\mu'} A(\mu - \beta(\mathbf{x}))A(\mu' - \beta(\mathbf{x}')) \\ &\quad \times m(\mathbf{x}, \mu)m(\mathbf{x}', \mu')d(\mu)d(\mu') - \gamma(\mathbf{m})\end{aligned} \qquad (8.81)$$

where $\gamma(m)$ denotes an as-yet-to-be-determined constraint on the synaptic strengths. To convert this energy function into a correlational hamiltonian, we form the ensemble average over the possible range of inputs, and sum over the cortical positions **r**. The resulting energy is

$$E = -\sum_{\mathbf{x},\mathbf{x}'} \tilde{I}(\mathbf{x}-\mathbf{x}') \sum_{\mu,\mu'} A(\mu - \beta(\mathbf{x}))A(\mu' - \beta(\mathbf{x}'))$$
$$\times m(\mathbf{x},\mu)m(\mathbf{x}',\mu')Q(\mu - \mu') - \gamma(\mathbf{m}) \quad (8.82)$$

where

$$\tilde{I}(\mathbf{x}-\mathbf{x}') = \sum_r I(\mathbf{r}-\mathbf{x})I(\mathbf{r}-\mathbf{x}') \quad (8.84)$$

is the effective cortical interaction function, and

$$Q(\mu - \mu') = \langle d(\mu)d(\mu') \rangle \quad (8.85)$$

is the correlation function. We now asumme that we have a step arbor function; that is, $A(\mathbf{x}) = 1$ if $|\mathbf{x}| \leq \mu_{\max}$, and 0 otherwise. The gradient descent dynamics for this hamiltonian is

$$\frac{dm(\mathbf{x},\mu)}{dt} = \begin{cases} \sum_{\mathbf{x}'} \sum_{\kappa(\mathbf{x}') < \mu_{\max}} I(\mathbf{x}-\mathbf{x}')m(\mathbf{x}',\mu)Q(\mu - \mu'), & \kappa(\mathbf{x}) \leq \mu_{\max} \\ 0, & \kappa(\mathbf{x}) > \mu_{\max} \end{cases} \quad (8.86)$$

where $\kappa(\mathbf{x}) = |\mathbf{x} - \beta(\mathbf{x})|$. The next step in solving this model is to express the synaptic weights in terms of the eigenfunctions of the correlation matrix. That is, we introduce the expansion

$$m(\mathbf{x},\mu) = \sum_{ln} a_{ln}(\mathbf{x})m_{ln}(\mathbf{x},\mu) \quad (8.87)$$

in terms of basis vectors defined by

$$\sum_{|\mu'-\alpha|} m_{ln}(\mathbf{x},\mu')Q(|\mu - \mu'|) = \lambda_{ln}m_{ln}(\mathbf{x},\mu - \alpha) \quad (8.88)$$

In a radially symmetric environment, the eigenfunctions can be written in the form

$$m_{ln}(\mathbf{x},\mu) = m_{ln}(\phi_{ln}(\mathbf{x}),\mu) = e^{i\phi_{ln}(\mathbf{x})} f_{ln}(\mu) \quad (8.89)$$

We now assume that all arbor functions have the same center point so that we may set $\alpha = \beta = 0$. This assumption produces a simplified dynamics termed the common input model by Shouval and Cooper. The hamiltonian for the common input model is

$$E = - \sum_{xx'ln} \tilde{I}(\mathbf{x} - \mathbf{x}')\lambda_{ln}a_{ln}(\mathbf{x})a_{ln}(\mathbf{x}')e^{i[\phi_{ln}(\mathbf{x}) - \phi_{ln}(\mathbf{x}')]} - \gamma(a) \quad (8.90)$$

We now simplify our model further by asuming that we are operating in a regime where $\lambda_{01} = \lambda_0$ and $\lambda_{11} = \lambda_1$ are the largest eigenvalues and that these eigenvalues are far larger than any of the others. Support for this assumption can be found in the studies by MacKay and Miller,[40] Liu and Shouval,[41] and Shouval and Liu.[42] We then have

$$E = - \sum_{rr'} \tilde{I}(\mathbf{r} - \mathbf{r}')\{\lambda_0 a(\mathbf{r})a(\mathbf{r}') + \lambda_1 b(\mathbf{r})b(\mathbf{r}')\cos[\phi(\mathbf{r}) - \phi(\mathbf{r}')]\} - \gamma(a, b) \quad (8.91)$$

If we now assume that the squares of the weights add to unity so that $a^2(\mathbf{r}) + b^2(\mathbf{r}) = 1$, then this model becomes equivalent to a nonisotropic Heisenberg spin model embedded in a two-dimensional lattice. If we further assume that $\lambda_1 \gg \lambda_0$ and that we can discard the terms multiplying λ_0 in Eq. (8.91), then this model reduces to the Cowan-Friedman or planar XY model with $\tilde{I}(\mathbf{r} - \mathbf{r}')$ as the coupling strengths. It may be noted the models of Linsker and Miller have a more complex dynamics than that of the common input model. In those models $\beta(\mathbf{x}) = \mathbf{x}$. Miller et al. has pointed out that their model can be related to phenomenological models of the Swindale type.

The information-processing activities by common input neurons may be summarized[34] as follows: For exclusive excitatory connections symmetry breaking does not occur, and all receptive fields have the same orientation selectivity. Inhibition affects both the organization and structure of the receptive fields. If there is sufficient inhibition, the network will develop orientation selective receptive fields even though the single-cell receptive fields are radially symmetric. The cortical cells self-organize into iso-orientation patches with pinwheel singularities. The character of the singularities so produced is sensitive to the symmetry of the receptive fields.

8.13 OBJECTIVE FUNCTION FORMULATION OF BCM THEORY

In examining the utility of the different statistical measures, it is useful to distinguish between information preservation (variance maximization) and classification (multimodality). In some instances the two goals coincide, but in other instances they will differ from one another. The situation may be best visualized in terms of another schematic plot of two-dimensional point-cloud data. Shown in Fig. 8.15 are two clusters of data points. We see that by projecting the data onto the first principal axis we maximize the variance. However, if we are interested in obtaining a maximum separation between

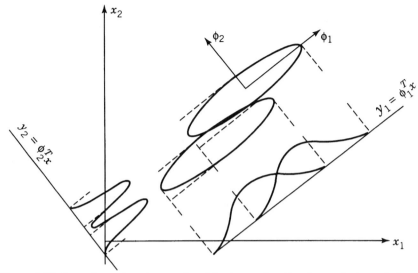

Figure 8.15. Two-dimensional example of feature extraction for a cluster pair showing the projections on the two principal axes. (From Fukunaga[94]. Reprinted with permission of Academic Press.)

clusters, then the projection onto the second principal axis is superior. Projection onto the second axis gives a smaller variance, but the separation between the two clusters is maximal and is therefore a more interesting projection from the viewpoint of classification.

8.13.1 Projection Pursuit

Projection pursuit is a method for finding the most interesting low-dimensional features of high-dimensional data sets. The term was introduced by Friedman and Tukey[86] to describe an exploratory technique for analyzing multivariate data sets. The objective is to find orthogonal projections that reveal interesting structure in the data.[87-90] In projection pursuit we select the most useful projections by local optimization over projection directions of some index of interestingness. Principal component analysis is a particular case of projection pursuit in which the index of interestingness is the proportion of total variance accounted for by the projected data.

The original purpose[86] of projection pursuit was to find interesting low-dimensional projections of high-dimensional point clouds by numerically maximizing an objective function, or projection index. High-dimensional spaces are inherently sparse. To illustrate this notion of sparseness, let us consider a ten-dimensional unit ball in which there is a uniform distribution of points. The radius of a ball enclosing just 3% of the points is $(0.03)^{0.10} = 0.70$. In situations such as this, the amount of training data needed to produce reasonable variance estimators is excessive. By finding interesting projections in the data, this "curse of dimensionality" may be reduced, if not avoided.[87] Most low-dimensional projections are approximately normal. One way of defining an interesting projection is to note that if classification is the purpose of the analysis of the data, then an interesting projection is one that departs from normalcy. In this view[87,90] it is better to seek deviations from normalcy that lie in the center of the distribution rather than those that occur in the wings.

In point-cloud data an interesting feature is one for which the data separates into meaningful clusters. However, not all data are of the cluster type. In the objective (energy) function formulation of BCM theory of Intrator,[91] Intrator et al.,[92] and Intrator and Cooper,[24] a feature is associated with each projection direction. We say that an input possesses a feature associated with that projection direction if the projection in that direction exceeds a threshold. Thus a one-dimensional projection may be interpreted as a single feature extraction. The goal in the approach is to find an objective (loss) function whose minimization produces a one-dimensional projection that is far from normal.

8.13.2 Objective Function Formulation of BCM Theory

We begin by redefining the threshold function Θ_M as the expectation of the square of the output activity,

$$\Theta_M = \langle c^2 \rangle = \langle (\mathbf{m}^T \mathbf{d})^2 \rangle \tag{8.92}$$

Next we represent the synaptic modification functions ϕ and by the simple analytic form

$$\phi(c, \Theta_M) = c^2 - c\Theta_M \tag{8.93a}$$

and

$$\hat{\phi}(c, \Theta_M) = c^2 - \frac{1}{2} c \Theta_M \tag{8.93b}$$

To find a projection index that emphasizes departures from normalcy that is multimodal, we must introduce polynomial moments higher than second. This can be accomplished by introducing a loss function that depends on the synaptic weights $\mathbf{m}$ and input $\mathbf{d}$ through an expression of the form

$$L_m(\mathbf{d}) = -\mu \int_0^{\mathbf{m}^T \mathbf{d}} \hat{\phi}(s, \Theta_M) ds = -\mu \left[\left(\frac{1}{3}\right)(\mathbf{m}^T \mathbf{d})^3 - \left(\frac{1}{4}\right) \langle (\mathbf{m}^T \mathbf{d})^2 \rangle (\mathbf{m}^T \mathbf{d})^2 \right] \tag{8.94}$$

where $\mu = \mu(t)$ is a learning rate that decays in time.

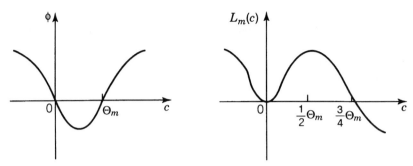

Figure 8.16. The modification and loss functions for a fixed synaptic weight vector and modification threshold. (From Intrator and Cooper[24]. Reprinted with permission of Elsevier Science, UK.)

We observe that the output activity c is the linear projection of **d** onto **m** and that the loss function reflects the neuron's decision whether to fire or not. Since the loss depends on both **m** and the stochastic process **d**, an appropriate choice for the projection index is one that minimizes the sum of the losses for all possible inputs. The desired synaptic weight vector, or parameter **m**, is the one that minimizes the mean loss, or risk.

If we assume that the modification threshold Θ_M and synaptic weight vector **m** are constant, then we can reexpress the loss function as

$$L_m(c) = -\mu c^2 \left(\left(\frac{c}{3} \right) - \left(\frac{\Theta_M}{4} \right) \right) \tag{8.95}$$

The synaptic modification and loss functions are plotted as a function of the neuron's output activity in Figs. 8.16a and 8.16b. As shown in Fig. 8.16a, the synaptic modification function is zero at the origin and at the modification threshold. It is negative for activity levels between these two points and is positive elsewhere. A examination of the plot for the loss function, Fig. 8.16b, reveals that the loss is small when the activity is negligible. The loss is also small when c is near the modification threshold, and it remains negative for above-threshold activities.

The projected distributions found by minimizing the risk are guaranteed to be multimodal. To see why this is so, we first note that the activity dependence of the loss function favors distributions to the right of the modification threshold. This behavior, together with the nonlinear dynamics of the modification threshold, Eq. (8.92), generates a projection direction for which the single-dimensional distribution differs from normalcy in its center. This happens because the modification threshold always adjusts itself so that part of the mass of the distribution is placed on each of its sides. As noted in Section 8.13.1, this manner of deviation, as contrasted with departures from normalcy in the wings of the distribution, is the preferred one.

Using Eq. (8.94), we find that the risk, or expected value of the loss, is equal to

$$R_m = \langle L_m(c) \rangle = -\mu \left\langle \left(\frac{1}{3}\right)(\mathbf{m}^T\mathbf{d})^3 - \left(\frac{1}{4}\right)\langle(\mathbf{m}^T\mathbf{d})^2\rangle(\mathbf{m}^T\mathbf{d})^2 \right\rangle$$
$$= -\mu \left[\left(\frac{1}{3}\right)\langle(\mathbf{m}^T\mathbf{d})^3\rangle - \left(\frac{1}{4}\right)\langle(\mathbf{m}^T\mathbf{d})^2\rangle^2 \right] \quad (8.96)$$

This risk is continuously differentiable, and we are able to minimize the risk by means of gradient descent with respect to m_i:

$$\frac{dm_i}{dt} = -\frac{\partial R_m}{\partial m_i} = \mu[\langle(\mathbf{m}^T\mathbf{d})^2 d_i\rangle - \langle(\mathbf{m}^T\mathbf{d})^2\rangle\langle(\mathbf{m}^T\mathbf{d})d_i\rangle]$$
$$= \mu[\langle c^2 d_i\rangle - \Theta\langle c d_i\rangle] \quad (8.97)$$
$$= \mu\langle \phi(m^T d, \Theta) d_i\rangle$$

This expression represents a slightly modified, deterministic version of the stochastic BCM modification equation. It may be concluded from an examination of the risk,[91] and from a recasting of the formalism into a correlation of activity framework,[92] that a BCM neuron is extracting third-order statistical correlates of the data. This would be a natural extension of principal component processing in the retina.

The above formulation can be easily extended to a nonlinear neuron by replacing the linear response $c = \mathbf{m}^T\mathbf{d}$ with the nonlinear form $c = \sigma(\mathbf{m}^T\mathbf{d})$, where σ is a sigmoidal function. This substitution generates a multiplicative factor, the derivative of the sigmoidal function evaluated at $\mathbf{m}^T\mathbf{d}$, into the above expressions for the loss and risk. This additional factor reduces the sensitivity to outliers, since the derivative of the sigmoidal function is small in those instances. The objective function formalism provides a general method for analyzing the stability of the fixed points and for studying the evolution of the cells under differing noise and patterned input situations.[24]

8.14 REFERENCES

1. Oppenheim, R. W. (1985). Naturally occurring cell death during neural development. Trends Neurosci., **8**, 487–493.
2. Cowan, W. M., Fawcett, J. W., O'Leary, D. D. M., and Stanfield, B. B. (1984). Regressive events in neurogenesis. Science, **225**, 1258–1265.
3. Crespo, D., O'Leary, D. D. M., and Cowan, W. M. (1985). Changes in the numbers of optic nerve fibers during late prenatal and postnatal development in the albino rat. Dev. Brain Res., **19**, 129–134.

4. O'Leary, D. D. M., Fawcett, J. W., and Cowan, W. M. (1986). Topographic targeting errors in the retinocollicular projection and their elimination by selective ganglion cell death. J. Neurosci., **6**, 3692–3705.
5. Garraghty, P. E., and Kaas, J. H. (1992). Dynamic features of sensory and motor maps. Curr. Opin. Neurobiol., **2**, 522–527.
6. Gilbert, C. D., and Wiesel, T. N. (1992). Receptive field dynamics in adult primary visual cortex. Nature, **356**, 150–152.
7. Jacobs, K. M., and Donoghue, J. P. (1991). Reshaping the cortical motor map by unmasking latent intracortical connections. Science, **251**, 944–947.
8. Hess, G., and Donoghue, J. P. (1994). Long-term potentiation of horizontal connections provides a mechanism to reorganize cortical motor maps. J. Neurophysiol., **71**, 2543–2547.
9. Cline, H. T., Debski, E. A., and Constantine-Paton, M. (1987). N-methyl-D-aspartate receptor antagonist desegregates eye-specific stripes. Proc. Nat. Acad. Sci. USA, **84**, 4342–4345.
10. Cline, H. T., and Constantine-Paton, M. (1989). NMDA receptor antagonists disrupt the retinotectal topographic map. Neuron, **3**, 413–426.
11. Kleinschmidt, A., Bear, M. F., and Singer, W. (1987). Blockage of "NMDA" receptors disrupts experience-dependent plasticity of kitten striate cortex. Science, **238**, 355–358.
12. Bear, M. F., Kleinschmidt, A., Gu, Q., and Singer, W. (1990). Disruption of experience-dependent synaptic modifications in striate cortex by infusion of an NMDA receptor antagonist. J. Neurosci., **10**, 909–925.
13. Bienenstock, E. L., Cooper, L. N., and Munro, P. W. (1982). Theory for the development of neuron selectivity: Orientation specificity and binocular interaction in visual cortex. J. Neurosci., **2**, 32–48.
14. Barlow, H. B. (1985). Cerebral cortex as model builder. In D. Rose and V. G. Dobson (eds.), *Models of the Visual Cortex*. New York: Wiley, pp. 37–46.
15. Anderson, J. A. (1970). Two models for memory organization using interacting traces. Math. Biosci., **8**, 137–160.
16. Anderson, J. A. (1972). A simple neural network generating an interactive memory. Math. Biosci., **14**, 197–220.
17. Kohonen, T. (1972). Correlation matrix memories. IEEE Trans. Comput., **C21**, 353–359.
18. Cooper. L. N. (1973). A possible organization of animal memory and learning. In B. Lundquist and S. Lundquist (eds.), *Proceedings of the Nobel Symposium on Collective properties of Physical Systems*. New York: Academic Press, pp. 252–264.
19. Malsberg, C. von der (1973). Self-organization of orientation sensitive cells in the striate cortex. Kybernetik, **14**, 85–100.
20. Pérez, R., Glass, L., and Shlaer, R. (1975). Development of specificity in the cat visual cortex. J. Math. Biol., **1**, 275–288.
21. Nass, M. M., and Cooper, L. N. (1975). A theory for the development of feature detecting cells in visual cortex. Biol. Cybern., **19**, 1–18.
22. Cooper, L. N., and Scofield, C. L. (1988). Mean-field theory of a neural network. Proc. Nat. Acad. Sci. USA, **85**, 1973–1977.
23. Clothiaux, E. E., Bear, M. F., and Cooper, L. N. (1991). Synaptic plasiticity in visual cortex: Comparison of theory with experiment. J. Neurophysiol., **66**, 1785–1804.
24. Intrator, N., and Cooper, L. N. (1992). Objective function formulation of the BCM theory of visual cortical plasticity: Statistical connections, stability conditions. Neural Networks, **5**, 3–17.

25. Cooper, L. N., Liberman, F., and Oja, E. (1979). A theory for the acquisition and loss of neuron specificity in visual cortex. Biol. Cybern., **33**, 9–28.
26. Oja, E. (1982). A simplified neuron model as a principle component analyzer. J. Math. Biol., **15**, 267–273.
27. Sanger, T. D. (1989). Optimal unsupervised learning in a single-layer linear feedforward neural network. Neural Networks, **2**, 459–473.
28. Linsker, R. (1988). Self-organization in a perceptual network. IEEE Computer, **88**, 105–117.
29. Linsker, R. (1986a). From basic network principles to neural architecture: Emergence of spatial-opponent cells. Proc. Nat. Acad. Sci. USA, **83**, 7508–7512.
30. Linsker, R. (1986b). From basic network principles to neural architecture: Emergence of orientation-selective cells. Proc. Nat. Acad. Sci. USA, **83**, 8390–8394.
31. Linsker, R. (1986c). From basic network principles to neural architecture: Emergence of orientation columns. Proc. Nat. Acad. Sci. USA, **83**, 8779–8783.
32. Miller, K. D., Keller, J. B., and Stryker, M. P. (1989). Ocular dominance column development: Analysis and simulation. Science, **245**, 605–615.
33. Miller, K. D. (1994). A model for the development of simple receptive fields and the ordered arrangement of orientation columns through activity-dependent competition between on- and off-center inputs. J. Neurosci., **14**, 409–441.
34. Shouval, H., and Cooper, L. N. (1996). Organization of receptive fields in networks with Hebbian learning: The connection between synaptic and phenomenological models. Biol. Cybern., **74**, 439–447.
35. Kirkwood, A., Dudek, S. M., Gold, J. T., Aizenman, C. D., and Bear, M. F. (1993). Common forms of synaptic plasticity in the hippocampus and neocortex *in vitro*. Science, **260**, 1518–1521.
36. Bear, M. F., Cooper, L. N., and Ebner, F. F. (1987). A physiological basis for a theory of synapse modification. Science, **237**, 42–48.
37. Mioche, L., and Singer, W. (1989). Chronic recordings from single sites of kitten striate cortex during experience-dependent modifications of receptive-field properties. J. Neurophysiol., **62**, 1℄–197.
38. Kammen, D. M., and Yuille, A. L. (1988). Spontaneous symmetry-breaking energy functions and the emergence of orientation selective cortical cells. Biol. Cybern., **59**, 23–31.
39. Yuille, A. L., Kanen, D. M., and Cohen, D. S. (1989). Quadrature and the development of orientation selective cortical cells by Hebb rules. Biol. Cybern., **61**, 183–194.
40. MacKay, D. J. C., and Miller, K. D. (1990). Analysis of Linsker's application of Hebbian rules to linear networks. Network, **1**, 257–297.
41. Liu, Y., and Shouval, H. (1995). Localized principal components of natural images—an analytic solution. Network, **5**, 317–325.
42. Shouval, H., and Liu, Y. (1996). Principal component neurons in a realistic visual environment. Network, **7**, 501–515.
43. Law, C. C., and Cooper, L. N. (1994). Formation of receptive fields in realistic visual environments according to the Bienenstock, Cooper and Munro (BCM) theory. Proc. Nat. Acad. Sci. USA, **91**, 7797–7801.
44. Koch, C., and Zador, A. (1993). The function of dendritic spines: devices subserving biochemical rather than electrical compartmentalization. J. Neurosci., **13**, 413–422.
45. Müller, W., and Conner, J. A. (1991). Dendritic spines as individual neuronal compartments for synaptic Ca^{2+} responses. Nature, **354**, 73–76.

46. Guthrie, P. B., Segal, M., and Kater, S. B. (1991). Independent regulation of calcium revealed by imaging dendritic spines. Nature, **354**, 76–80.
47. Zador, A., Koch, C., and Brown, T. H. (1990). Biophysical model of a Hebbian synapse. Proc. Nat. Acad. Sci. USA, **87**, 6718–6722.
48. Gold, J. I., and Bear, M. F. (1994). A model of dendritic spine Ca^{2+} concentration exploring possible bases for a sliding synaptic modification threshold. Proc. Nat. Acad. Sci. USA, **91**, 3941–3945.
49. Bailey, C. H., Chen, M., Keller, F., and Kandal, E. R. (1992). Serotonin-mediated endocytosis of apCAM: An early step of learning-related synaptic growth in *Aplysia*. Science, **256**, 645–649.
50. Mayford, M., Barzilai, A., Keller, F., Schacher, S., and Kandal, E. R. (1992). Modulation of an NCAM-related adhesion molecule with long-term synaptic plasticity in *Aplysia*. Science, **256**, 638–644.
51. Lüthl, A., Laurent, J.-P., Figurov, A., Muller, D., and Schachner, M. (1994). Hippocampal long-term potentiation and neural cell adhesion molecules L1 and NCAM. Nature, **372**, 777–779.
52. Cremer, H., et al. (1994). Inactivation of the N-CAM gene in mice results in size reduction of the olfactory bulb and deficits in spatial learning. Nature, **367**, 455–459.
53. Sheppard, A., Wu, J., Rutishauer, U., and Lynch, G. (1991). Proteolytic modification of neural cell adhesion molecule (NCAM) by the intracellular proteinase calpain. Biochem. Biophys. Acta, **1076**, 156–160.
54. Dudek, S. M., and Bear, M. F. (1992). Homosynaptic long-term depression in area CA1 of hippocampus and effects of N-methyl-D-aspartate receptor blockade. Proc. Nat. Acad. Sci. USA, **89**, 4363–4367.
55. Malinow, R., Schulman, H., and Tsien, R. W. (1989). Inhibition of postsynaptic PKC or CaMKII blocks induction but not expression of LTP. Science, **245**, 862–866.
56. Malenka, R. C., Kauer, J. A., Perkel, D. J., Mauk, M. D., Kelly, P. T., Nicoll, R. A., and Waxman, M. N. (1989). An essential role for postsynaptic calmodulin and protein jinase activity in long-term potentiation. Nature, **340**, 554–557.
57. Mulkey, R. M., Herron, C. E., and Malenka, R. C. (1993). An essential role for protein phosphatases in hippocampal long-term depression. Science, **261**, 1051–1055.
58. Bliss, T. V. P., and Collingridge, G. L. (1993). A synaptic model of memory: Long-term potentiation in the hippocampus. Nature, **361**, 31–39.
59. Bear, M. F., and Malenka, R. C. (1994). Synaptic plasticity: LTP and LTD. Curr. Opin. Neurobiol., **4**, 389–300.
60. Griffith, L. C., Wang, J., Zhong, Y., Wu, C.-F., and Greenspan, R. J. (1994). Calcium/calmodulin-dependent protein kinase II and potassium channel subunit Eag similarity affect plasticity in *Drosophila*. Proc. Nat. Acad. Sci. USA, **91**, 10044–10048.
61. Silva, A. J., Stevens, C. F., Tonegawa, S., and Wang, Y. (1992). Deficient hippocampal long-term potentiation in α-calcium-calmodulin kinase II mutant mice. Science, **257**, 210–206.
62. Silva, A. J., Paylor, R., Wehner, J. M., and Tonegawa, S. (1992). Impaired spatial learning in α-calcium-calmodulin kinase II mutant mice. Science, **257**, 206–212.
63. Mayford, M., Wang, J., Kandal, E. R., and O'Dell, T. J. (1995). CaMKII regulates

the frequency-response function of hippocampal synapses for the production of both LTD and LTP. Cell, **81**, 891–904.
64. Bach, M. E., Hawkins, R. D., Osman, M., Kandal, E. R., and Mayford, M. (1995). Impairment of spatial but not contextual memory in CamKII mutant mice with a selective loss of hippocampal LTP in the range of the θ frequency. Cell, **81**, 905–915.
65. Pettit, D. L., Perlman, S., and Malinow, R. (1994). Potentiated transmission and prevention of further LTP by increased CaMKII activity in postsynaptic hippocampal slice neurons. Science, **266**, 1881–1885.
66. Hanson, P. I., and Schulman, H. (1992). Neuronal Ca^{2+}/calmodulin-dependent protein kinases. Ann. Rev. Biochem., **61**, 559–601.
67. Michelson, S., and Schulman, H. (1994). CaM Kinase: A model for its activation and dynamics. J. Theor. Biol., **171**, 281–290.
68. Miller, S. G., and Kennedy, M. B. (1986). Regulation of brian type II Ca^{2+}/calmodulin-dependent protein kinase by autophosphorylation: A Ca^{2+}-triggered molecular switch. Cell, **44**, 861–870.
69. Molloy, S. S., and Kennedy, M. B. (1991). Autophosphorylation of type II Ca^{2+}/calmodulin-dependent protein kinase in cultures of postnatal rat hippocampal slices. Proc. Nat. Acad. Sci. USA, **88**, 4756–4760.
70. Meyer, T., Hanson, P. I., Stryer, L., and Schulman, H. (1992). Calmodulin trapping by calcium-calmodulin-dependent protein kinase. Science, **256**, 1199–1202.
71. Lisman, J. (1989). A mechanism for Hebb and anti-Hebb processes underlying learning and memory. Proc. Nat. Acad. Sci. USA, **86**, 9574–9578.
72. Huang, Y.-Y., Colino, A., Selig, D. K., and Malenka, R. C. (1992). The influence of prior synaptic activity on the induction of long-term potentiation. Science, **255**, 730–733.
73. Kirkwood, A., Rioult, M. G., and Bear, M. F. (1966). Experience-dependent modification of synaptic plasticity in visual cortex. Nature, **381**, 526–528.
74. Kirkwood, A., Dudek, S. M., Gold, J. T., Aizenman, C. D., and Bear, M. F. (1993). Common forms of synaptic plasticity in the hippocampus and neocortex *in vitro*. Science, **260**, 1518–1521.
75. Kirkwood, A., and Bear, M. F. (1994). Hebbian synapses in visual cortex. J. Neurosci., **14**, 1634–1645.
76. Field, D. J. (1987). Relations between the statistics of natural images and the response properties of cortical cells. J. Opt. Soc. Am., **A4**, 2379–2394.
77. Ruderman, D. L., and Bialek, W. (1994). Statistics of natural images: Scaling in the woods. In J. D. Cowan, G. Tesauro, and J. Alspector (eds.), *Advances in Neural Information Processing Systems*, vol 6. San Francisco: Morgan Kaufmann.
78. Ruderman, D. L. (1994). The statistics of natural images. Network, **5**, 517–548.
79. Hancock, P. J. B., Baddeley, R. J., and Smith, L. S. (1992). The principal components of natural images. Network, **3**, 61–70.
80. Linsker, R. (1990). Designing a sensory processing system: What can be learned from principal components analysis. In *Intern. Joint Conf. Neural Networks*, vol. 2. Applications Track. Hillside, NJ: Laurence Erlbaum, pp. 291–297.
81. Cowan, J. D., and Friedman, A. E. (1991). Simple spin models for the development of ocular dominance columns and iso-orientation patches. In R. P. Lippman, J. E. Moody, and D. S. Touretzky (eds.), *Advances in Neural Information Processing Systems*, vol. 3. San Mateo: Morgan Kaufmann, pp. 26–31.
82. Swindale, N. (1980). A model for the formation of orientation columns. Proc. Roy. Soc. Lond., **B208**, 243–264.

83. Blasdel, G. G., and Salama, G. (1986). Voltage-sensitive dyes reveal a molular organization in monkey striate cortex. Nature, **321**, 579–585.
84. Bonhoeffer, T., and Grinvald, A. (1991). Iso-orientation domains in cat visual cortex are arranged in pinwheel-like patterns. Nature, **353**, 429–431.
85. Blasdel, G. G. (1992). Orientation selectivity preference and continuity in monkey striate cortex. J. Neurosci., **12**, 3139–3161.
86. Friedman, J. H., and Tukey, J. W. (1974). A projection pursuit algorithm for exploratory data analysis. IEEE Trans. Comput., **C-23**, 881–889.
87. Huber, P. J. (1985). Projection pursuit. Ann. Statist., **13**, 435-475.
88. Friedman, J. H. (1987). Exploratory projection pursuit. J. Am. Statist. Assoc., **82**, 249–266.
89. Jones, M. C., and Sibson, R. (1987). What is projection pursuit? (with discussion). J. Roy. Statist. Soc., **A150**, 1–36.
90. Hall, P. (1989). On polynomial-based projection indices for exploratory projection pursuit. Ann. Statist., **17**, 589–605.
91. Intrator, N. (1990). A neural network for feature extraction. In D. S. Touretzky and R. P. Lippman (eds.), *Advances in Neural Information Processing Systems*, vol. 2. San Mateo: Morgan Kaufmann, pp. 719–726.
92. Intrator, N., Bear, M. F., Cooper, L. N., and Paradiso, M. A. (1993). Theory of synaptic plasticity in visual cortex. In M. Baudry, R. F. Thompson, and J. L. Davis (eds.), *Synaptic Plasticity: Molecular, Cellular and Functional Aspects*. Cambridge: MIT Press.
93. Bear, M. F., and Cooper, L. N. (1990). Molecular mechanisms for synaptic modification in the visual cortex: Interaction between theory and experiment. In M. A. Gluck and D. E. Rumelhart (eds.), *Neuroscience and Connectionist Theory*. Lawrence Erlbaum.
94. Fukunaga, K. (1990). *Introduction to Statistical Pattern Recognition*, 2nd edn. San Diego: Academic Press.

9

RHYTHMS AND SYNCHRONY

9.1 BIOLOGICAL RHYTHMS AND SYNCHRONY

9.1.1 Nonlinear Dynamics

In the last chapter we began our examination of dynamic cooperativity by exploring experience-dependent synaptic modification. We studied the stability properties of the synaptic strengths in the BCM model by scrutinizing the convergence of the trajectories, representing different initial conditions, to fixed points in phase space. An important ingredient in the model was the nonlinear threshold for synaptic modification, which permitted the system to dynamically adapt to the mean activity level. We then made some tentative connections between the theoretical model, in particular, the dynamic threshold and the biochemical substrate that supports modifications. At the subsynaptic level we encountered coupled receptors and a dynamic modulation by calcium and protein networks. The subsynaptic systems in turn have a cooperative dynamics that serves to intimately link together several levels of organization.

In this chapter we will consider ensembles of elements undergoing oscillatory and other forms of rhythmic behavior. Steady oscillatory states of a dynamic system are represented in phase space by limit cycles. A limit cycle oscillator has a preferred amplitude and wave form to which it will return subsequent to being given any small perturbation. Thus limit cycle oscillators are self-sustaining oscillators, describing closed trajectories in phase space that are stable against small perturbations. Limit cycle oscillations do not occur in systems that are linear or conservative. For instance, a simple pendulum if perturbed will establish a new orbit, whereas a limit cycle oscillator, which has a dissipative mechanism that removes excess energy and a source that restores it, will maintain an orbit's periodicity.

A distinguishing characteristic of nonlinear dynamic systems is their richness of response properties. A linear system will either oscillate or not, and it will do one or the other over its entire parameter range. Nonlinear systems such as model neurons exhibit different response characteristics as their physical parameters are varied. For example, it is not uncommon to encounter steady state behaviors of a given cell corresponding to bistability, plateau behavior, and rhythmic oscillations, depending on the parameter regimes. Change in electrophysiological parameters can be initiated by internal and/or external signals. This property plus the existence of a multiplicity of steady state behaviors endow these systems with their dynamic adaptive abilities.

9.1.2 Excitable Membranes

The subject matter of this chapter, rhythms and synchrony, is a continuation of our exploration of nonlinear dynamics far from equilibrium started in the last chapter with the topic of synaptic plasticity. A large number of pioneering contributions have contributed to the evolution of this field, and we can give only the briefest of sketches of this body of work. One of the early efforts on this subject was that of van der Pol[1,2] who modeled the beating of a heart using a pair of coupled relaxation oscillators. Van der Pol's model of heartbeat was merged with the Hodgkin-Huxley[3] equation by FitzHugh[4,5] and by Nagumo[6] to yield an elegant model of neural excitability and limit cycle oscillations. This modeling approach was advanced further by Morris and Lecar[7] who replaced the linear dependence on membrane voltage in the second of the two FitzHugh-Nagumo equations with a more detailed conductance-based expression.

The term excitable membrane is often used to describe members of a diverse family of dynamic cooperative systems capable of propagating pulses and waves of activity over long distances. Spatially organized as systems of repeating regenerative units, these structures propagate activity through nearest-neighbor interactions. They have a stable rest state, are capable of damping out small perturbations, and respond strongly to above-threshold stimuli. Action potentials in nerve fibers and pulsatile responses of smooth muscle are two examples of propagation in excitable membranes. The works by Hodgkin, Huxley, and Katz[3,8] describing the electrodynamics of the squid giant axon, which culminated with the Hodgkin-Huxley equations, mark the beginning of computational neuroscience.

The extension of electrodynamic modeling of active nerve fibers to considerations of passive membrane properties that promote the propagation and spread of impulses along axons, and support the integration of postsynaptic potentials, was initiated by Hodgkin and Rushton.[9] Their effort was inspired by the mathematical theory developed in the last century for telegraphy by Kelvin and Heaviside. Subsequently cable theory and compartmental models of neurons have been developed into mathematical and computational tools by Rall, Rinzel, and Shepherd.[10-13] Recent efforts in this area have focused on dendritic spines. As discussed in the last chapter, these structures are the main locus of excitatory synaptic input and must play an important role in synaptic plasticity.

9.1.3 Population Oscillations

Another striking property of nonlinear dynamic systems is their ability to synchronize their motions when coupled together. The question of how synchrony can arise in a system of oscillatory units, each moving with its own native frequency, was first sucessfully addressed by Winfree.[14] In his approach Winfree simplified the dynamics, treating the assembly in a weak coupling limit where each oscillator remains close to its limit cycle and any amplitude variations can be neglected. The oscillatory units in the resulting phase-coupled system synchronize their motions through influences of the collective rhythms of all other units. These cooperative effects are expressed through the mean-field phase. This mean-field phase is another counterpart to the Weiss molecular field in an Ising spin system which we encountered in Chapter 3. The synchrony that can be achieved by such as system depends on the width of the frequency distribution relative to the strength of the coupling. If that spread is excessive, or equivalently, if the coupling is too weak, synchrony cannot emerge. In these model systems there is a phase transition from a incoherent to a coherent phase at some critical coupling strength (frequency width). The frequency distribution playing the role of the temperature, and the alignment of oscillator phases serves as a counterpart to the spin alignment of an Ising system. A rigorous mathematical treatment of this problem was subsequently provided by Kuramoto[15,16] and by Kuramoto and Nishikawa.[17] In their mathematical treatments of large populations of rotators under mean-field couplings, we encounter spontaneous (temporal) symmetry breaking and a complex order parameter.

There are both strong similarities and differences between these dynamical systems far from equilibrium and the equilibrium properties of Ising models. One of the new elements is the emergence of local clustering in place of global clustering; another is the appearance of traveling waves in chains of oscillators. Motivated by a desire to understand the rhythmic activity produced by central pattern generators for locomotion, Cohen, Holmes, and Rand[18] and Kopell and Ermentrout[19] investigated the properties of chains of oscillators coupled to one another through nearest-neighbor interactions involving their phases. An important observation[19] is that properties of the coupling interaction such as symmetry have a role in determining the nature of the steady states generated by the system. This observation is in contrast with the notion of universality, in which the detailed character of the hamiltonian is viewed as unimportant, that was invoked in Chapter 3 in connection with critical phenomena and the renormalization group.

Oscillatory patterns can be generated in several ways. Some neurons are endowed with intrinsic electrophysiological properties that promote rhythmic firing, and these cells serve as pacemakers. Others do not but still have intrinsic properties that support rhythmicity when driven by pacemakers. More interesting, cells that are not intrinsically oscillatory may become so as a consequence of networks properties. A key development in the field was the finding by Wilson and Cowan[20-21] that limit cycle phenomena can arise in populations of excitatory and inhibitory neurons reciprocally coupled to one another. A variety of cooperative phenomena are exhibited by these model networks, including oscillations, bifurcations, and hysteresis.[20-22]

9.1.4 Neural Rhythms and Synchrony

The central nervous system exhibits many prominent forms of synchronous activity. Electroencephalographic (EEG) recordings of oscillatory wave patterns in the brain are a well-established tool for identifying abnormal conditions, and various stages of sleep and arousal. Delta waves, oscillations associated with deep sleep and abnormal function, are of low frequency in the range 0.1 to 4 Hz. Theta-wave oscillations, 6 to 7 Hz, are found in components of the limbic system such as the hippocampus. Alpha waves, rhythmic neural activity in the 10-Hz range, are observed during drowsiness and relaxed activity, and spindle activity associated with the early stages of quiescent sleep is observed in the 7- to 14-Hz range. The detection of these low-frequency, large-amplitude oscillations using macroelectrodes supports the idea that we are observing synchronized oscillations among large populations of cells. Repetitive firing at high frequencies are of low amplitude and appear in EEG recordings as irregular fluctuations in the beta-band from 15 to 30-Hz, and in the gamma-band from 30 to 60 Hz. These correlated activity patterns correspond to arousal and attentive states.

Recently developed techniques using simultaneous multielectrode recording devices provide us with a far more detailed picture of cortical activity patterns at high frequency. Repetitive firing patterns in the 40-Hz range were first observed some time ago by Adrian[23] and later by Freeman[24] in studies of the olfactory bulb. Synchronized, repetitive firing activity in the visual cortex of the cat was discovered more recently by Gray et al.[25,26] and by Eckhorn et al.[27] using arrays of microelectrodes. The rhythmic activity patterns seen by these researchers may be best described as irregular oscillations, or synchronized bursting, spread over a broad frequency range spanning the gamma-band. These high-frequency oscillations are visually evoked but not stimulus locked, and they are accompanied by oscillations in the membrane potential which imply their synaptic origin.[28] Temporally correlated activity has been observed between cells located in a particular cortical column,[26,27] in separate columns in a given cortical area,[25,27,29] in columns in different cortical areas,[27,30] and across hemispheres.[31] Finally, another form of high-frequency synchrony, hippocampal sharp waves spanning a broad range of frequencies in the vicinity of 200 Hz has been observed in the hippocampus by Buzsáki et al.[32] and Ylinen et al.[33] These activity patterns are associated with states of awake immobility.

The cells exhibiting high-frequency synchronous activity are capable of rapidly organizing and reorganizing themselves on time scales as short as a few tens of milliseconds. This aspect was demonstrated in experiments by Gochin et al.,[34] where rapid changes in functional connectivity were observed in rat dorsal cochlear nucleus. These modulations of the effective connectivity took place during the presence of tonic bursts. The functional coupling of cells through temporally correlated activity is stimulus initiated and behavioral context dependent. The dynamic couplings may extend across multiple cortical regions at a number of different frequencies and operate across several time scales. The first point was shown by Ahissar et al.[35] and by Vaadia et al.[36] in experiments with awake monkeys, and the second by Bressler et al.[37] also in studies in the monkey. The discovery of waves of synchronous activity moving across the retina were discussed in Chapter 5. Traveling waves of synchronous active clusters of cells have also been observed in hippocampal tissue slices by Miles, Traub, and Wong.[38]

Multiple steady states, bifurcations, clustering, and propagating waves of synchrony are characteristic properties of a nonlinear dynamic system. The temporally correlated firing patterns occurring in the sensory, motor, and higher cortical areas may be utilized for a variety of purposes. For example, correlated firing patterns may perform a number of instructional functions during development. As already mentioned, rhythmic discharges have been found by Meister et al.[39] to propagate across the retina at regular time intervals. This simple form of temporal correlation is thought to contribute to the refinement of the topographic maps in the mammalian nervous system. Another function may be that of selective gain control. Temporally correlated signaling was observed in a study by Sillito et al.[40] of the reciprocal pathway from layer VI of the visual cortex to the LGN. In this instance, feedback projections induce a synchronized firing in the LGN relay neurons that may promote a selective elevation of the convergent thalamic output to the stimulus responsive layer IV cortical cells. It is clear from our exploration of synaptic plasticity that correlated firings of an assembly of neurons convergent on a common postsynaptic target are effective mechanisms for activating NMDA receptors and selectively enhancing the synaptic efficiency of common target cells. The presence of widespread synchronous activity in the hippocampus is suggestive, since this region is associated with forms of learning and memory.

It has been argued for some time that timing is important. It is a major element in Abeles's[41] notion of synfire chains, collections of synchronously firing neurons linked together to form the nodes of a string of sequentially active sets of cells. It is also central to Malsburg's[42,43] assembly coding hypothesis. In assembly coding membership in a particular cell constellation is signaled by temporally correlated firing rather then by an elevated mean firing rate. Individual neurons participate at various times in different assemblies distributed across the various cortical areas, representing disparate objects. According to the cell assembly hypothesis, neurons that encode attributes that belong together are integrated, or bound, together by their synchronous firing through cooperative processes mediated by the network connectivity. It has been suggested that temporal correlations serve as a mechanism for attention, recall, and feature (attribute) integration. The idea that temporally correlated firing serving as an attention "searchlight" is due to Treisman and Gelade[44] and has been elaborated upon further by Crick.[45] The use of synchronous activity as a mechanism for recall, another binding problem, was advanced by Damasio.[46] In his model records of previously activated patterns of synchronous activity are recalled through the use of convergence zones, sets of cells that project back to the neurons in the early feature-processing regions.

Receptive fields and topographic maps are not static structures but instead are dynamic entities able to encode sensory information spatially and temporally. For example, rats use their facial whiskers to make repetitive contacts with their surroundings in a manner similar to our use of fingertips to probe shape and texture. Information from the periphery of the rat's somatosensory system is fed to the ventral posterior medial thalamus (VPM). If we examine the stimulus response of single neurons in the VPM, we find[47] that the receptive fields are large and overlapping and that the spatial locations of the receptive fields tend to shift from caudal-most to rostral-most whiskers in the face over the first 35 ms of poststimulus time. We may infer from these observations that the representation of the face in the awake, behaving rat is dynamic and distributed.

Cortical, thalamic, and brainstem neurons in the somatosensory system exhibit synchronous oscillations in the 7- to 12-Hz range that are distinct from the spindling oscillations discussed earlier. These oscillations begin as traveling waves of synchronous activity in the cortex and spread to the thalamus and then to the brainstem just prior to the onset of rhythmic whisker twitching. The oscillations do not occur in all parts of the system but instead appear in structures such as the VPM that exhibit dynamic and distributed somatosensory maps. As noted by Nicolelis et al.,[47,48] the overall picture is one of the distributed coding of spatiotemporal information by assemblies of cells linked to one another by networks of feedforward and feedback connections.

9.2 OUTLINE OF THE CHAPTER

We will begin our tour in Section 9.3 by deriving the entrainment condition for a pair of coupled rotators. In systems of rotators, a phase-pulling mechanism synchronizes their motions. We will explore the mean-field scheme of Kuramoto and Nishikawa[17] for globally coupling a population of such oscillators. We will then look at a random pinning model that generates a first-order phase transition in place of the second-order phase transition of Winfree[14] and Kuramoto-Nishikawa. We then turn to nearest-neighbor couplings and examine the several forms of local synchronization in these systems. We conclude our tour of phase oscillators in Section 9.4 with a second examination of macroscopic clustering.

In the next portion of the chapter, we will look at rhythmic behavior and synchrony in populations of model neurons. As noted in our first section, neurons and neural ensembles are equipped with a rich variety of response characteristics over the allowable ranges of their physiological parameters. Phase plane methods are an important tool for analyzing these complex response properties. We will precede our main discussion of dynamic cooperativity with a short overview in Section 9.5 of phase plane methods such as linear stability and nullcline analysis, the Poincaré-Bendixson theorem, and the Hopf bifurcation theorem. Examples of their use will appear throughout the remainder of this chapter.

One of the prototypic models of rhythmic firing is that of two populations of cells, one excitatory and the other inhibitory, reciprocally coupled to one another. As mentioned before, population models of this form, introduced by Wilson and Cowan, exhibit multiple stable states, hysteresis, and limit cycle behavior in response to different stimuli. We will derive the coarse-grained equations of motion for models of the Wilson-Cowan form in Section 9.6. The discovery of synchronized bursting and oscillatory responses in the cortex has inspired a number of models of repetitive firing. We will examine the further developments of the Wilson-Cowan approach in this direction in Section 9.7. These extensions include a mean-field model of global coupling that is reduced to a system of coupled rotators, which synchronize their motions by phase pulling, and a nearest-neighbor coupling scheme that is used to construct a repetitive firing model in which the influences of delay couplings can be studied.

A theme common to all cortical models of oscillations and synchrony is that these repetitive firing patterns are an intrinsic property of cortical networks. Temporally correlated firing patterns arise naturally in networks characterized by recurrent excitation and global feedback inhibition, given the appropriate input and physiological parameters. The significance of the irregular oscillations in the gamma band is thought to reside in the synchrony rather than in the oscillations. This point is highlighted by models that produce synchronous bursting without oscillations. We will briefly discuss one approach of this type in Section 9.7, which focuses on the common network properties that support synchrony. We will then discuss hippocampal oscillations and synchrony, where we encounter similar network organizational principles. In the last part of Section 9.7, we will discuss feature integration from the viewpoint of Malsburg's assembly coding hypothesis. We will observe that this cooperative mechanism for feature binding or temporal tagging is analogous to the spatial tagging in the Markov random field approach to feature integration of Chapter 6. Finally, we will examine some insights into temporal tagging provided by models characterized by parallel networks of reciprocal connections.

In the next two sections we will study how the ionic currents self-organize in a membrane to produce a variety of stable firing states. We will begin this exploration with a brief review in Section 9.8 of the Hodgkin-Huxley equations.

We will follow this preamble with an examination of the phase plane characteristics of the Morris-Lecar model of membrane excitability and oscillations. Network considerations will be introduced by means of a model developed by Somers and Kopell[49,50] of relaxation oscillators coupled to one another through nearest-neighbor interactions. We will find that global synchronization is possible in this model, in contrast to our earlier finding for coupled rotators. We will next look at spindle waves in the thalamocortical system. We will start in Section 9.9 with the cellular properties that support spindling and then turn to the network mechanisms. Lastly, we will explore intracellular calcium oscillations. This phenomenon will provide us in Section 9.10 with an opportunity to examine how cellular automata may be used to model limit cycle behavior.

9.3 PHASE-COUPLED OSCILLATORS

9.3.1 Mutual Synchronization

Spontaneous, or mutual, synchronization is a cooperative process in which a population of interacting nonlinear oscillators synchronize themselves to a common frequency. As mentioned in the introductory remarks, mutual entrainment is a basic property of a nonlinear dynamic structure, and can be encountered in a host of physical, chemical, and biological systems. The examples given in Chapter 1 such as lasers, networks of heart pacemaker cells, and large populations of chirping crickets and flashing fireflies are just a few of the many instances of this self-organizational principle.

In this section, and throughout this chapter, we will consider the behavior of limit cycle oscillators. The question we intend to answer is what will happen in a large population of similar, limit cycle oscillators, each with its own native frequency, when these elements are weakly coupled together through interactions that communicate phase information from one member of the population to the others. Let us now define our problem in mathematical terms. Our starting point is Eq. (8.2), which we now rewrite in a way that makes explicit its form as a set of coupled first-order differential equations

$$\frac{d\mathbf{X}_i}{dt} = \mathbf{F}_i(\mathbf{X}_i) + \sum_{j=1}^{N} \mathbf{G}_{ij}(\mathbf{X}_i, \mathbf{X}_j) \qquad (9.1)$$

As before, the $\mathbf{X}_i$ are phase space coordinates. These equations can describe an assembly of N coupled limit cycle oscillators. In this situation, the coupling terms G_{ij} are assumed to be small, and the dependence of $\mathbf{F}$ on the oscillator index i is weak, since we are assuming similar oscillators. Under these conditions it can be demonstrated[16,22] that the only variable of importance is the phase, and Eq. (9.1) can be simplified to

$$\frac{d\theta_i}{dt} = \omega_i + \sum_{j=1}^{N} \Gamma_{ij}(\theta_i - \theta_j) \qquad (9.2)$$

where the ω_i are the oscillator native frequencies and the θ_i are the oscillator phases. A particularly simple form for the coupling interactions is obtained if we assume that each oscillator interacts with the remaining $N - 1$ oscillators in the same manner. One way of accomplishing this is to define the couplings to be of the rotator form

$$\Gamma_{ij}(\theta) = \frac{K}{N} \sin \theta \qquad (9.3)$$

where K is a positive coupling strength and N is a normalization factor. We will shortly examine what happens in a rotator model with mean-field couplings, due to Kuramoto and Nishikawa.[17] We will then reexamine this problem in a nearest-neighbor lattice model. Before discussing the mean-field and lattice approaches, it is instructive to examine entrainment in a system consisting of a single pair of phase-coupled oscillators.

9.3.2 Entrainment in a Rotator Model

For the case of a system consisting of two similar limit cycle oscillators our equations of motion are

$$\begin{aligned} \frac{d\theta_1}{dt} &= \omega_1 + \left(\frac{K}{2}\right) \sin(\theta_2 - \theta_1) \\ \frac{d\theta_2}{dt} &= \omega_2 + \left(\frac{K}{2}\right) \sin(\theta_1 - \theta_2) \end{aligned} \qquad (9.4)$$

To see how entrainment occurs, suppose that the phase difference $\theta_1 - \theta_2$ is positive; that is, the phase of first oscillator is ahead of that of the second. Then the coupling term in the first equation will be negative so that $d\theta_1/dt < \omega_1$ and the first oscillator will slow down. Similarly, if the phase of the first oscillator is behind that of the second, it will speed up. The same considerations hold for the second oscillator. This mechanism for achieving synchrony, called *phase pulling*, has the property that the force exerted by one oscillator on the other is the same at all points along the limit cycle trajectory, and the amount of attraction depends on the coupling strength and the phase difference.

Let us introduce a phase difference variable $\phi = \theta_1 - \theta_2$. We then obtain

$$\frac{d\phi}{dt} = \omega_1 - \omega_2 - K \sin \phi \qquad (9.5)$$

As illustrated in Fig. 9.1, phase locking can be achieved if and only if

$$|\omega_1 - \omega_2| \leqslant K \tag{9.6}$$

This is the basic requirement for phase-locking, and it states that the difference in native frequencies of the pair of oscillators must not be too large relative to the coupling strength. Thus limit cycle oscillators must be similar in order to phase lock. If the requirement, Eq. (9.6), is satisfied, then we have entrainment in which oscillators possessing different natural frequencies start to oscillate at the same frequency as a result of their weak coupling.

Let us rescale Eq. (9.5) by introducing the variables $\tau = Kt$ and $\kappa = \phi/K$. Then our dynamic equation becomes $d\phi/dt = \kappa - \sin\phi$. As shown in Fig. 9.1, if $\kappa = 0$, then there is a stable fixed point at the origin, and both oscillators become synchronized. If the native frequencies of the two oscillators differ from one another, but the condition for entrainment is satisfied, namely $0 < \kappa < 1$, then the stable fixed point moves to some positive value, and the two units eventually become phase-locked, oscillating with a constant phase difference. Finally, if the entrainment condition, Eq. (9.6), is not satisfied, there are no fixed points, and phase-locking does not occur.

9.3.3 Mean-Field Model

We next discuss the generalization of the two-oscillator system to a population of limit cycle oscillators in which each oscillator is coupled to every other oscillator. Winfree first noted that populations of such oscillators will undergo a phase transition to a fully synchronized phase at a critical coupling strength. We will examine the behavior of this globally coupled system using a rotator model with mean-field coupling.[17] In the mean-field oscillator model, each unit interacts with every other oscillator in the assembly with the same coupling

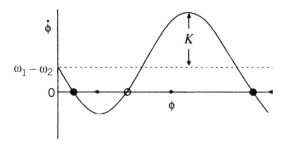

Figure 9.1. Mutual entrainment of two coupled rotators. There are two solutions, a stable fixed point (filled circle) and an unstable fixed point (open circle). The stable fixed point is approached asymptotically for all initial conditions. (From Strogatz and Mirollo[51]. Reprinted with permission of IOP Publishing Limited.)

strength K. The equation of motion is of the form

$$\frac{d\theta_i}{dt} = \omega_i + \frac{K}{N} \sum_{j=1}^{N} \sin(\theta_j - \theta_i) \quad (9.7)$$

We assume that the native frequencies ω_i are symmetrically distributed about some mean frequency, ω_0. This frequency distribution may be written

$$g(\omega) = \frac{1}{N} \sum_{j=1}^{N} \delta(\omega - \omega_j) \quad (9.8)$$

The frequency distribution may be simplified by introducing new phase variables ψ_i as

$$\psi_i = \theta_i - \omega_0 t \quad (9.9)$$

In this rotating coordinate system, the frequency distribution may be taken to have zero mean and unit variance. Upon replacing ω_i with $\omega_i - \omega_0$, our equations of motion become

$$\frac{d\psi_i}{dt} = \omega_i - \frac{K}{N} \sum_{j=1}^{N} \sin(\psi_i - \psi_j) \quad (9.10)$$

As we have seen in the last subsection, if the strength of the coupling is sufficiently large relative to the phase differences, the cooperativity embodied in the couplings will serve to align the phases of the oscillators.

We now introduce an order parameter that serves as a mean-field for the oscillator community. It is given by the expression

$$r \exp(i\Theta) = \frac{1}{N} \sum_{j=1}^{N} \exp(i\psi_j) \quad (9.11)$$

This order parameter Θ represents the average phase of the rotator system. The modulus r provides a measure the amount of synchrony. A value $r = 0$ corresponds to an absence of phase-locking, while $r = 1$ denotes complete phase-locking. If we multiply both sides of this equation by $\exp(-i\psi_i)$ and decompose this relationship into its real and imaginary parts, we see that

$$r \sin(\Theta - \psi_i) = \frac{1}{N} \sum_{j=1}^{N} \sin(\psi_j - \psi_i) \quad (9.12)$$

Inserting this equality back into our earlier mean-field expression yields the following form of the mean-field dynamics for a phase-coupled population of oscillators

$$\frac{d\psi_i}{dt} = \omega_i - Kr\sin(\psi_i - \Theta) \qquad (9.13)$$

This system undergoes a phase transition at a critical value of the ratio of coupling strength to the width of the frequency distribution. This synchronization phenomenon is strikingly similar to the ferromagnetic phase transitions studied in Chaper 3. In place of spontaneous spatial symmetry breaking in the Ising system, we have spontaneous temporal symmetry breaking. The population of oscillators breaks into two subpopulations. One of the subpopulations is composed of oscillators whose phases satisfy the entrainment condition and undergo motions that are synchronized. The second subpopulation contains those oscillators that fail to satisfy the requirement for phase-locking, and therefore remain unsynchronized.

9.3.4 The Random Pinning Model

The smooth onset of ordering in the mean-field model is similar to that of a second-order phase transition in a spin system. This situation is altered of we append a random pinning term to our equations of motion. The random pinning field attempts to pin each oscillator phase θ_i at a random setting α_i. This pinning force is opposed by an attractive interaction between phases of the usual form so that we have the dynamical equation

$$\frac{d\theta_i}{dt} = \sin(\alpha_i - \theta_i) + \frac{K}{N}\sum_{j=1}^{N}\sin(\theta_j - \theta_i) \qquad (9.14)$$

This model resembles an ensemble of coupled spins in a local random magnetic field, and was studied by Strogatz et al.[52] and by Mirollo and Strogatz.[53] The hamiltonian for this system is

$$E = -\sum_{i=1}^{N}\cos(\alpha_i - \theta_i) - \frac{K}{2N}\sum_{i=1}^{N}\sum_{j=1}^{N}\cos(\theta_i - \theta_j) \qquad (9.15)$$

By treating this as a system at zero temperature, we may use gradient descent to find a local minimum of the hamiltonian. Specifically, we have

$$\frac{d\theta_i}{dt} = -\frac{\partial E}{\partial \theta_i} \qquad (9.16)$$

At the energy minimum we have a stable steady state. The character of this state is determined by the balance between random pinning encouraged by the $\cos(\alpha_i - \theta_i)$ term and phase alignment promoted by the $\cos(\theta_i - \theta_j)$ term. The relative importance of these two processes is controlled by the coupling strength K. For $K = 0$, we have random pinning, and for $K \to \infty$, we have phase coherence. Unlike the phase-coupled system discussed in the last subsection, the present system undergoes a discontinuous, first-order phase transition at the critical coupling stength K_c.

In order to facilitate our study of the phase transition behavior, we make an infinite-range coupling approximation. This simplification allows us to use a continuum approximation for the discrete dynamical system, and in doing so we replace Eq. (9.14) with the expression

$$\frac{d\theta_\alpha}{dt} = \sin(\alpha - \theta_\alpha) + \frac{K}{2\pi} \int_0^{2\pi} \sin(\theta_\beta - \theta_\alpha) d\beta \tag{9.17}$$

The states of the system are continuous periodic function that relate (map) α to θ_α. In the continuum approximation the order parameter, Eq. (9.11), becomes

$$r \exp(i\Theta) = \frac{1}{2\pi} \int_0^{2\pi} \exp(i\theta_\alpha) d\alpha \tag{9.18}$$

and our mean-field equation is

$$\frac{d\theta_\alpha}{dt} = \sin(\alpha - \theta_\alpha) + Kr \sin(\Theta - \theta_\alpha) \tag{9.19}$$

The critical states are the mappings of α to θ_α for which $d\theta_\alpha/dt = 0$. We may simplify our system by transforming to a coordinate frame for which Θ is zero. Then the critical states must satisfy the condition

$$\sin(\alpha - \theta_\alpha) - u \sin \theta_\alpha = 0 \tag{9.20}$$

where we have introduced a parameter $u = Kr$, and $\alpha \in (0, 2\pi)$. For $u = 0$, the solutions to Eq. (9.20) are $\theta_\alpha = \alpha$ and $\theta_\alpha = \alpha + \pi$. For $u > 0$, we can proceed by rewriting $\sin(\alpha - \theta_\alpha)$ as a complex exponential. We then find that

$$\exp(i\theta_\alpha) = \frac{u + \exp(i\alpha)}{|u + \exp(i\alpha)|} \tag{9.21}$$

and therefore

$$\frac{u}{K} = \frac{1}{2\pi} \int_0^{2\pi} \frac{u + \exp(i\alpha)}{|u + \exp(i\alpha)|} d\alpha \equiv f(u) \tag{9.22}$$

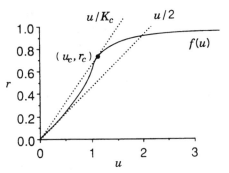

Figure 9.2. Graph of f(u) versus u as defined by Eq. (9.17). Dashed lines correspond to jump bifurcations at $K = K_c$ and $K = K_T = 2$. (From Mirollo and Strogatz[53]. Reprinted with permission of the Society for Industrial and Applied Mathematics.)

We now use the auxiliary function $f(u)$ to find the self-consistent values of the order parameter r as the coupling strength K is varied. The graphical solution obtained by locating the intersections of the curves $r = f(u)$ and u/K is shown in Fig. 9.2. We observe that the two curves intersect tangentially at $K_c = 1.486$ and $K_T = 2$. These points mark the locations of jump bifurcations. The changes in the order parameter as the coupling strengths are increased corresponding to the stable critical states are displayed in Fig. 9.3. We see that if K is small, the system will evolve from an initial state to a stable critical state having $r = 0$. The system will remain in this incoherent condition until $K = K_T$, when it will jump to a coherent stable state ($r \sim 1$) where it will remain until K is lowered to K_c. Thus the system exhibits jump bifurcations and hysteresis as the coupling strength is systematically varied. These properties are characteristic of a first-order phase transition.

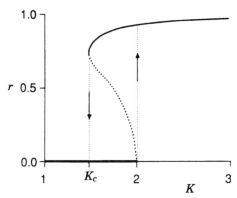

Figure 9.3. The coherence order parameter r versus the coupling strength K for the critical states of the system. Heavy lines denote locally stable states, and thin broken lines represent unstable states. The system exhibits jump bifurcations and hysteresis. (From Mirollo and Strogatz[53]. Reprinted with permission of the Society for Industrial and Applied Mathematics.)

9.3.5 Lattice Model

In Chapter 3 we studied the Ising model. We observed that the Weiss molecular field approximation utterly failed to describe the critical properties of the Ising system in one dimension. The mean-field approximation did better in two dimensions and was a rather good approximation to the exact Ising lattice in three dimensions. In other words, as the dimensionality of the lattice system increased, the mean-field behavior of an infinite-range system approached that of a nearest-neighbor system. This equivalence does not hold for constellations of coupled rotators. We have just treated constellations of globally coupled rotators; we will now turn to the lattice, or nearest-neighbor coupled system. The local form of the phase-coupling interaction was studied by Sakaguchi, Shinomoto, and Kuramoto[54] and by Strogatz and Mirollo,[51,55] and we will follow their treatments.

Our starting point is the lattice form of Eq. (9.1), namely

$$\frac{d\theta_i}{dt} = \omega_i - K \sum_{j \in N_i} \sin(\theta_i - \theta_j) \tag{9.23}$$

where the coupling sum is over the nearest neighbors of oscillator i. We first generalize the two-oscillator case discussed in Section 9.3.1 to a linear chain of N oscillators coupled through nearest-neighbor interactions. We impose free boundary conditions so that there is only one coupling term for the unit at each of the end points of the chain. The result is the system of equations

$$\frac{d\theta_1}{dt} = \omega_1 - K \sin(\theta_1 - \theta_2)$$

$$\frac{d\theta_i}{dt} = \omega_i - K \sin(\theta_i - \theta_{i-1}) - K \sin(\theta_i - \theta_{i+1}) \tag{9.24}$$

$$\frac{d\theta_N}{dt} = \omega_N - K \sin(\theta_N - \theta_{N-1})$$

In the model the native frequencies ω_i are quenched random variables. To establish the analogue entrainment condition, Eq. (9.6), for a chain of N-coupled oscillators, let us assume that we have phase-locking. This means that $d\theta_i/dt = d\theta_j/dt$ for all i and j. If we add together our N equations under this assumption, we get

$$\frac{d\theta_i}{dt} = \bar{\omega} = \frac{1}{N} \sum_{k=1}^{N} \omega_k \tag{9.25}$$

If we then replace each of the time derivatives of the phase variables with the average native frequency and add together the first j equations, we find that

$$j\bar{\omega} = \sum_{k=1}^{j} \omega_k + K \sin(\theta_{j+1} - \theta_j) \tag{9.26}$$

Equation (9.26) can be rewritten as

$$K \sin \phi_j = X_j \tag{9.27}$$

where

$$X_j = \sum_{k=1}^{j} (\omega_k - \bar{\omega}) \tag{9.28}$$

and $\phi = \theta_j - \theta_{j+1}$. The desired entrainment condition is

$$\max_{1 \leq j \leq N} |X_j| \leq K \tag{9.29}$$

This expression is similar to our previous expressions for entrainment. However, the required condition, Eq. (9.29), is given in terms of the quantity X_j defined by Eq. (9.28) as a cumulative sum of differences of frequencies. If we examine the probability for phase-locking in linear chains where N is large, we observe that the coupling strength required for phase-locking scales as $O(N^{1/2})$. For any fixed value of the coupling strength, the probability for phase-locking in a linear chain approaches zero as the number of oscillators increases. Thus large-scale phase-locking in linear chains of rotators coupled through nearest-neighbor interactions is impossible.

We may summarize the one-dimensional case by the statement that for fixed couplings a chain of oscillators cannot form synchronized clusters of a macroscopic size. In higher dimensions large clusters of synchronized oscillators may form, but these constellations necessarily have a spongelike spatial structure.[50] That is, the domains of synchronization will be irregular in shape and contain holes. An example of domain formation in a two-dimensional oscillator lattice is presented in Fig. 9.4. This result generalizes to higher dimensions.[51] Thus the phase-locked behavior of a nearest-neighbor system differs from that of a mean-field constellation regardless of the dimension. Unlike the Ising lattice examined in Chapter 3, a globally coupled rotator system interacting through its mean-field must be regarded as an entity distinct from a nearest-neighbor lattice.

9.3.6 Frequency Plateaus

Further insight into nearest-neighbor systems can be obtained by examining chains of nearest-neighbor oscillators possessing a uniform gradient in native

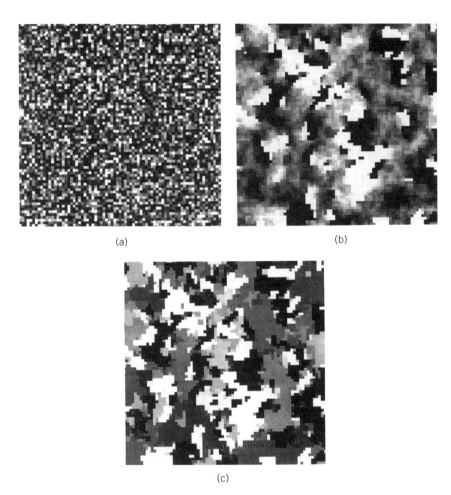

Figure 9.4. Formation of clusters of synchronized rotators in a two-dimensional lattice. Grey levels denote oscillator frequency in the range 0 (black) to 1 (white). (a) Initial distribution of native frequencies. (b) Coupling-modified frequencies at time $t = 10^3$, and (c) at $t = 10^4$. (From Strogatz and Mirollo[51]. Reprinted with permission of IOP Publishing Limited.)

frequencies from one element to another.[54-56] If the coupling term is an odd function of the phase variable as in $\sin \theta$ coupling, then a phase-locked solution exists provided that the differences in natural frequencies between pairs of oscillators is small. This phase-locked solution corresponds to the case where all oscillators move with the same frequency and each pair of oscillators has a fixed (in time) phase difference. For larger frequency gradients just beyond the point where phase-locking is lost, we encounter frequency plateaus. In a frequency plateau the oscillators are synchronized to a common frequency, but the phase relation between each oscillator pair is not necessarily constant in time. If we sufficiently increase the frequency gradient, the system develops a large amplitude stable limit cycle corresponding to a fragmentation of the chain into two frequency plateaus. Additional frequency plateaus form at still larger gradients. In a given regime in moderate-sized chains, as the coupling strength is raised, the subchains of locally synchronized clusters will increase in length. However, unlimited increases in coupling strength lead to oscillator death. Several different mechanisms can produce this breakdown in phase-locking in chains of rotators coupled to one another through nearest-neighbor interactions.[57]

The general case where no particular symmetry is assumed for the coupling was studied by Kopell and Ermentrout.[22] They found that the symmetry properties will strongly influence the critical frequency for phase-locking and will affect the scaling behavior as the frequency gradient is increased. For odd symmetry phase-locking is lost sooner than if a non-odd component is present in the coupling interaction. One of the goals of the studies[20,22] of frequency plateaus was to model central pattern generators for locomotion in fish. An important consideration in the analysis of the possible mathematical models is whether the couplings are diffusive or synaptic. In the context of phase oscillators, these terms have the following meaning: In diffusive coupling, if a pair of oscillators are in the same state, that is, have the same phase, their coupling interaction is zero. In the case of synaptic couplings, the interactions do not vanish when both oscillators are in the same state. The assumption of diffusive couplings requires the presence of a pacemaker to produce a local inhomogeneity that initiates the wave, whereas the assumption of synaptic couplings leads to a model that produces constant speed traveling waves without pacemakers.

9.4 CLUSTERING OF GLOBALLY COUPLED PHASE OSCILLATORS

We now return to the case of global coupling. So far we have considered a rather simple type of coupling interaction. A more general class of globally coupled oscillators is represented by phase equations of the form

$$\frac{d\theta_i}{dt} = \omega + f(\theta_i) - \frac{1}{N} \sum_{j=1}^{N} g(\theta_j) \qquad (9.30)$$

for $i = 1, \ldots, N$, and where f and g are periodic functions. In the absence of interactions, the motion of each oscillator obeys the equation of motion $d\theta_i/dt = \omega + f(\theta_i)$, and its trajectory in phase space converges to either a fixed point or to a limit cycle. In Eq. (9.30) we treat the interactions in a mean-field approximation so that each oscillator interacts equally with every other oscillator in the population, with $K_{ij} = 1/N$ for all i and j, with $i \neq j$. The general properties of this system was investigated by Golomb et al.[58] The functions f and g were modeled in their study by Fourier expansions, in which the first three terms in f and the leading term in g were retained. The trajectories for the coupled spatiotemporal system exhibited several types of behavior as the parameters of the Fourier expansions were varied. These included fixed points and several forms of limit cycle phenomena, the most notable of which was the formation of macroscopic spatial clusters of synchronized oscillators. We now briefly sketch the model and describe their results.

We start by defining the single-phase distribution function

$$P_N(\theta, t) = \frac{1}{N} \sum_{i=1}^{N} \delta(\theta - \theta_i) \tag{9.31}$$

which will be used by us to help characterize the asymptotic behavior of the system when we consider the influences of stochastic noise. We will consider those situations where N is large. The quantity $P(\theta, t)$ defined in the limit where N becomes large as

$$P(\theta, t) = \lim_{N \to \infty} P_N(\theta, t) \tag{9.32}$$

obeys the continuity equation

$$\frac{\partial}{\partial t} P(\theta, t) + \frac{\partial}{\partial \theta} [\{\tilde{\omega} + f(\theta)\} P(\theta, t)] = 0 \tag{9.33}$$

where $\tilde{\omega}(t)$ is defined through the relation

$$\tilde{\omega} = \omega - \int_0^{2\pi} P(\theta', t) g(\theta') d\theta' \tag{9.34}$$

The distribution function $P(\theta, t)$ must be nonnegative at all times, must satisfy the periodicity requirement

$$P(2\pi, t) = P(0, t) \tag{9.35}$$

and obey the normalization condition

$$\int_0^{2\pi} P(\theta, t) d\theta = 1 \tag{9.36}$$

This system of globally coupled phase oscillators can exhibit fixed point, limit cycle, nonperiodic, and quasi-periodic behavior at long times. One of the classes of asymptotic states of the system are the homogeneous fixed points. In these situations all oscillators have a constant phase θ_0 that satisfies the relation $\omega + f(\theta_0) - g(\theta_0) = 0$. These asymptotic states are the simplest attracting fixed points of the system. Among the various types of limit cycle phenomena are homogeneous limit cycles, clusters, and steady states. In homogeneous limit cycles, all oscillators evolve in time with a common phase θ that obeys the equation $d\theta/dt = \omega + f(\theta) - g(\theta)$. The most interesting cases are those for which the initially spatially uniform system fragments into several macroscopically large clusters, each of which is composed of fully synchronized oscillators. The distribution function for a cluster state consists of a sum of N_{cl} delta functions, defined in terms of their mutual phases Θ_k that satisfy the equations of motion

$$\frac{d\Theta_k}{dt} = \omega + f(\Theta_k) - \sum_{j=1}^{N_{cl}} \varepsilon_j g(\Theta_k) \tag{9.37}$$

where ε_j is the normalized fraction of oscillators in the jth cluster. Another class of limit cycle states are the stationary distributions in which the phases of the oscillators are uniformly distributed over the interval 0 to 2π. Macroscopic clusters are absent, and each oscillator obeys the equation of motion $d\theta/dt = \tilde{\omega} + f(\theta)$. In this steady state condition all oscillators have the same frequency $\tilde{\omega}$, and the phase difference between any pair of oscillators evolves in time. The distribution function in a steady state is

$$P(\theta) = \frac{v}{\tilde{\omega} + f(\theta)} \tag{9.38}$$

and

$$\tilde{\omega} = \omega - \int_0^{2\pi} \frac{vg(\theta)}{\omega + f(\theta)} d\theta \tag{9.39}$$

We now incorporate stochastic noise into our dynamic system. As was done in Chapter 7, we add a noise term $\eta_i(t)$ to our equations of motion

$$\frac{d\theta_i}{dt} = \omega + f(\theta_i) - \frac{1}{N}\sum_{j=1}^{N} g(\theta_j) + \eta_i(t) \tag{9.40}$$

which is of a Gaussian form with zero mean

$$\langle \eta_i(t) \rangle = 0 \tag{9.41}$$

and autocorrelation

$$\langle \eta_i(t)\eta_j(t') \rangle = 2T\delta_{ij}\delta(t-t') \tag{9.42}$$

We then introduce the noise-averaged distribution function

$$P(\theta, t) = \left\langle \lim_{N \to \infty} \frac{1}{N} \sum_{i=1}^{N} \delta(\theta - \theta_i) \right\rangle \qquad (9.43)$$

This probabilistic quantity obeys the Fokker-Planck equation

$$\frac{\partial}{\partial t} P(\theta, t) + \frac{\partial}{\partial \theta}[\{\tilde{\omega} + f(\theta)\} P(\theta, t)] = T \frac{\partial^2 P(\theta, t)}{\partial \theta^2} \qquad (9.44)$$

The stationary distribution $P_s(\theta)$ is a solution to the equation

$$(\tilde{\omega} + f(\theta)) P_s(\theta) - T \frac{\partial P_s(\theta)}{\partial \theta} = v \qquad (9.45)$$

subject to the positivity and periodicity conditions. We can determine the frequency $\tilde{\omega}$ and the quantity v using Eq. (9.34) and the normalization requirement, Eq. (9.36).

We may define an order parameter in a manner analogous to rotator model expression given in Eq. (9.11). For our more general case we take as our order parameter

$$Z(t) = \frac{1}{N} \sum_{j=1}^{N} e^{i\theta_j} = \int_0^{2\pi} P(\theta, t) e^{i\theta} d\theta \qquad (9.46)$$

The magnituide of the time dependence of the order parameter $Z(t)$ is given by another order parameter Z_{rms}, representing the root-mean-square average distance between a point on the limit cycle and its center

$$Z_{rms} = \left[\frac{1}{t_p} \int_0^{t_p} |Z(t) - \bar{Z}|^2 dt \right]^{1/2} \qquad (9.47)$$

where the mean value of Z is

$$\bar{Z} = \frac{1}{t_p} \int_0^{t_p} Z(t) dt \qquad (9.48)$$

The response properties of this system of phase oscillators may be summarized as follows: At low noise levels the behavior of the system at $T = 0$ determines its behavior for $T > 0$. We discuss two cases. First, if the Fourier parameters are chosen so that the stationary distribution is stable and attracting at $T = 0$, then it will remain so for $T > 0$. Second, if the parameters of the model are chosen so that the system is in a cluster-state regime at $T = 0$, the stationary distribution will be unstable at $T = 0$. It will remain unstable up to a critical noise level T_c. The system will undergo a second-order phase transition at this critical value. Above T_c the stationary state will be the global attractor. These response properties are illustrated in Fig. 9.5 for the case where there are two macroscopic cluster states. In the first panel we observe that distribution, which has two delta-function peaks at $T = 0$, is smeared out by the addition of noise ($T > 0$). As the noise level is increased to values of T exceeding T_c, the distribution function becomes smeared out over all phases. A general observation is that in all parameter regimes the addition of stochastic noise stabilizes the steady state attractor. The asymptotic behavior of the order parameter Z as T is increased is shown in the second section, and the evolution of the order parameter Z_{rms} near T_c is plotted in the third part of Fig. 9.5. In examining the behavior of Z_{rms} near T_c, we find that

$$Z_{rms} \propto (T - T_c)^{1/2} \tag{9.49}$$

That is, the order parameter vanishes as T approaches T_c from below with critical exponent $1/2$. We may conclude by noting[53] that the convergence of the system to (partially homogeneous) cluster states represents the spontaneous breaking of the spatial symmetry of the globally coupled phase oscillator system. The grouping of oscillators into clusters is random, and this degeneracy can be broken by external imputs possessing a spatial structure that overlaps that of a specific cluster. Thus the oscillatory system may single out and enhance spatial structure in input patterns.

Cluster states have been observed in other nonlinear systems characterized by global couplings among their elements and appear to be a general feature of such systems. For example, clusters of synchronized elements has been found by Kaneko[59,60] in studies of globally coupled networks of chaotic elements. In general, these sytems will evolve through a succession of phases as their nonlinearity in increased. These system will start in a coherent phase having a single synchronized attractor, pass through an ordered phase characterized by formation of a small number of clusters, then a partially ordered phase having coexisting attractors some with large clusters and others with small ones, and finally a turbulent phase. A clustering is characterized by the number of clusters k and a partitioning $\{k, N_1, \ldots, N_k\}$ specifying the number of elements in each cluster. In the partially ordered phase the partitioning complexity possesses a treelike structure reminiscent of the ultrametricity of the Sherrington-Kirkpatrick spin glasses discussed in Chapter 4.

9.5 PHASE SPACE

9.5.1 Linear Stability Analysis

In a linear stability analysis we study the behavior of a system as its undergoes small excursions in the vicinity of a fixed point. Let us consider a system

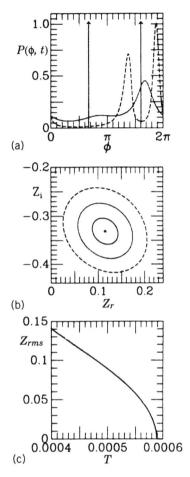

Figure 9.5. Behavior of the globally coupled oscillator system in the presence of noise. The critical temperature is $T_c = 5.886 \times 10^{-4}$. (a) At $T = 0$ the system is in a two-cluster state. These are denoted by the delta functions. The dashed curve represents the distribution function at a nonzero temperature $T < T_c$, and the solid curve denotes the distribution function at a temperature $T > T_c$. (b) Asymptotic trajectories of the order parameter at several values of the temperature. As the critical temperature is approached from below, the size of the orbits decrease. The solid circle denotes the trajectory at a temperature just above the critical value. (c) The order parameter Z_{rms} versus temperature. (From Golomb et al.[58]. Reprinted with permission of the American Physical Society.)

described by the equations of motion

$$\frac{dx}{dt} = F_1(x, y)$$
$$\frac{dy}{dt} = F_2(x, y) \tag{9.50}$$

in the vicinity of a fixed point (x_0, y_0), where $F_1(x_0, y_0) = 0 = F_2(x_0, y_0)$. We assume that in the vicinity of this fixed point we may approximate the nonlinear system by a linear one. To obtain the linearized form of the dynamics equations, we expand the right-hand side of the equations of motion in a Taylor's series about the fixed point, retaining terms up to first order in $(x - x_0)$ and $(y - y_0)$. The result for the first of the two equations is

$$\dot{x} F_1(x_0, y_0) + \frac{\partial F_1(x_0, y_0)}{\partial x}(x - x_0) + \frac{\partial F_1(x_0, y_0)}{\partial y}(y - y_0)$$

or

$$\dot{x} = \frac{\partial F_1(x_0, y_0)}{\partial x} x + \frac{\partial F_1(x_0, y_0)}{\partial y} y \tag{9.51}$$

with the understood meaning that we carry out the indicated partial differentiations and then evaluate the resulting expressions at the fixed point, and in writing the second equality, we have dropped constant terms, since they do not influence the dynamics. Similar results holds for the other equation. Thus the linearized version of the equations of motion can be written in matrix notation as

$$\begin{pmatrix} \dot{x} \\ \dot{y} \end{pmatrix} = \begin{pmatrix} \frac{\partial F_1(x_0, y_0)}{\partial x} & \frac{\partial F_1(x_0, y_0)}{\partial y} \\ \frac{\partial F_2(x_0, y_0)}{\partial x} & \frac{\partial F_2(x_0, y_0)}{\partial y} \end{pmatrix} \begin{pmatrix} x \\ y \end{pmatrix} = J \begin{pmatrix} x \\ y \end{pmatrix} \tag{9.52}$$

where J is the Jacobian matrix of the linearized system.

An evaluation of the eigenvalues of the Jacobian, made by solving the characteristic equation,

$$\det(J - \lambda I) = 0 \tag{9.53}$$

provides information about the stability properties of the steady states, away from complicated cases where nonlinear terms are important. In brief, we find that the steady state is stable (an attractor) if both eigenvalues have negative real parts; it is unstable (a repellor) if both eigenvalues have positive real parts, and it is a saddle if one eigenvalue has a positive real part and the other has a negative one. If the eigenvalues are real, the steady states are nodes; otherwise, they are inward or outward winding spirals, except for the case where the real part vanishes. In that case we have a neutral center (concentric circles). If we write the characteristic equation as

$$\lambda^2 - b\lambda + c = 0 \tag{9.54}$$

TABLE 9.1 Classification of steady states

Steady state	b	c	$b^2 - 4c$
Unstable node	+	+	+
Saddle point		−	+
Stable node	−	+	+
Unstable spiral	+	+	−
Neural center	0	+	−
Stable spiral	−	+	−

then we may classify the types of steady state in terms of the signs of b, c, and $b^2 - 4c$ which determines whether the eigenvalues are real or complex. A resulting, slightly overspecified, enumeration is given in Table 9.1.

9.5.2 Nullclines

The locus of points for which $F_1(x, y) = 0$ is termed the x *nullcline*, and the locus of points for which $F_2(x, y) = 0$ is called the y *nullcline*. An examination of the crossings of the nullclines, the shape of the nullclines, and and flow in their vicinity provides useful information on the underlying dynamics. With respect to the crossings we note that the intersections of the two curves in phase space satisfy the conditions $dx/dt = 0$ and $dy/dt = 0$ and therefore give the steady states of the system. A first step in identifying the flow of trajectories in phase space is to determine the flow near the nullclines where the analysis is simple. As an example of the aforementioned ideas, including information provided by the shape of the nullclines, let us consider the van der Pol[1,2] oscillator described by the equation of motion

$$\ddot{x} - \mu(1 - x^2)\dot{x} + x = 0 \tag{9.55}$$

This equation may be cast into a set of first-order differential equations as in Eqs. (9.1); specifically

$$\dot{x} = \mu[y - F(x)]$$
$$\dot{y} = -\left(\frac{1}{\mu}\right)x \tag{9.56}$$

with

$$F(x) = \left(\frac{1}{3}\right)x^3 - x \tag{9.57}$$

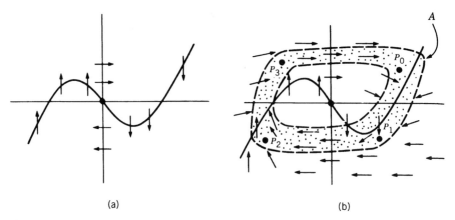

Figure 9.6. Flow along a cubic nullcline: (*a*) Direction of the flow along the *y* and *x* nullclines; (*b*) trapping of trajectories flowing into an annular region A. Starting at point P_0 a trajectory will flow toward point P_1, and then toward points P_2 and P_3, and back to P_0. (From Edelstein-Keshet[61]. Reprinted with permission of McGraw-Hill.)

We observe the following. The *x* nullcline, given by the relation $y = F(x)$, has the cubic form illustrated in Fig. 9.6*a*. The *y* nullcline, given by the expression $x = 0$, coincides with the *y*-axis. This system has one fixed point, located at the origin $x = 0$, $y = 0$, where the two nullclines cross one another.

Let us examine the flow along these nullclines. First, we note that the flow is vertical along the *x* nullcline where dx/dt is zero and is horizontal along the *y* nullcline where dy/dt is zero. Next, from the second of these equations we see that when $x > 0$, *y* decreases, and along the *y* nullcline *x* and $F(x)$ are both zero. As a result, when moving near the *y* nullcline, *x* will increase when $y > 0$ and *x* will decrease when $y < 0$. The flow in the vicinity of the *x* and *y* nullclines is depicted in Fig. 9.6*a*. The flow of a trajectory into the annular region about the nullclines, as illustrated in Fig. 9.6*b*, will become trapped in that part of phase space. Application of the Poincaré-Bendixson theorem (Section 9.5.3) to this situation tells us that there is a limit cycle in this region whose trajectory is depicted in Fig. 9.7*a*.

Unlike the rotators discussed previously, motion along the limit cycle trajectory involves two time scales, a fast horizontal movement and slow vertical motion. In this picture we assume that the parameter $1/\mu$ is small. When *y* is near $F(x)$, namely near the *x* nullcline, both dx/dt and dy/dt vary gradually, and the movement is slow. When the trajectory departs from the cubic nullcline dy/dt is large, and the horizontal movement is fast. This situation is shown in Fig. 9.7*a*, and the time scales and wave form for the van der Pol oscillator are presented in Fig. 9.7*b*. Systems of equations such as those for the van der Pol oscillator describe a form of repetitive behavior known as *relaxation oscillations*.

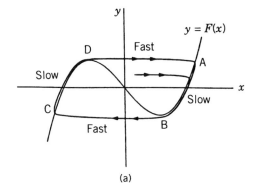

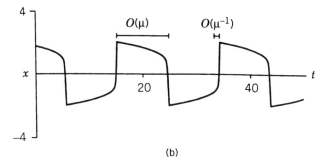

Figure 9.7. (*a*) Limit cycle trajectory and the accompanying fast and slow timescales. (*b*) Wave form of a van der Pol oscillator. (From Strogatz[62]. Reprinted with permission of Addison-Wesley Longman, Inc.)

The identification of a cubic nullcline with a limit cycle oscillatory phenomenon is a representative result. In general, limit cycle behavior will be encountered whenever there are nullclines having shapes resembling an N or S. In the remainder of this chapter, we will encounter several additional examples of nullclines with these shapes. We will examine the properties of coupled relaxation oscillators when we discuss synchronization in the visual cortex (Section 9.7) and when we use cellular automata to model calcium waves and oscillations (Section 9.10).

9.5.3 Poincaré-Bendixson Theorem

The Poincaré-Bendixson theorem describes two situations in which limit cycles form. It states that if for some time $t > t_0$ a trajectory is bounded in some region of phase space, and that region does not enclose any stable steady states, then that trajectory is either a closed periodic orbit (or limit cycle) or approaches a closed periodic orbit as t progressively increases. The theorem establishes that a limit cycle will form if (1) the flow is trapped in a particular region containing a repelling unstable node or spiral, or (2) flow is trapped in an annular region. In the first case, flow near the outer boundary is trapped and directed inward, and flow from the vicinity of the enclosed unstable state is directed outward. In the second case, flow is captured within an annular ring that excludes any stable steady states lying within the inner boundary.

9.5.4 Hopf Bifurcation

Another tool in analyzing the evolution of a nonlinear system from one type of stability to another as a control parameter is varied is the Hopf bifurcation theorem. This theorem describes the appearance of a limit cycle about any steady state that undergoes a transition from a stable to an unstable focus as the parameter of interest is varied. In this approach we identify the stable steady states and then perform a linear stability analysis in the vicinity of those states to find eigenvalues of the Jacobian matrix.

Let us again consider a two-dimensional system with a stable steady state. For this case the system can have either two real eigenvalues that are both negative or a pair of eigenvalues that are complex conjugates of one another. For this system to lose stability as some parameter is varied, one or both eigenvalues must move into the right half of the complex plane as that parameter is changed. When this happens, the node or stable spiral changes into either an unstable node or unstable spiral surrounded by a limit cycle marking the onset of repetitive or oscillatory behavior. For the supercritical form of Hopf bifurcation, the newly born limit cycle grows slowly starting with a small size and increase gradually with a frequency $\omega \propto \text{Im } \lambda$. For subcritical Hopf bifurcation, the situation is more complex, and large-scale oscillations appear in place of the small-scale repetitive activity typical of supercritical Hopf bifurcation. Further discussions of phase plane methods are given by Edelstein-Keshet.[61]

9.6 POPULATION DYNAMICS IN THE WILSON-COWAN MODEL

Our objective in this next section is to derive a set of coupled nonlinear equations that describes the dynamics of a local population of excitatory and inhibitory neurons. The solution of this set of equations will show us that global temporal order can arise from locally random interactions. The ensuing phase plane analysis of the response properties will reveal simple, multiple hysteresis, and limit cycle behavior for the model system. We will find that the frequency of the limit cycle activity in this model depends monotonically on the stimulus intensity. The assumptions of the model, due to Wilson and Cowan,[18,19] are quite gentle. First, we assume that a population of excitatory and inhibitory cells in close spatial proximity undergoes random interactions. Second, we regard the population as being sufficiently dense to ensure that there is at least one path, either direct or through an interneuron, linking any two cells. In this model we neglect spatial details and emphasize the temporal dynamics.

There are two dynamic variables: $E(t)$, the proportion of excitatory cells firing per unit time at time t, and $I(t)$, the corresponding proportion of inhibitory neurons. The resting state is the zero activity state, $E(t) = 0 = I(t)$, measured relative to the fraction of cells that fire spontaneously. We assume that the values of $E(t)$ and $I(t)$ at time $t + \tau$ are equal to the fractions of the neurons in the populations that are (1) not refractory and (2) receive at least threshold excitation at time t. We proceed by treating first the refractory property and then the thresholded excitation.

We denote the refractory periods for the excitatory and inhibitory cells as τ_E and τ_I. The proportion of the excitatory cells that are refractory is

$$\int_{t-r_E}^{t} E(t')dt'$$

and therefore the fraction of cells that are not refractory (i.e., sensitive) is

$$1 - \int_{t-r_E}^{t} E(t')dt'$$

Similar expressions hold for the inhibitory cells. We must next construct an expression for the average level of excitation in a cell of each population. We assume that a given cell sums its inputs and that the resulting stimulation decays in time as decribed by the function $\alpha(t)$. The average amount of excitation generated in an excitatory cell at time t is

$$\int_{-\infty}^{t} \alpha(t-t')[c_1 E(t') - c_2 I(t') + P(t')]dt'$$

The constants c_1 and c_2 are connectivity coefficients representing the average number of excitatory and inhibitory synapses per cell, and $P(t)$ is the external input to the excitatory population. Similarly the average excitation in an inhibitory cell at time t is

$$\int_{-\infty}^{t} \alpha(t-t')[c_3 E(t') - c_4 I(t') + Q(t')]dt'$$

with connectivity coefficients c_3 and c_4, and external input $Q(t)$.

If we assume that the sensitivity of the cells is not correlated with their level of excitation, then we can construct a pair of equations governing the dynamics of the two populations of cells by multiplying together the two terms. The desired expressions for the activities are

$$E(t+\tau_E) = \left[1 - \int_{t-r_E}^{t} E(t')dt'\right] S_E\left(\int_{-\infty}^{t} \alpha(t-t'')[c_1 E(t'') - c_2 I(t'') + P(t'')]dt''\right) \quad (9.58)$$

and

$$I(t+\tau_I) = \left[1 - \int_{t-r_I}^{t} I(t')dt'\right] S_I\left(\int_{-\infty}^{t} \alpha(t-t'')[c_3 E(t'') - c_4 I(t'') + Q(t'')]dt''\right) \quad (9.59)$$

where $S_E(\cdot)$ and $S_I(\cdot)$ are sigmoidal response functions for the excitatory and inhibitory neurons, respectively.

Our next step is to simplify these integral expressions by introducing moving temporal averages. This is done by replacing temporal variables by their mean values. Specifically, we define the temporal mean of a variable $f(t)$ over the time interval s as

$$\bar{f}(t) = \frac{1}{s}\int_{t-s}^{t} f(t')dt' \quad (9.60)$$

We now use this definition to replace the integrals of the dynamic variables appearing in Eqs. (9.58) and (9.59) by temporal mean values

$$\int_{t-r_E}^{t} E(t')dt' \leftrightarrow r_E \bar{E}(t) \quad (9.61)$$

where r_E is the absolute refractory period and

$$\int_{-\infty}^{t} \alpha(t-t')E(t')dt' \leftrightarrow k_E \bar{E}(t) \quad (9.62)$$

where k_E is a constant. Coarse-grained replacements analogous to Eqs. (9.61) and (9.62) hold for the inhibitory activity as well. The assumption in this procedure is that rapid fluctuations in the dynamic variables may be neglected. The resulting equations in this approximation are

$$\begin{aligned} E(t+\tau_E) &= [1 - r_E \bar{E}(t)]S_E(k_E c_1 \bar{E}(t) - k_E c_2 \bar{I}(t) + k_E P(t)) \\ I(t+\tau_I) &= [1 - r_I \bar{I}(t)]S_I(k_I c_3 \bar{E}(t) - k_I c_4 \bar{I}(t) + k_I Q(t)) \end{aligned} \quad (9.63)$$

We now expand the left-hand side of each of these last two equations in a

Taylor's series about $\tau_E = 0$ and $\tau_I = 0$, respectively. Upon retaining terms up to first order, we obtain the desired dynamic equations for the two populations of cells

$$\tau_E \frac{d\bar{E}(t)}{dt} = -\bar{E}(t) + [1 - r_E \bar{E}(t)]S_E(k_E c_1 \bar{E}(t) - k_E c_2 \bar{I}(t) + k_E P(t)) \quad (9.64)$$

and

$$\tau_I \frac{d\bar{I}(t)}{dt} = -\bar{I}(t) + [1 - r_I \bar{I}(t)]S_I(k_I c_3 \bar{E}(t) - k_I c_4 \bar{I}(t) + k_I Q(t)) \quad (9.65)$$

Small final adjustments must be made to the above to ensure that $E = 0 = I$ is a steady state solution when $P(t) = 0 = Q(t)$, namely when there are no external inputs, and to guarantee that the resting state is, in fact, stable. This is accomplished by transforming the sigmoidal response functions so that $S_E(0) = 0 = S_I(0)$ and by modifying the refractory terms. The final form of the Wilson-Cowan dynamic equations is

$$\tau_E \frac{dE(t)}{dt} = -E(t) + [k_E - r_E E(t)]S_E(c_1 E(t) - c_2 I(t) + P(t)) \quad (9.66)$$

$$\tau_I \frac{dI(t)}{dt} = -I(t) + [k_I - r_I I(t)]S_I(c_3 E(t) - c_4 I(t) + Q(t)) \quad (9.67)$$

The equations for the E and I nullclines can be obtained by introducing the inverses S_E^{-1} and S_I^{-1} of the sigmoidal response functions:

$$c_2 I = c_1 E - S_E^{-1}\left(\frac{E}{k_E - r_E E}\right) + P \quad (9.68)$$

and

$$c_3 E = c_4 I + S_I^{-1}\left(\frac{I}{k_I - r_I I}\right) - Q \quad (9.69)$$

It is noteworthy that negative feedback (nonzero c_2 and c_3) between the two populations is required if we are to have nontrivial nullclines. The phase plane plot for the case where $P = 0$ and $Q = 0$ (no external input) is shown in Fig. 9.8a. We observe that the E nullcline possesses a kink, or an N (or S) shape. This kink makes possible the rich spectrum of response characteristics such as multiple stable states, hysteresis, and oscillations.

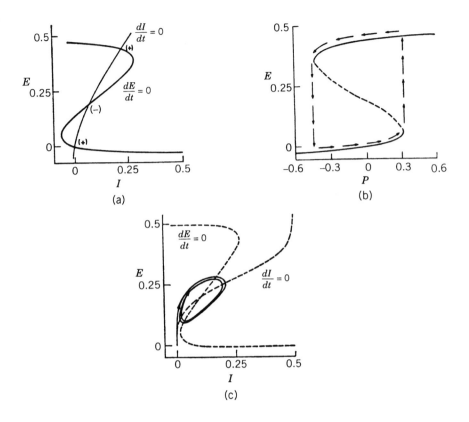

Figure 9.8. Steady state, hysteresis, and limit cycle behavior in the Wilson-Cowan model. (a) Phase plane portrait showing the nullclines for a system with three steady states; two of these are stable (+) and one is unstable (−). (b) An example of a system with a single hysteresis loop; stable states are indicated by the solid lines and unstable states by the dashed line. (c) Phase plane portrait showing the nullclines and ensuring limit cycle trajectory for a constant excitatory stimulus. (From Wilson and Cowan[18]. Reprinted with permission of the Biophysical Society.)

The sigmoidal response function may be expressed as

$$S(x) = \frac{1}{1 + \exp\{-a(x-b)\}} - \frac{1}{1 - \exp\{ab\}} \tag{9.70}$$

which satisfies the requirement that $S(0) = 0$. For this function the parameter b, fixes the position of maximum slope and the parameter a determines the magnitude of that slope. A sufficient condition for generating an N-shaped E nullcline is that the slope of the E nullcline is positive at the inflection point for S_E^{-1}. This will happen provided that $c_1 > 9/a_E$, where a_E is the slope parameter for $S_E(x)$. The three steady state solutions, corresponding to the intersections of the E and I nullclines, is illustrated in Fig. 9.8a. Two of the solutions are stable, while the third is unstable. The effects of the stimuli P and Q to the excitatory and inhibitory cells is to translate the nullclines parallel to the phase plane axes. Hysteresis phenomena will occur, as can be seen Fig. 9.8b, where we plot E against P for a fixed Q. Similar phenomena can be generated by varying Q while P is held constant.

The solutions illustrated in Figs. 9.8a and 9.8b are not the only ones possible. Larger sets of stable states and more complex hysteresis loops can be generated by other conditions on the c's and a's. We may conclude from a phase plane anslysis of Eqs. (9.66) and (9.67) that localized aggregates of cells may be excited to high levels, and these levels are maintained by a lower excitation than was needed to trigger the movement. Multiple stable states may occur, with transitions between them forming multiple hysteresis loops.

Limit cycle oscillations can arise in response to a constant stimulation. The condition that must be met for these oscillations to develop is that there is a single steady state that is unstable. A single unstable steady state will arise when the nullclines intersect at a location in the vicinity of an inflection point of the response function. The trajectories will remain within a closed region of phase space, and the condition on the intersection of the nullclines will produce limit cycle activity in accordance with the Poincaré-Bendixson theorem. Displayed in Fig. 9.8c is a phase plane portrait showing the intersection of the two nullclines in an unstable steady state, and the resulting limit cycle trajectory. The dependence of the limit cycle activity upon P and Q is interesting. There is threshold for generating the limit cycle oscillations, and once produced, their frequency will be a monotonic-increasing function of the stimulus intensity. If stimuli to the excitatory and inhibitory populations are independently alterable, then the combined population may exhibit either limit cycle or hysteresis activity, switching between different steady states in a manner dependent on the relationships between the two sets of stimuli.

9.7 OSCILLATIONS AND SYNCHRONY IN THE VISUAL CORTEX AND HIPPOCAMPUS

In this next two subsections we will expand on our model of population dynamics to explore how limit cycle oscillations may arise in a single cortical column in response to external stimuli. We will begin by reformulating the dynamic equations in a macroscopic mean-field theory. We will find that oscillations are intrinsic to a column and may arise in densely interconnected clusters of excitatory and inhibitory cells. We will find that these oscillations may be described by the phases of limit cycle oscillators. We will then briefly sketch the results of a study of the influences of delay couplings on synchronization and desynchronization in a nearest-neighbor approach. As will be made clearer in this section, the nearest-neighbor coupled oscillators function as relaxation oscillators.

9.7.1 Mean-Field Model of Cortical Oscillations

We start by modeling a cortical column as a cluster of N_E excitatory and N_I inhibitory neurons. Each cell in the cluster is characterized by its mean firing rate $E_i(t)$, $i = 1, \ldots, N_E$, or $I_j(t)$, $j = 1, \ldots, N_I$. We assume, as shown in Fig. 9.9a, that each neuron is coupled to all others in a column, and write the dynamic equations for the activities as

$$\frac{dE_i}{dt} = -E_i + S\left(a_E\left(\frac{1}{N_E}\sum_{j=1}^{N_E} u_{ij}E_j - \frac{1}{N_I}\sum_{j=1}^{N_I} v_{ij}I_j - \theta_i^E + p_i\right)\right) \quad (9.71)$$

where $i = 1, \ldots, N_E$ and

$$\frac{dI_i}{dt} = -I_i + S\left(a_I\left(\frac{1}{N_E}\sum_{j=1}^{N_E} w_{ij}E_j - \frac{1}{N_I}\sum_{j=1}^{N_I} z_{ij}I_j - \theta_i^I\right)\right) \quad (9.72)$$

where $i = 1, \ldots, N_I$. In these equations, u_{ij}, v_{ij}, w_{ij}, and z_{ij} are cell-cell couplings, θ_i^E and θ_i^I are thresholds, p_i are external inputs to the excitatory cells, there are no external inputs to the inhibitory neurons, and the sigmoidal response function is taken to be $S(x) = (1 + \exp(-x))^{-1}$.

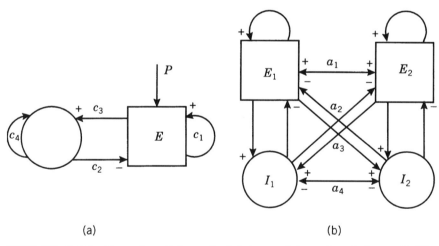

Figure 9.9. Construction of oscillator units: (a) A population of excitatory cells (E) is coupled to a population of inhibitory cells (I); couplings are labeled as c_1 to c_4 with (+) denoting excitatory couplings and (−) inhibitory ones. (b) Two oscillator units are linked together; couplings between oscillators are denoted as a_1 to a_4. The intrinsic couplings are as in panel (a), but have been omitted from the figure to maintain clarity. (From Schuster and Wagner[63]. Reprinted with permission of Springer-Verlag.)

The Wilson-Cowan couplings are now identified with the mean values of the cell-cell couplings:

$$c_1 = \frac{1}{N_E^2} \sum_{ij} u_{ij}, \quad c_2 = \frac{1}{N_E N_I} \sum_{ij} v_{ij}$$
$$c_3 = \frac{1}{N_E N_I} \sum_{ij} w_{ij}, \quad c_4 = \frac{1}{N_I^2} \sum_{ij} z_{ij}$$
(9.73)

The mean activities, thresholds, and external input may be defined in a similar manner:

$$E(t) = \frac{1}{N_E} \sum_i E_i(t), \quad I(t) = \frac{1}{N_I} \sum_i I_i(t)$$
$$\theta^E = \frac{1}{N_E} \sum_i \theta_i^E, \quad \theta^I = \frac{1}{N_I} \sum_i \theta_i^I$$
(9.74)

and

$$P = \frac{1}{N_E} \sum_i P_i \qquad (9.75)$$

If we sum Eqs. (9.71) and (9.72) over the index i, we obtain

$$\frac{dE}{dt} = -E + \frac{1}{N_E} \sum_{i=1}^{N_E} S(x_i^E) \qquad (9.76)$$

$$\frac{dI}{dt} = -I + \frac{1}{N_I} \sum_{i=1}^{N_I} S(x_i^I) \qquad (9.77)$$

We now expand the the sigmoidal response function in each equation in a Taylor's series about the mean values of each term in its argument

$$S(x_i) = S(\langle x \rangle) + S'(\langle x \rangle) \delta x_i + \left(\frac{1}{2}\right) S''(\langle x \rangle)(\delta x_i)^2 \qquad (9.78)$$

In the above the symbols x_i^E and x_i^I denote the arguments to the sigmoidal function given in Eqs. (9.71) and (9.72). Each term in the argument may be decomposed into its mean value plus a fluctuation about that mean; for example, $u_{ij} = c_1 + \delta u_{ij}$. The first term in the Taylor's expansion gives

$$\frac{dE}{dt} = -E + S(a_E(c_1 E - c_2 I - \theta^E + P)) \qquad (9.79)$$

$$\frac{dI}{dt} = -I + S(a_I(c_3 E - c_4 I - \theta^I)) \qquad (9.80)$$

The second term in the Taylor's expansion gives no contribution, and the third term is $O(1/N^{1/2})$. Equations (9.79) and (9.80) are equivalent to the Wilson-Cowan results for time intervals $\tau_1 = 1, \tau_E = 1$, and refractory period $r = 0$. As found by Wilson and Cowan, a stable fixed point characterizes the system at low-stimulus intensities, and there is a transition to limit cycle activity once the stimulus intensity increases beyond a critical value. The amplitude and frequency of the limit cycle activity is plotted as a function of the stimulus intensity in Figs. 9.10a and 9.10b.

This model can be developed further by coupling together a number of excitatory-inhibitory clusters. For instance, the linking of two units in the manner illustrated in Fig. 9.9b gives rise to the dynamics equations

$$\frac{dE_i}{dt} = -E_i + S(A_i^E + \eta a_E U_j) \tag{9.81}$$

$$\frac{dI_i}{dt} = -I_i + S(A_i^I + \eta a_I V_j) \tag{9.82}$$

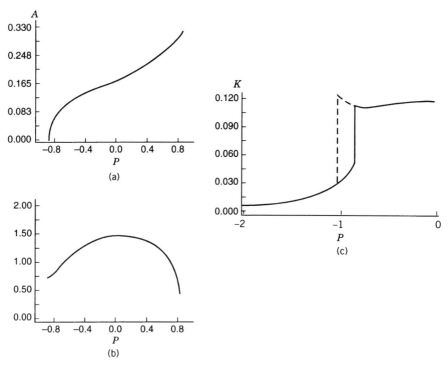

Figure 9.10. Properties of the limit cycle oscillators: (a) Plot of the amplitude A as a function of the input stimulus P. (b) Changes in the frequency of the limit cycle as a function of the input stimulus. (c) Hysteresis in the activity-dependent coupling K of the passive oscillator as a function of the input stimulus; the input to the active oscillator is held fixed. In panel (c) the solid line denotes the response for increasing input, while the dashed line shows the dependence for decreasing input. (From Schuster and Wagner[63]. Reprinted with permission of Springer-Verlag.)

where $i,j = 1, 2$ and $i \neq j$. The first term in the response function takes into account the couplings within a single column. The second term, which now appears in the sigmoidal function handles the interactions between the two cluster oscillators presented schematically in Fig. 9.9b:

$$U_j = a_1 E_j - a_2 I_j, \quad V_j = a_3 E_j - a_4 I_j \tag{9.83}$$

$$A_i^I = a_I(c_3 E_i - c_4 I_i - \theta^I) \tag{9.84}$$

and

$$A_i^E = a_E(c_1 E_i - c_2 I_i - \theta^E + P_i) \tag{9.85}$$

The starting point for reducing the above set of coupled equations to a phase coupled dscription is (1) to assume that the couplings η are weak allowing for a first-order expansion in this coupling strength,

$$S(A_i^E + \eta a_I U_j) = S(A_i^E) + \eta a_E S'(A_i^E) U_j \tag{9.86}$$

and similarly for the inhibitory activity, and (2) to introduce the phase variables ϕ_i through the deviations of the activities from the unstable fixed points (E_{i0}, I_{i0}),

$$\Delta E_i = E_i - E_{i0} \approx r_i \cos \phi_i \tag{9.87}$$

and

$$\Delta I_i = I_i - I_{i0} \approx r_i \sin \phi_i \tag{9.88}$$

The expansion with respect to the couplings produces couplings terms that are sensitive to the external inputs to the two oscillators. The dependence of the coupling strength $K = a_E \eta S'(A_i^E)$ on the input P into one of the oscillators is displayed in the right-hand panel of Fig. 9.10. Enhancements of an order of magnitude can be produced if the two coupled columns are each active. The reduced equations of motion for the coupled columns are

$$\begin{aligned}\dot{\phi}_1 &= \omega_1 - K_{12} \sin(\phi_1 - \phi_2) \\ \dot{\phi}_2 &= \omega_2 - K_{21} \sin(\phi_2 - \phi_1)\end{aligned} \tag{9.89}$$

where $\omega_{1,2}$ are the oscillator frequencies and K_{ij} are the couplings proportional to η. The procedure for generating the phase equations is given by Schuster and Wagner.[63,64]

9.7.2 Delay Connections and Nearest-Neighbor Interactions

Another aspect of stimulus-dependent oscillator assembly operation is the effects of temporal delays in the connections. We may study the influences of delay connections in a simplified oscillator unit obeying the following equations of motion

$$\tau_0 \frac{dE(t)}{dt} = -\alpha_E E(t) + c_2 S(I(t-\tau)) + P(t) + N_E(t) \qquad (9.90)$$

$$\tau_0 \frac{dI(t)}{dt} = -\alpha_I I(t) + c_3 S(E(t-\tau')) + N_I(t) \qquad (9.91)$$

In this unit the c_1 and c_4 connections have been eliminated. An excitatory unit is connected to an inhibitory unit with strength c_3 and temporal delay τ', and the inhibitory unit makes a reciprocal connection to the excitatory unit with strength c_2 and delay τ. This nearest-neighbor coupling model is illustrated in Fig. 9.11. As before, P denotes an external stimulus to the excitatory unit, and we have added white noise inputs N_E and N_I and damping constants α_E and α_I. The sigmoidal response function S is of the form

$$S(x) = \frac{1}{1 + \exp(-a(x-\theta))} \qquad (9.92)$$

where a is the slope parameter and θ is the threshold. This model was used by König and Schillen,[65] and by Schillen and König,[66] to study the effects of excitatory delay connections in promoting synchronization and desynchronization within a single oscillator unit and in two-dimensional layers of oscillator units undergoing nearest-neighbor interactions with one another.

In the investigation by König and Schillen, coupling delays were taken to be on the order of an oscillator's intrinsic delay. They observed that when the time delays are either too small or too large a system of two coupled units will relax to a stable fixed point determined by the coupling parameters and

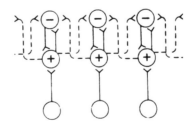

Figure 9.11. Nearest-neighbor coupling scheme illustrating the coupling of an excitatory unit of an oscillator to the inhibitory units of the neighboring oscillators using delay connections. (From König and Schillen[65]. Reprinted with permission of MIT Press.)

external input. There is a broad range of delays in the vicinity of 4 to 5 ms for which the system will exhibit stable limit cycle behavior. Similar results were found when the model was extended to a two-dimensional layer of oscillatory units. They note that these systems will either relax to a fixed point or exhibit limit cycle oscillations with a frequency that is independent of the magnitude of the input activity.

Desynchronization was promoted in the study by Schillen and König by adding a second set of delay connections operating between next nearest neighbors. The addition of this second set of delay connections gives rise to nonzero phase lags and frustration. In contrast to the desynchronizing effects of noise, desynchronizing delay connections actively dephase different neighboring assemblies, influence stimulus responses in a specific way, and provide a mechanism for segregating partially overlapping stimuli.

9.7.3 Burst Synchronization

In the investigation of Bush and Douglas[67] the mechanisms responsible for burst synchronization was studied in a network composed of excitatory pyramidal and inhibitory basket (smooth) neurons. In this network LGN neurons projected to bursting pyramidal cells that make excitatory connections to each other and to a common pool of basket cells. The basket cells made inhibitory feedback connections to the pyramidal cells. In the Bush-Douglas model the pyramidal and smooth cells were endowed with a fairly complete set of electrophysiological properties. Active conductances had Hodgkin-Huxley-like dynamics (to be discussed in the next section), and passive properties were consistent with data from intracellular recordings for these classes of cells. The electrophysiological properties were allowed to vary from one pyramidal cell to another in a manner that produced a broad range of burst frequencies. In this circuit we find that there is a rapid onset of synchronous bursting with randomly varying interburst intervals.

Synchronous bursting in the Bush-Douglas system was studied analytically by Koch and Schuster.[68] In their investigation they simplified the Bush-Douglas model to one containing all-to-all excitatory binary (McCulloch-Pitts) neurons plus a single global inhibitor. This model was shown to be capable of generating burst synchronization without frequency locking. This model demonstrates that the neural circuitry functions as a coincidence detector, amplifying coincident input signals to produce a coherent response, with recurrent excitation playing the main role. The importance of inhibition was brought out in studies of oscillations in the piriform and visual cortices by Wilson and Bower.[69] They found that inhibition, and particularly mutual inhibition, improves frequency locking and determines the frequency of the oscillatory firing pattern.

To summarize, in model networks in the visual cortex there two types of excitatory synaptic input. First, we have afferent excitatory input from the LGN to the excitatory pyramidal cells. Second, these cells make both short-

and long-range connections with one another. The pyramidal cells also make excitatory connections with a pool of inhibitory interneurons (basket cells). These interneurons make reciprocal inhibitory connections with excitatory cells, and with each other (mutual inhibition). The network of inhibitory cells may be replaced with a common feedback inhibition. The operation of this network is as follows: The short-range cortical reexcitation plus geniculate input gives rise to repetitive firing that amplifies coincident events. The long-range connections establishes a rough global coherence in the firing pattern. The global inhibition establishes a robust synchrony by reducing phase lags to zero within a few cycles of the network, and the frequency of the repetitive firing is determined by the time course (delays) within the inhibitory circuitry.

Experimentally we find that excitatory cortico-cortico synapses are the predominant form of synaptic connection in the neocortex. Most of these connections are between neighboring cells, but long-ranging tangential connections are also present. There is also evidence for distributed pools of inhibitory interneurons with the required fast time course to support synchrony. The theoretical finding is that these neural circuits evolve to either fixed points, as discussed in the last chapter, or to some form of synchronous bursting or oscillatory behavior given the appropriate LGN input and physiological parameters. A number of distinct mechanisms are subsumed within this general theme of recurrent excitation plus inhibition, and we will encounter some additional examples of these in our discussions of hippocampal synchrony (Section 9.7.4, below) and thalamocortical spindling (Section 9.9).

9.7.4 Rhythmic Population Oscillations in the Hippocampus

Before discussing hippocampal synchrony, let us recall some of our findings from Chapter 8. In our study of use-dependent synaptic modification, we observed that changes in the efficiency of synaptic transmission are initiated by spatially and temporally correlated afferent input. The chain of events leading to synaptic modification begins with the activation of AMPA receptors by the correlated excitatory input. The changes in membrane permeability brought on by the AMPA receptor activation produces a depolarization that relieves a voltage-dependent Mg^{2+} block from NMDA receptors, thereby allowing Ca^{2+} to enter the postsynaptic cell. The Ca^{2+} entry then triggers the processes leading to synaptic modification. Two forms of synaptic plasticity, LTP and LTD, are encountered in the hippocampus, and this part of the central nervous system is thought to be a locus of short-term learning and memory. Given these observations, it is significant that large populations of cells in the hippocampus can fire synchronously. Temporally correlated patterns of activity such as those produced by synchronous rhythms are an effective means of altering synaptic transmission. The large-scale synchrony may, in some as yet not fully understood manner, promote synaptic modification and information storage.

The composition of the hippocampus seems ideally suited to support large-scale cooperative events. First, we find that hippocampal cells extend out widely arborizing axon collaterals. These collaterals provide the connectivity to generate recurrent excitation in networks of pyramidal cells. Second, GABAergic interneurons are present. Three types of interneurons have been found[70] in the hippocampus. These interneurons, basket, bistrified and axo-axonic, have distinctive axonal and dendritic morphologies, establish their synaptic contacts at different places on the target cell, and have different postsynaptic kinetics. Each of these types of GABA (γ-amino-butyric acid) releasing interneuron establishes between 6 and 12 contacts with a postsynaptic principal neuron. Their inhibitory postsynaptic potentials (IPSPs) have short latencies and fast rise times. The characteristics of the IPSPs are consistent with the timing required for recurrent inhibition. Third, cells in this region are capable of repetitive bursting as a consequence of their intrinsic electrophysiological properties. Specifically, the membrane potential of single pyramidal cells can oscillate in the 4- to 10-Hz range. Intrinsically oscillatory cells in the entorhinal cortex projecting to hippocampal neurons can also drive cells into 4- to 10-Hz oscillations. Thus the hippocampus exhibits several different types of rhythmicity and has a number of possibly redundant mechanisms for inducing collective responses.

Synchronous activity in the hippocampus includes theta rhythms in the just mentioned 4- to 10-Hz range, carbachol-driven rhythmic population oscillations in the 4- to 10-Hz range, high-frequency activity in a broad 30- to 100-Hz band, and sharp waves in another broad frequency range spanning 200 Hz. The amplitude and frequency of these rhythmic bursting events are dependent on both the intrinsic properties of the cells, and the connectivity and strength of excitatory and inhibitory synapses. The carbachol-driven oscillations induced by disinhibition are generated by recurrent excitation among a network of intrinsically bursting cells. They have properties that differ from those of the *in vivo* theta rhythms.[71] The 40-Hz oscillations are a collective behavior of the network of inhibitory interneurons in the hippocampus.[72] Mutual inhibition plays a key role in generating this form of oscillatory activity. Another source of high-frequency rhythmicity are the intrinsic 40-Hz oscillatory interneurons.[73] The hippocampal sharp waves observed in pyramidal cells in the CA1 region of the hippocampus are initiated by synchronously bursting cells in the CA3 region. The sharp waves seen in the EEG recordings are indicative of a transient depolarization of apical dendrites of pyramidal cells receiving input from Schaffer collaterals. These waves are accompanied by spindle-shaped fast oscillatory synaptic potentials and by the synchronized bursting of interneurons during periods of oscillatory synaptic activity.[32] Works[74,75] on establishing how these modes of oscillatory activity influence synaptic plasticty have revealed a possible priming function, selective gating, and highly plastic oscillatory states.

Rhythmic population oscillations resembling those observed experimentally in hippocampal slice preparations can be generated by model networks characterized by recurrent excitation, sparse but strong synaptic connections, and intrinsic bursting cells as their elements. A variety of synchronous behaviors can be produced by turning inhibition on or off, and more generally by varing the synaptic strengths and connectivity. The firing patterns, most notably traveling waves that propagate across the model slices, and the formation of large and small clusters of synchronously bursting cells, resemble those generated in one- and two-dimensional arrays of coupled rotators.

These characteristics of hippocampal rhythmicity were elucidated in a series of experimental and computational studies by Traub et al.,[76] by Miles, Traub, and Wong,[38] and by Traub, Miles, and Wong.[77] In their first study,[76] a hippocampal network was constructed from two hundred excitatory neurons arranged in a two-dimensional array, and ten inhibitory neurons were distributed uniformly across the array. Physiological parameters and connectivities were chosen so that the model pyramidal neurons developed intrinsic bursts and, when organized in networks of sparse but strong excitatory synaptic connections, had the ability to propagate bursts from cell to cell. Inhibitory neurons were of two kinds, one with a fast time course and the other with a slow one. When fast inhibition is present the bursting neurons self-organize into clusters of synchronously firing cells. Different cells participate in each cluster. A cluster is formed when one or two cells bursting synchronously recruit others by means of recurrent excitation. The situation changes when fast inhibition is blocked to create a form of epileptic synchrony. Now most of the cells in the population burst coherently. This collective bursting is followed by a silent period and then by partially synchronized bursting. As is the case with inhibition present synchrony is spread by recurrent excitation. Cluster formation is limited by the refractoriness of the participants and by the slow inhibition.

In further studies of population oscillations in hippocampal slices, we encounter constant speed traveling waves of activity. To explore large-scale response properties observed in EEG recordings, Miles et al.[38] and Traub et al.[77] constructed arrays containing 9000 model excitatory cells and 900 model inhibitory neurons according to the principles described above. The connectivity properties of the networks were neither global all-to-all nor strictly nearest neighbor, but were closer to the latter than to the former. The connections were exponentially distributed in space. The probability that a cell at position x_1 established contact with a neuron at position x_2 decreased exponentially with distance $|x_1 - x_2|$ at a rate determined by space constants for excitatory and inhibitory units.

Pyramidal cells established excitatory synaptic contacts with each other and with inhibitory cells, and inhibitory neurons made synaptic connections with one another and with pyramidal cells. The connection densities were sparse. Each excitatory cell contacted 22 neurons; 20 of these were excitatory cells, and 2 were inhibitory units. Each inhibitory neuron communicated with 220 cells; 200 were excitatory cells, and 20 were inhibitory neurons. Thus a typical cell received 20 excitatory synaptic inputs and 20 inhibitory inputs. Excitatory cells were also assigned a temporal delay varied with distance. Typically within a region about an excitatory unit containing half its outgoing connections, the probability that a particular cell was contacted was about 1%. The propagation of synchronous firing is shown in Fig. 9.12. In this figure we observe a traveling wave of synchronous bursting propagating at constant velocity across the hippocampal slice. During each rhythmic event, some neurons fire while many others receive correlated synaptic input. The oscillations observed in EEG recordings are indicative of the synchronous synaptic input. Thus local couplings in the hippocampal slice are able to coordinate the cells' activities, producing coherent oscillations in the synaptic potentials over distances large (up to 10 mm) compared to the space constants.

In a recent modeling study by Traub et al.[114] networks of inhibitory interneurons were shown to be capable of entraining large populations of excitatory pyramidal cells to which they are linked. More specifically, they demonstrated that excitatory pyramidal cells belonging to spatially separated clusters will synchronize their discharges in the gamma frequency range when inhibitory interneurons to which they are linked fire two rapid spikes or doublets such that the time interval (on the order of 5 ms) between the spikes equals the transition time between clusters. Evidence of interneuron doublet firing in conjunction with oscillations in the gamma frequency range was provided by Traub et al. in a complementary set of experiments done in CA1 region of rat hippocampal slices. The doublet mechanism was found to be capable of promoting synchrony with zero or near-zero phase lags over long distances.

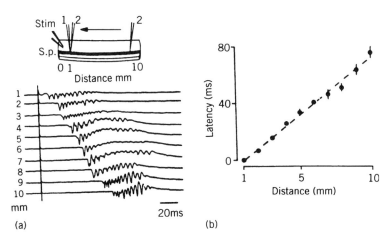

Figure 9.12. Spread of synchronous firing in a hippocampal slice. (*a*) Recordings of field potentials from multiple sites at distances up to 10 mm from the stimulus. (*b*) Linear propagation of the population wave. The conduction velocity given by the linear fit to the data is 0.12 m/s. (From Miles, Traub, and Wong[38]. Reprinted with permission of the American Physiological Society.)

This rapid spiking mechanism allows us to address an issue posed by the König-Schillen delay line study. In the model of Section 9.7.2 the authors assumed that the local circuit time constant matched the delay time between neighboring local circuits. They presumed that the time delay between pyramidal cell firing and excitation of neighboring interneurons matched the conduction delay between separate groups of cells. The doublet mechanism alleviates the need for slow conduction in the local circuit, since the doublet interval of 4 to 5 ms can function as the local circuit time constant.

9.7.5 Feature Integration

The processing of early visual information such as contours, depth, motion, and color is distributed, in parallel, among large populations of neurons in different cortical areas such that each population encodes a particular aspect of the visual scene. It is an open question how these attributes are integrated to produce an unambiguous representation of the component features and objects in the scene. The issue of how attributes are integrated to produce a segmentation of the scene into its component surfaces and a segregation of objects from their backgrounds is known as the *binding problem*. One possible mechanism for the reintegration of features is through a convergence of low-level inputs into a small number of higher-level neurons called *grandfather* or *cardinal cells* located in object-specific cortical areas. This hierarchical process, and the reasoning supporting it, has been described in detail by Barlow.[78] Another mechanism for integrating features is through *assembly coding*, that is, through flexible associations of large numbers of simultaneously active neurons.

According to the assembly coding hypothesis, neurons that encode attributes that belong together are bound together by their synchronous firing. Synchronized firing should occur in different columns within a given cortical area to encode local information, and in different cortical areas to bind features. The experimental studies of Gray et al.,[25,26] Eckhorn et al.,[27] Livingston et al.,[29] and Engel et al.[30,31] provide us with evidence that clusters of synchronously discharging cells form within one or more columns located in a particular cortical area and within columns located in different cortical regions and hemispheres. These results tell us that at least some of the conditions for assembly coding are satisfied. The theoretical investigations demonstrate that there are a variety of mechanisms supporting the formation of clusters of rhythmically firing oscillatory or bursting units, and show that cluster formation is an intrinsic property of networks of model neurons.

In assembly coding, temporal clusters form and reform in response to input signals. If there are multiple objects in a scene, then there are several co-active clusters. Individual neurons belong at various times to different assemblies and can rapidly associate into a functional group while at the same time disassociating from a different functional group. The experimental evidence of Vaadia et al.[35] is of a rapid coalescence of cells into functional groups and of their dissociation from competing groups. They find no evidence for an elevation in the mean firing rate in these processes, consistent with another tenet of assembly coding that synchrony alone signifies membership in a temporal cluster.

As noted in Chapter 1, the organization and reorganization of neural elements into different functional clusters is encountered[79] in lower organisms such as *Tritonia*, where the computational function performed by a circuit can change in response to alterations in environmental conditions. For instance, the *Tritonia* escape swim circuit contains a pattern generator mode that produces the alternating, dorsal-ventral bursting responsible for escape swimming. The swim circuit functionally reconfigures itself when changing from a reflexive withdrawal state to the pattern generator mode.[79]

There are a number of requirements that must be satisfied if integration mechanisms based on assembling coding are to segregate objects from their backgrounds and segment scenes. The requirements for forming perceptual clusters are embodied in the Gestalt laws of proximity, continuity, similarity, and common motion (fate). Briefly, these are guidelines that characterize which elements should belong to a particular grouping of synchronously active cells. The rules tell us that cells that encode features that (1) are spatially contiguous, (2) have similar and smoothly varying attributes such as texture, color, and position (depth), and (3) share the same motion, should be bound together in a synchronous cluster. Experimental evidence that synchronous clusters form in the visual cortex in a manner consistent with the Gestalt laws has been provided by Engel et al.[80]

We have encountered processes of this type previously in our discussion in Chapter 6 of integration using Markov random fields. In generalizing the line process of Geman and Geman, we introduced the idea of integrating early vision modules by assigning specific labels to groups of pixels with common sets of intrinsic properties such as texture, motion, color, and depth. The analogue of that procedure in a neural system is the formation of clusters of cells whose membership is signaled by temporal correlations. Assembly coding and MRF-based integration-by-labeling are self-organizing processes that reinforce and improve the integration of features from one iteration to the next and are robust against noise. Both use multiple cues in an iterative manner to resolve conflicting or ambiguous situations arising from single cues, and both maintain a distributed representation of the stimulus. Outputs in one area are used in another area in its own operations and are then reentered back into the original area.

A number of mechanisms may help drive these processes of dynamic and context-dependent modulation of temporally correlated firing. One possible mechanism is that of reciprocal reentrant connections between groups excitatory and inhibitory neurons distributed across feature encoding areas. In reentrant signaling, cells that belong to a neuronal group in parallel, repeatedly, and in register exchange signals with cells in other groups. The reentrant signals drive the integration process, promoting the elimination of conflicting and ambiguous responses within a single area, permitting outputs in one area to be used in another area and enabling the reuse of outputs back into the originating feature encoding region. In an initial study using reentrant signaling, Finkel and Edelman[81] demonstrated that reentrant circuits are able to integrate features distributed among different cortical area and synthesize responses to illusory contours. In a second study by Sporns et al.,[82] reentrant signaling was shown to lead to the formation of clusters of cells oscillating in phase with one another, and in a later study by Sporns et al.,[83] the model was shown to be capable of solving problems in figure-ground segregation.

In two other studies by Sompolinsky, Golomb, and Kleinfeld[84] and by Malsburg and Buhmann,[85] the complex neural dynamics was treated using simplified networks of phase oscillators. In these studies, as well as in the study by Sporns et al.,[83] visual cortical areas were built from feature selective cells arranged topographically into cortical columns. The connectivity of these networks consisted of short-range intracolumnar connections linking together cells that have overlapping receptive fields, and long-range connections coupling together cells in separate spatial locations that encode similar attributes. In the model of Sompolinsky et al.,[84] networks of dense short-range interconnections within local clusters of cells were treated in a mean-field approximation. Long-range coherence was described in terms of interactions between the average phases of the locally coherent clusters.

The stimulus presentations in the above-mentioned studies were modeled after the experimental findings of Gray et al.,[25] Gray and Singer,[26] and Eckhorn et al.,[27] and the objective was to reproduce their results. The stimuli used by Sompolinsky et al.[84] consisted of a pair of oriented short bars and a single oriented long bar, where "short" and "long" refer the lengths of the bars relative to the size of the receptive fields of the cells. Schillen and König[65,66] used two short bars and a long bar as their stimuli, and also examined motion/position coherency, while Sporns et al.[84] investigated orientation and motion coherency. The consistency with respect to the Gestalt laws, of the segmentation and figure-ground segregation through assembly coding, was discussed by Malsburg and Buhmann[85] and by Sporns et al.[83]

Assembly coding has been identified with gamma-band rhythmicity. An argument supporting this identification has been presented by Singer.[86] He notes that the time required by the visual system for the segmentation of a visual scene, as revealed in psychophysical studies, is on the order of 100 to 200 ms. Thus rhythmic firings in the alpha- and beta-bands are too slow for feature binding. An upper limit on the useful periodicity is provided by the coupling delay considerations of Schuster and Wagner,[63,64] Schillen and Konig,[65,66] Sporns et al.,[83] and Sompolinsky et al.[84] In their studies we find that synchrony with zero phase lags can be established for coupling delays up to about one-quarter of the average periodicity. For coupling delays outside this range synchrony with zero phase lags may not be possible. For physical delays of about 1 ms per millimeter of length of slow conducting axon, we find that oscillations in the 20–25-ms range are maximal if we wish to maintain zero phase lags over distances of 5–6 mm. Therefore the gamma-band represents a compromise between the dual requirements of rapid binding and transmission time limitations. Its irregularities (high noise levels) may promote rapid desynchronization.

Pacemaker cells provide another mechanism for synchronizing the firing of populations of cells distributed across functional columns in the visual cortex. As noted previously gamma-band activity in the visual cortex is characterized by high-frequency burst discharges in single-cells. Gray and McCormick[115] have identified a class of pyramidal cells which fire rapid bursts of action potentials in the gamma-frequency range. These neurons, termed chattering cells, exhibit intraburst spiking at frequencies as high as 800/s. Intracellular recordings reveal that chattering is intrinsic properties of the cells. Chattering cells are well positioned anatomically to promote synchrony over long distances. These neurons, found in layers II and III, extend axon collaterals horizontally and ramify within layers I, II, III, and IV in addition to projecting out of the local cortical region.

We end this section by noting that fast spontaneous cortical rhythms are generated in brain-active states upon waking and during REM sleep in humans and other mammals. The fast synchronous activity originates in the brainstem, spreads to the thalamus, and from there propagates to the cortex. In contrast to the low-frequency rhythmicity characteristic of sleep, the pools of neurons synchronized upon arousal by the brainstem-thalamus-cortex activating system are spatially limited and conform to the notion of a local cluster. As pointed out by Steriade et al.,[87] the fast rhythmicity produced upon waking may contribute to the background activity of the brain, facilitating synchronization and potentiating responses to signals. In a study by Munk et al.,[88] the arousal process was found to facilitate synchronous responses among distributed pools of cortical neurons to visual stimuli. Significantly the activation-system-generated stinulation did not alter the specificity of the neuronal responses to visual input. Thus the activation system may dynamically influence which responses are selected for higher-level integration, such as sensorimotor integration, and may play an important role in attentional processes.

9.8 NEURAL EXCITABILITY AND OSCILLATIONS

Neurons in the central nervous system are themselves complex nonlinear dynamic systems. In these cells, the differing voltage- and ligand-gated ionic conductances self-organize to produce stereotypic firing patterns such as single spiking, bursting, and rhythmic oscillations. The cells have an internal communication system and are influenced by a host of neurotransmitters and neuromodulators. The sensitivity to these biochemical signaling agents allows the neurons to adapt to changes in the internal and external environments through modifications in their physiological parameters which result in alterations in their firing patterns.

9.8.1 The Hodgkin-Huxley Equations

The construction by Hodgkin and Huxley[3] of a mathematical model of the time course of sodium and potassium permeability changes, produced in response to depolarization of the nerve membrane, marks the beginning of computational neuroscience. The development of their model was preceded by detailed studies of the the flow of electric current through the surface membrane of the squid giant axon. The basic elements of the Hodgkin-Huxley model are a set of ionic currents separated into contributions from sodium ions (I_{Na}), potassium ions (I_K), and the leakage of other ionic species such as chlorine (I_L). These currents are related to the voltage-dependent membrane permeabilities expressed in terms of the ionic conductances, g_{Na}, g_K, and g_L, and the displacements of the membrane potential V from contribution to the resting membrane potential from the particular ion species. The relationship between current, conductance, and voltage is, from Ohm's law,

$$I_x = g_x(V - V_x) \tag{9.93}$$

where x represents the ion type under consideration, $g_x = 1/R_x$ for resistance R_x, and V_x is a reference potential. The net ionic current, assuming that the circuit elements operate in parallel, is given as the sum of the potassium, sodium, leakage, capacitive, and any external applied or synaptic currents. The capacitive current I_C is, from Faraday's law,

$$I_C = C\frac{dV}{dt} \tag{9.94}$$

where C is the capacitance, and the resulting circuit equation is

$$C\frac{dV}{dt} = -g_K(V - V_K) - g_{Na}(V - V_{Na}) - g_L(V - V_L) + I \tag{9.95}$$

where I is the applied current.

To model the opening and closing of the sodium and potassium channels in response to depolarization of the nerve membrane, Hodgkin and Huxley introduced three gates. As depicted in Fig. 9.13, the sodium channel is controlled by two voltage-dependent gates, the fast m-gate and the slow h-gate. At rest, the m-gate is closed and the h-gate is open. The m-gate opens quickly when the membrane is depolarized allowing for the entry of sodium. The h-gate then closes slowly to subsequently block the sodium inflow. The potassium channel is controlled by the n-gate. This gate is closed at rest. It opens after a delay in response to depolarization to permit the outflow of potassium.

The next step in the construction of the dynamic model is to describe the rise and fall in the sodium conductance, and the rise in the potassium conductance, following depolarization. This was done using dimensionless

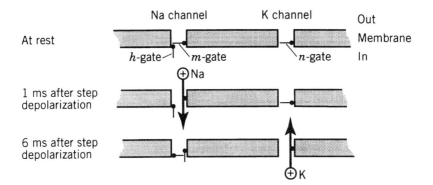

Figure 9.13. Schematic of the opening and closing of the sodium and potassium channels in the Hodgkin-Huxley model. (From Kuffler, Nicholls, and Martin[89]. Reprinted with permission of Sinauer Associates, Inc., MA.)

variables m, h, and n associated with the ion gates. Hodgkin and Huxley used power law approximations to the observed sigmoidal time course of the increase in sodium and potassium conductances for a given voltage step. Their formal model for the sodium channel was

$$g_{\mathrm{Na}} = \bar{g}_{\mathrm{Na}} m^3 h \tag{9.96}$$

with an associated dynamics governed by the equations

$$\frac{dm}{dt} = \alpha_m(V)(1-m) - \beta_m(V)m \tag{9.97}$$

$$\frac{dh}{dt} = \alpha_h(V)(1-h) - \beta_h(V)h \tag{9.98}$$

In the above, $\bar{g}_{\mathrm{Na}}$ is a constant, and α_m, α_h, β_m, and β_h are voltage-dependent rate constants, and the variables m and h assume values between 0 and 1. The product $m^3 h$ may be interpreted as the fraction of open sodium channels. The corresponding model for the potassuim channel is

$$g_{\mathrm{K}} = \bar{g}_{\mathrm{K}} n^4 \tag{9.99}$$

with a dynamics given by the analogous expression

$$\frac{dn}{dt} = \alpha_n(V)(1-n) - \beta_n(V)n \tag{9.100}$$

As before, $\bar{g}_{\mathrm{K}}$, α_n and β_n are constants, and n^4 is the number of open potassium channels.

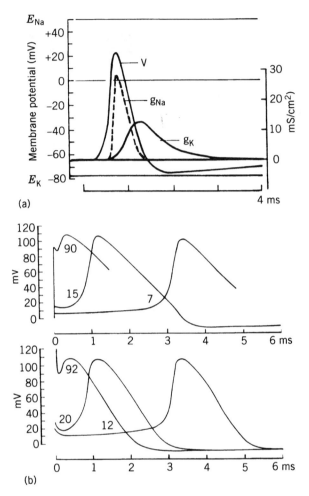

Figure 9.14. Time course of changes in membrane potential and ion conductances. (a) Action potential (V) and sodium and potassium conductance variations based on results of voltage clamp measurements. (b) Comparison of theoretical (upper) and measured (lower) action potentials initiated by brief shocks to a squid giant axon at three different intensities. (From Kuffler, Nicholls, and Martin.[89] Reprinted with permission of Sinauer Associates, Inc., MA.)

Equations (9.95), (9.97), (9.98), and (9.100) comprise a family of four coupled ordinary differential equations. This system of equations described the formation of regenerative action potentials that propagate down axons without attenuation. The action potential, and underlying variations in sodium and potassium conductances, measured in voltage clamp experiments is presented in Fig. 9.14. Also shown in this figure are comparisons of experimental membrane action potentials with those predicted by the Hodgkin-Huxley equations.

Rhythmicity is generated in neurons through a cooperative and sequential activation of a set of conductances. For example, rhythmic oscillations can be initiated[90] in inferior olivary neurons by an increase in a voltage-gated calcium conductance that allows for the inflow of calcium. The transient increase in intracellular calcium activates a calcium-dependent potassium conductance that leads to a hyperpolarization of the membrane followed by a rebound low-threshold calcium conductance. This last step reinstates the cycle and sustains the oscillatory activity.

There are many different kinds of neuron in the central nervous system. Each cell class possesses a distinctive morphology, connectivity, and electrophysiological properties.[91] We have earlier noted that there exist layer 4 interneurons that are able to oscillate at high frequencies near 40-Hz. These cells have extensive local axon collaterals and may contribute to the columnar synchronization discussed in the last section. The high-frequency oscillations are generated in these interneurons by a persistent sodium conductance and a delayed rectifier potassium conductance, activated sequentially, rather than by calcium-dependent conductances, which appear to be absent in these cells.[67]

In the model of Traub et al.,[76] five active conductances were used together with a passive leakage conductance to generate bursting. The active conductances were sodium, delayed rectifier potassium, calcium, calcium-dependent potassium, and an M-current potassium. Cells that fired repetitively were modeled by eliminating the calcium-related conductances and the M-current, and modifying the potassium conductance. In bursters, the potassium conductance was given the form $g_K = \bar{g}_K n^4 y$, where y is an inactivation variable. For repetitive firing, y was set equal to unity. The bursting pyramidal neurons in the Bush-Douglas[67] calculations were modeled in a similar manner. Bursting was promoted by the delayed rectifier and calcium-dependent potassium conductances. Also included were spike and persistent sodium conductances, a calcium conductance, an A conductance, and a leakage conductance. Inhibitory neurons had somewhat different spiking characteristics and were modeled by retaining only the large and fast delayed rectifier and sodium conductances.

9.8.2 The Morris-Lecar Model

The first application of phase plane methods to the Hodgkin-Huxley equations was made by FitzHugh. In his first[4] of two studies, FitzHugh simplified the dynamics by noting that the variables V and m change more rapidly than h and n. A study of the reduced, $V-m$ phase plane portrait revealed that there were three stable states. Two of these were stable nodes, and the third was a saddle point. A $Vmhn$ phase portrait was then reconstructed by considering small displacements of h and n from their resting values.

In his second[5] investigation, FitzHugh analyzed the dynamics further in a stereotypic model inspired in part by the van der Pol oscillator. In his simplified model there were two variables representing the excitation and recovery of the membrane. The inclusion of an applied current term in the equation for the excitation variable enabled FitzHugh to examine the evolution of the system in the phase plane as a function of the strength of the stimulus. For small values of the stimulating current, he found little change in the behavior of the system. However, for larger stimuli, the system traced out a trajectory through the region of phase space delineated by the N-shaped nullcline for the excitation variable. This large-scale movement represents the firing of an action potential. For still larger stimulating currents, the system underwent a transition to limit cycle behavior. This transition corresponds to the onset of repetitive firing of the membrane. In all cases removal of the stimulating current resulted in a return to rest conditions.

The work by FitzHugh was followed by a similar model by Nagumo et al.,[6] and the second-order dynamic system analyzed by these authors is commonly referred to as the FitzHugh-Nagumo equations. In this section we will examine a hybrid of the Hodgkin-Huxley and FitzHugh-Nagumo approaches developed by Morris and Lecar.[7] Their model is a third-order family of dynamic equations

$$C\frac{dV}{dt} = -g_K N(V - V_K) - g_{Ca} M(V - V_{Ca}) - g_L(V - V_L) + I \quad (9.101)$$

$$\frac{dM}{dt} = \lambda_M(V)(M_\infty(V) - M) \quad (9.102)$$

$$\frac{dN}{dt} = \lambda_N(V)(N_\infty(V) - N) \quad (9.103)$$

In these equations there are contributions to the current balance from calcium, potassium, leakage, and applied currents. The quantity g_{Ca} is the calcium ion conductance, V_{Ca} is the equilibrium potential for calcium ions, and M and N are the fractions of open Ca^{2+} and K^+ channels. The constants $M_\infty(V)$ and $N_\infty(V)$ represent the fractions of open calcium and potassium channels in the steady state, and the multiplicative factors $\lambda_M(V)$ and $\lambda_N(V)$ are rate constants for the opening of these channels. The forms for steady state channel fractions and rate constants are, on the basis of statistical arguments,

$$\begin{aligned} M_\infty(V) &= \left(\frac{1}{2}\right)\left\{1 + \tanh\left[\frac{V - V_1}{V_2}\right]\right\} \\ \lambda_M(V) &= \bar{\lambda}_M \cosh\left[\frac{V - V_1}{2V_2}\right] \\ N_\infty(V) &= \left(\frac{1}{2}\right)\left\{1 + \tanh\left[\frac{V - V_3}{V_4}\right]\right\} \\ \lambda_N(V) &= \bar{\lambda}_N \cosh\left[\frac{V - V_3}{2V_4}\right] \end{aligned} \quad (9.104)$$

and V_{1-4}, $\bar{\lambda}_M$, and $\bar{\lambda}_N$ are parameters.

In the Morris-Lecar model calcium replaces sodium as a rapid activating inward current. We will study repetitive behavior in this model using a reduced second-order set of equations based on the observation that the Ca^{2+} system operates on a much faster time scale than the K^+ system. As a consequence, g_{Ca} is instantaneously in the steady state at any time so that $M = M_\infty(V)$. Under these conditions the above set of equations can be replaced by the V, N reduced second-order system

$$C\frac{dV}{dt} = -g_K N(V - V_K) - g_{Ca}M_\infty(V)(V - V_{Ca}) - g_L(V - V_L) + I \quad (9.105)$$

and

$$\frac{dN}{dt} = \lambda_N(V)(N_\infty(V) - N) \quad (9.106)$$

Before examining limit cycle behavior, it is worthwhile to consider the case where there is a single stable rest state and the system generates action potentials when a brief threshold-exceeding current pulse is supplied. The time courses of the variations in voltage are shown for several different current pulses in Fig. 9.15a. The second part of Fig. 9.15 shows the phase portrait of the system for this parameter regime. The nullclines are given by the expressions

$$V(N) = \frac{I + g_L V_L + g_{Ca}V_{Ca}M_\infty + g_K V_K N}{g_L + g_{Ca}M_\infty + g_K N} \quad (9.107)$$

and

$$N(V) = N_\infty(V) \quad (9.108)$$

The phase plane trajectories corresponding to the various current pulses are shown along with the nullclines in Fig. 9.15b. These plots show the voltage displacements produced by the current pulse leading to trajectories that execute a loop in phase space and return to the rest state.

We now consider the case of limit cycle behavior. The V and N nullclines plotted in the first part of Fig. 9.16 intersect once to give a single steady state. We would like to establish the conditions for the onset of oscillatory behavior by using the Poincaré-Bendixson theorem. To accomplish this objective, we first show that the trajectories remain bounded in a region of the phase plane. Then we find the conditions leading to the absence of stable steady states inside that bounded domain. The establishment of a bounded domain is straightforward. First, N remains bounded between 0 and 1 by definition. Next, we note

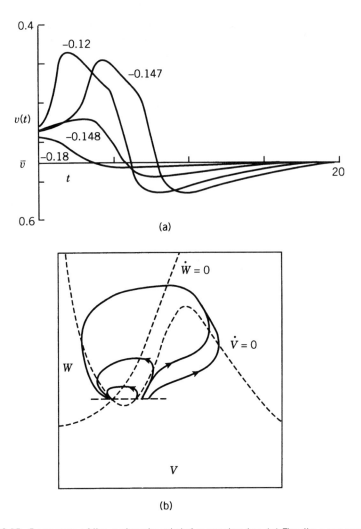

Figure 9.15. Response of the system to a brief current pulse. (*a*) The time courses of the voltage changes for four different stimuli. (*b*) Phase plane plots of the four trajectories. The variable N discussed in the text is plotted in its reduced form as w. The variable V is also plotted in its reduced form as v. (From Rinzel and Ermentrout[92]. Reprinted with permission of MIT Press.)

that $V_{Ca} > V_1$ and $V_K < V_L$. Thus V is bounded by the limiting potentials

$$V_{min} = \frac{I + g_L V_L + g_K V_K}{g_L + g_K} \tag{9.109}$$

and

$$V_{max} = \frac{I + g_L V_L + g_{Ca} V_{Ca}}{g_L + g_{Ca}} \tag{9.110}$$

We now apply linear stability analysis to the behavior of the system in the vicinity of the steady state. The eigenvalues λ that describe the stability properties of the linearized system are solutions of the characteristic equation, Eq. (9.54), with the coefficients b and c given by

$$b = \left(\frac{\partial F_1}{\partial V} + \frac{\partial F_2}{\partial N}\right)_{(V_o, N_o)}$$

$$c = \left(\frac{\partial F_1}{\partial V}\frac{\partial F_2}{\partial N} - \frac{\partial F_2}{\partial V}\frac{\partial F_1}{\partial N}\right)_{(V_o, N_o)}$$

(9.111)

The two coefficients are equal to the sum and product of the eigenvalues, respectively. If we denote the two eigenvalues as λ_+ and λ_-, then we have $\lambda_+ + \lambda_- = b$, and $\lambda_+ \cdot \lambda_- = c$. The condition for limit cycle oscillations is that $b > 0$ and $c > 0$. These inequalities will ensure that both eigenvalues will be positive if the eigenvalues are real and that the eigenvalues will have positive real part if they are complex. Upon substituting the expressions for the functions F_1 and F_2, and evaluating the requisite partial derivatives, we obtain the conditions that must be satisfied by the conductance parameters in order to generate oscillatory behavior, namely

$$g_{Ca}\left(\frac{\partial M_\infty}{\partial V}\right)_{S_o}(V_{Ca} - V_{S_o}) > g_L + g_K N_{S_o} + g_{Ca} M_\infty(V_{S_o}) + C\lambda_N(V_{S_o}) \quad (9.112)$$

and

$$g_{Ca}\left(\frac{\partial M_\infty}{\partial V}\right)_{S_o}(V_{Ca} - V_{S_o}) < g_L + g_K N_{S_o} + g_{Ca} M_\infty(V_{S_o}) + g_K\left(\frac{\partial N_\infty}{\partial V}\right)_{S_o}(V_{S_o} - V_K)$$

(9.113)

The spiral trajectory shown in the first part of Fig. 9.16 approaches a stable limit cycle as the applied current is increased from 0 to 300. The steady state, located at crossing of the nullclines inside the limit cycle, is an unstable node. A plot of the real and imaginary parts of the eigenvalue λ_+ of the linearized system as determined from the characteristic equation is shown in the second part of Fig. 9.16 for different values of the applied current. The negative eigenvalue λ_- traces out a mirror reflection of the positive eigenvalue. The firing property predicted by the Morris-Lecar model is that of an onset of voltage oscillations as the applied current is raised above a threshold followed by a diminishing of repetitive firing at higher magnitudes of the current. The range of currents leading to limit cycle oscillations is shown in the eigenvalue plot.

A variety of firing properties manifest themselves as the conductances are varied over their parameter domains. Shown in Fig. 9.17 are sketches of how

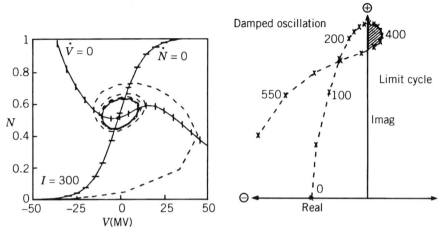

Figure 9.16. (a) Phase plane portrait showing the nullclines in a parameter regime that produces a limit cycle. The solid point at the intersection of the nullclines represents an unstable node. The bars attached to the nullclines show the direction a trajectory must cross. (b) Plot of the changes in the real and imaginary parts of the eigenvalue of the linearized system as the current in the barnacle giant muscle fiber is increased. The curve denotes the positive root. The curve for the negative root is a mirror reflection across the real axis. (From Morris and Lecar[7]. Reprinted with permission of the Biophysical Society.)

the system will respond as g_{Ca} and g_K are systematically varied while the remaining parameters are held constant. In the first of these stability diagrams, we observe several regions of monostability along with domains of damped and limit cycle oscillations. In the region of damped oscillations, the eigenvalues are complex with negative real parts. The system develops an oscillatory response to a transient current which then decays to a constant voltage. In the second part of the figure, the applied current has been set equal to zero. There are two regions of stability. In the monostable domain, there is a single stable node. In the bistable region, there are three stable states; two of these are stable nodes, and the third is a saddle point. The system can exhibit bistable behavior in this portion of the parameter manifold. A more detailed analysis of the Morris-Lecar model has been presented by Rinzel and Ermentrout.[92]

9.8.3 Waves and Synchrony in Systems of Relaxation Oscillators

In Section 9.7 we discussed the synchronization in interconnected populations of excitatory and inhibitory neurons. In the Schuster-Wagner[63,64] model we studied arrays of oscillators in a global mean-field approach which led to a description in terms of phase-coupled oscillators that synchronize by means of phase-differences. In their study Schuster and Wagner noted that their arrays of phase-coupled oscillators failed to synchronize when linked together through nearest-neighbor couplings. In the König-Schillen[65,66] model oscil-

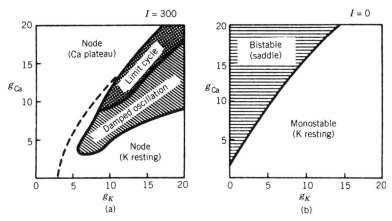

Figure 9.17. Stability diagrams delineating the types of steady state behavior as a function of g_{Ca}, g_K, and I. (a) $I = 300$: Limit cycles occur for those combinations of the parameters giving an unstable fixed point. (b) $I = 0$: Oscillations cannot occur. There are two stability regimes. In one (*unshaded*) the system has a stable node; in the other (*shaded*) the system exhibits bistability. (From Morris and Lecar[7]. Reprinted with permission of the Biophysical Society.)

lators operating in a different manner were able to synchronize when coupled to one another through local, nearest-neighbor interactions. We now discuss synchronization in global and local coupling schemes further. To understand the above-mentioned results, we must first distinguish between relaxation and rotator regimes of nonlinear oscillators and then examine the contrasting synchronization mechanisms used by arrays of units operating as relaxation oscillators and those functioning as rotators.

Relaxation oscillators were mentioned earlier in Section 9.5.2 where we examined the limit cycle trajectory and wave form for the van der Pol oscillator operating in the strong nonlinear regime. We found that the motion of the oscillator was characterized by the presence of two alternating time scales. These rather different time scales appear as slow vertical and fast horizontal jump stages in the limit cycles, and they give rise to a distinctive sawtooth-shaped wave form. These properties may be contrasted with those of the van der Pol oscillator in the weakly coupled regime. Whenever the coefficient of the nonlinear term is small, the limit cycle trajectories are more circular, and differences between slow and fast stages of a period are largely absent. The resulting wave form for the oscillator is far more sinusoidal.

An instructive example of this multiplicity is the Morris-Lecar model. The parameter λ_N appearing in the second equation in the reduced second-order Morris-Lecar system governs the rate of conductance changes in the potassium channels. Altering this intrinsic property of our model neuron can shift the wave form from a sawtooth to a sinusoidal shape. Shown in Fig. 9.18 are the wave forms in the relaxation and rotator regimes, calculated by taking λ_N in Eq. (9.78) to be first small and then large. In the relaxation regime we have a

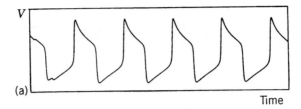

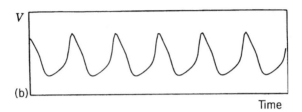

Figure 9.18. Morris-Lecar wave forms in the (a) relaxation ($\lambda_N = 0.02$) and rotator ($\lambda_N = 0.33$) regimes. (From Somers and Kopell[49]. Reprinted with permission of Springer-Verlag.)

situation analogous to that encountered for the van der Pol oscillator. We have alternating fast and slow stages, and in the latter the trajectory hugs either the outer or inner branch of the v nullcline.

The key observation, due to Somers and Kopell,[49,50] is that arrays of coupled relaxation oscillators ultilize a different mechanism to synchronize their motions in the phase plane from that used by groups of coupled rotators. In place of phase-pulling used by oscillators coupled by means of phase differences, relaxation oscillators employ a method of synchronization called *fast threshold modulation*. To facilitate our discussion of fast threshold modulation, we consider the Morris-Lecar model describing the ith oscillator in a coupled array

$$\bar{\lambda}_N \frac{dv_i}{dt} = -g_{Ca} m_\infty(v_i)(v_i - 1) - g_K w_i(v - v_K) - g_L(v_i - v_L) + I_{ext}(v_i, w_i, v_j) \tag{9.114}$$

and

$$\frac{dw_i}{dt} = \frac{1}{\tau_w(v_i)}(w_\infty - w_i) \tag{9.115}$$

These equations have been written in a standard reduced form where the voltages have been scaled by V_{Ca}, the currents has been similarly scaled, and the time has been scaled by the capacitance. In the second equation we have

NEURAL EXCITABILITY AND OSCILLATIONS   393

taken $\lambda_N(v) = \lambda_N/\tau_w(v_i)$, and the scale factor λ_N appears on the left-hand side of the first equation through another scale adjustment. In the above, the external current contains a dependence on the fast variable of the nearest-neighbor oscillator. We assume that $I(v_j)$ has a sigmoidal shape that is aligned with the v_i nullcline so that the inner branch of the nullcline coincides with unsaturated portion of $I(v_j)$ and the outer branch with the saturated segment. This alignment of input current and v_i nullcline is depicted in Fig. 9.19.

We model the external input excitatory current to the ith cell in the array, assumed for simplicity to be linear, through a nearest-neighbor coupling term of the form

$$I_{ext} = -\frac{\alpha g_{Ca}}{2}(m_\infty(v_{i-1}) + m_\infty(v_{i+1}))(v_i - 1) \qquad (9.116)$$

The local minima and maxima, or knees, of a limit cycle trajectory serve as the thresholds for the fast jumps from the outer to the inner branches of the N-shaped nullclines, and vice versa. The positions of these knees are altered by the coupling interactions in the manner shown in Fig. 9.19. A pair of coupled relaxation oscillators will synchronize their phases rapidly in just a few cycles, for a considerable range of initial conditions, by modifying these thresholds for

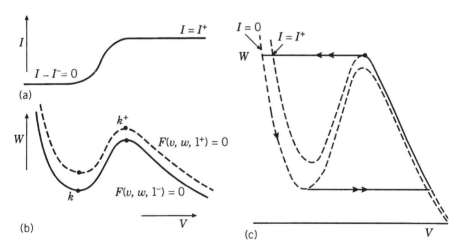

Figure 9.19. Alignment of the synaptic input with the v nullcline. (*a*) Plot of the sigmoidal response of the current to changes in potential. The nonconstant range of I is aligned with the middle branch of the v nullcline. (*b*) The shift in the v nullcline produced by a change in current. The nullclines are computed for the two limiting values of the current. The filled circles labeled as k and k^+ mark the location of the knees. (*c*) In fast threshold modulation the trajectory is followed by an oscillator when its onsets and offsets are synchronous with those of the oscillators that provide its input. (From Somers and Kopell[50]. Reprinted with permission of Elsevier Science, The Netherlands.)

fast jumps. A sufficient condition for synchronization is that the rate of change in the slow variable before the jump is less than that following the jump. This variation generates a compression of the phase difference of the two oscillators along a branch of the nullcline. The variation in the rate of change of the slow variable gives rise to a scalloped appearance when plotting the slow variable as a function of time. Fast threshold modification plus scalloping is able to compensate for differences in native frequences of up to 50% between pairs of coupled oscillators.

These results can be extended to arrays of relaxation oscillators coupled through nearest-neighbor interactions such as those given by Eq (9.116). The speed of synchronization is dependent on the extent of the scalloping in the slow variable. In systems where the nullcline for the slow variable is S-shaped as in the Morris-Lecar model, scalloping and the attendant phase compression will enhance the rate of synchronization. This rate will be slower in systems such as the van der Pol oscillator and the Fitzhugh-Nagumo model which lack an S-shaped nullcline for the slow variable. Finally, as noted by Cohen, Holmes, and Rand[20] and by Kopell and Ermentrout,[22] we find that rotators coupled to one another through nearest-neighbor synaptic interactions will tend to produce traveling waves when phase-locked.

9.9 SPINDLE WAVES

The low-frequency oscillations generated during sleep and arousal originate in the thalamocortical system. Several mechanisms contribute to the low-frequency rhythmicity. The intrinsic electrophysiological properties of thalamic neurons have been characterized by Jahnsen and Llinás.[87,88] In addition to the sodium and potassium conductances underlying the initiation of action potentials, several additional conductances endow thalamic neurons with the ability to function as a relay system and as single-cell oscillators. Thalamic neurons have a number of stable firing states. For example, if the membrane is hyperpolarized, the cells will oscillate at 5 to 6 Hz, and if the membrane is slightly depolarized, the neurons will oscillate at 9 to 11 Hz.

There are perhaps three or more distinct forms of low-frequency thalamo-cortical rhythmicity. Delta-band oscillations are observed in thalamic relay neurons in the frequency range from 0.5 to 4 Hz, spindling is seen at frequencies from 7 to 14 Hz, and as discussed earlier, oscillations are found in the range 7 to 12 Hz associated with dynamic maps[48] and perhaps with the 8- to 15-Hz triggered, behaviorally modulated short-term plasticity.[95] The spindle wave process, which we will examine in this section, depends on both intrinsic cell and network properties, and it provides us with an archetypal example of synchronized neural oscillations. *In vitro* waxing and waning oscillations occur at lower frequencies, from 0.5 to 3.2 Hz, while the higher frequency *in vivo* rhyhmicity depends on interactions with neurons in the thalamic reticular nucleus. With results of recent voltage clamp measurements, we have a characterization of the ionic conductances contributing to the waxing and waning oscillations. We will examine results of the recent modeling work by Destexhe, Babloyantz, and Sejnowski[96] that take into account the new experimental findings. We will then look at some investigations of how the network interactions modulate the firing properties of the relay neurons.

9.9.1 Ionic Mechanisms

Three ionic currents cooperate to produce spindling oscillations. The first of these is the mixed sodium/potassium current I_h, activated by hyperpolarization in the subthreshold range of membrane potentials. The second is the low-threshold calcium current I_T, and the third is the voltage-dependent potassium current I_{K2}. Therefore our starting point for a cellular model of spindle oscillations in thalamic neurons is a Hodgkin-Huxley scheme with

$$C\dot{V} = -g_L(V - V_L) + I_T + I_h + I_{K2} + I \qquad (9.117)$$

The quantity I is an applied external current, and for simplicity, the action-potential-generating sodium and potassium currents have not been included.

A four variable model for the low-threshold, or T-type, calcuim current I_T has been developed by Wang, Rinzel, and Rogawski.[97] It is of the form

$$I_T = -\bar{g}_{Ca} m^3 h (V - E_{Ca}) \qquad (9.118)$$

with

$$\begin{aligned}\dot{m} &= \lambda_m(V)(m_\infty(V) - m) \\ \dot{h} &= \alpha_1(V)[1 - h - d - K(V)h] \\ \dot{d} &= \alpha_2(V)[K(V)(1 - h - d) - d]\end{aligned} \qquad (9.119)$$

where m is the activation variable and h and d are the pair of inactivation variables. The second inactivation variable d describes the slow recovery of I_T from inactivation. The quantity $\bar{g}_{Ca}$ is the maximum conductance of the calcium current, and E_{Ca} is the calcium reversal potential. Our goal in this section is primarily to examine the new modeling features that appear when we try to understand thalamocortical oscillations. Details of the several models, their relationships to the voltage clamp data, the voltage dependences and values for all constants appearing in the above and the following ionic current models are given by Destexhe et al.

The new features introduced in the spindling model include new ways of treating the time course of the ionic currents and the inclusion of a dependence on the intracellular calcium concentration, which couples different ionic channel types. In the Morris-Lecar model the calcium and potassium channels were noninactivating. We now have multiple Hodgkin-Huxley channel gates describing slow and fast components of the currents. The regulation of ionic conductances by intracellular calcium concentration has been considered in a number of modeling efforts. For example, in a study of how intrinsic cellular properties (maximal conductances) are modulated by a neuron's electrical activity in order to maintain stability, LeMasson, Marder, and Abbott[98] used the intracellular calcium concentration as a measure of activity. In the spindling model, calcium which enters a cell through I_T regulates the hyperpolarization-activated inward current I_h.

The voltage clamp data show that I_h is a noninactivating current. It activates in the same subthreshold range of potentials as the low-threshold calcium current I_T. The model adopted for the hyperpolarization activated current I_h is

$$I_h = -\bar{g}_h S_1 F_1 (V - V_h) \tag{9.120}$$

with

$$\begin{aligned}\dot{S}_1 &= \alpha_S(V)(1 - S_1) - \beta_S(V)S_1 \\ \dot{F}_1 &= \alpha_F(V)(1 - F_1) - \beta_F(V)F_1\end{aligned} \tag{9.121}$$

In the above system there are two activation gates, fast (F) and slow (S). During activation F_1 opens rapidly, while S_1 opens more slowly. The dependence of the current on the product of S_1 and F_1 in Eq. (9.120) ensures that the activation kinetics is determined mainly by S_1. During inactivation the situation is reversed. Since F_1 closes rapidly, the product form guarantees that the inactivation kinetics is determined primarily by F_1.

Two possible mechanisms for the waxing and waning oscillations were presented by Destexhe et al. Both are strongly dependent on the kinetic properties of I_h. The first involves regulation of I_h by the intracellular calcium concentration. The second mechanism entails interactions among the three currents, I_h, I_T, and the slow, depolarization-activated outward current I_{K2}. In modeling the regulation of I_h by intracellular calcium, Destexhe et al. assume that the calcium ions directly affect I_h channels. This regulatory influence is depicted schematically in Fig. 9.20a. A second pair of gate variables, F_2 and S_2, representing the fractions of fast and slow gates bound to calcium, are added, and the kinetic equations are modified to reflect the additional variables. The model is completed by describing the influx of Ca^{2+} through I_T channels and the efflux of Ca^{2+} by means of active pumps.

Waxing and waning oscillations are initiated when the progessive hyperpolarization of the membrane reaches the point where I_T becomes deinactivated. In the Ca^{2+}-regulatory model, the binding of calcium to the I_h channels induces concentration-dependent shifts of the activation function of I_h toward more positive membrane potentials, as shown in Fig. 9.20b, producing a gradual depolarization during the oscillatory phase. This depolarization

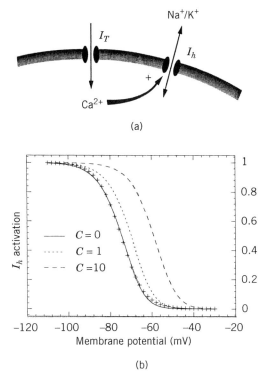

Figure 9.20. Ca^{2+} and voltage-dependent activation of I_h. (a) Illustration of the coupling of the I_T and I_h channels. The low-threshold Ca^{2+} current I_T lets Ca^{2+} enter the cell. These ions bind to the mixed Na^+/K^+ channel I_h and modify its voltage dependent properties. (b) The shift in the voltage dependence of the current toward positive membrane potentials produced by direct binding of Ca^{2+}. The curve labeled $C = 0$ represents the response properties at the resting intracellular calcium concentration. The curves corresponding to higher concentrations are labeled accordingly. (From Destexhe, Babloyantz, and Sejnowski[96]. Reprinted with permission of the Biophysical Society.)

eventually prevents I_T from activating, thereby dampening the oscillations. By this means, the slow oscillations wax and wane as a result of the interactions among the two subthreshold currents and the regulation of I_h by intracellular calcium. Figures 9.21a and 9.21b illustrate how the firing patterns change as a function of the maximum conductance $\bar{g}_h$ and as a function of the time constant k_2^{-1} of the intracellular Ca^{2+} binding to I_h. We see, for example, that the length of the silent phase and the duration of the oscillatory phase are proportional to the binding time constant. The length of the silent phase increases with increasing $\bar{g}_h$, while the duration of the oscillatory phase decreases with increasing $\bar{g}_h$.

The dynamical states of the system can be studied as a function of the slow calcium gating variable S_2. When this is done, we find that the system is in either a stable resting state or undergoing limit cycle oscillations. The

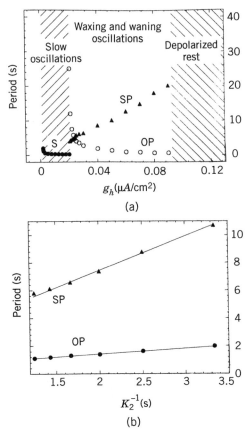

Figure 9.21. Dependence of the period of the waxing and waning oscillations on the maximal conductance of I_h and the time constant of Ca^{2+} binding to I_h channels. The length of the silent phase (SP) and the duration of the oscillatory phase (OP) are shown as a function (a) the maximal conductance $\bar{g}_h$ and (b) the time constant k_2^{-1}. (From Destexhe, Babloyantz, and Sejnowski[96]. Reprinted with permission of the Biophysical Society.)

transition form one to the other at a critical S_2 is a subcritical Hopf bifurcation. In waxing and waning, S_2 oscillates sequentially driving the system into resting and slow oscillatory branches with a pronounced hysteresis.

The results found using the mechanism of Ca^{2+}-regulation of I_h appear to be in closer agreement with the experimental data than the model based on interactions among I_h, I_T, and I_{K2}. In the latter case the frequencies of the oscillations are higher, from 10 to 14 Hz, and therefore appear to be in poorer agreement with the experimental observations. Similar transitions among stable resting states, limit cycle oscillations, and hysteresis loops are found in a phase plane analysis assuming the interacting current mechanism.

9.9.2 Network Mechanisms

The synchronous spindle waves that arise during sleep and arousal are generated through reciprocal interactions between thalamocortical relay neurons and GABAergic cells in thalamic reticular nucleus (nRt) or perigeniculate nucleus (PGN). There are two essential sets of network-related findings in the studies by Steriade, McCormick, and Sejnowski,[99] by von Krosigk, Bal, and McCormick,[100] and by Wang and Rinzel.[101] First, we acquire some insights[99,100] into the manner by which the reciprocal network connections initiate the sequences of events that generate spindling at the requisite frequencies. Second, we gain some understanding[101] of the network mechanism that promotes synchronization of these oscillations among population of neurons.

With regard to the initiation of *in vivo* spindling, we observe that thalamic relay neurons undergo widespread inhibition when they receive input from a critical number of activated PGN GABAergic neurons. The activation of the PGN neurons results in a generation of IPSPs in thalamic relay neurons that terminates the inactivation of the low threshold Ca^{2+} spike, allowing these relay cells to generate rebound Ca^{2+} spikes and associated bursts of action potentials. The rebound Ca^{2+} spikes in turn depolarize PGN neurons, thereby activating low-threshold Ca^{2+} spiking and leading to bursting in the PGN cells. Spindling is initiated by the shifting voltage dependence of I_h as the intracellular Ca^{2+} concentration is reduced and is terminated when the voltage dependence of I_h shifts back as the intracellular Ca^{2+} concentration is increased. Thus the Ca^{2+} related currents enhance and then decrement the spindle waves thereby generating the observed pattern of waxing and waning in the 7- to 14-Hz range.

Let us now consider network mechanisms that promote synchronization. This isssue was investigated by Wang et al.,[97] who found that mutual inhibition can synchronize the discharges provided that cells in the thalamic reticular nucleus possess a T-type calcium current and a leakage current. The full model describing the dynamics of a network of identical, mutually inhibiting neurons studied by Wang and Rinzel[101] contains T-type calcium, leakage, and synaptic currents:

$$C\frac{dV_i}{dt} = -\bar{g}_T m_\infty^3 h_i (V_i - V_{Ca}) - g_L(V_i - V_L) - \sum_{i,j} J_{ij} s_j g_{syn}(V_i - V_{syn}) \quad (9.122)$$

with

$$\frac{dh_i}{dt} = -\frac{\phi}{\tau_h(V)}(h_i - h_\infty(V))$$

$$\frac{ds_i}{dt} = -k_r s_i - s_\infty(V)(s_i - 1) \quad (9.123)$$

In the above, J_{ij} are the elements of the connectivity matrix, and s_i is the postsynaptic conductance denoting the fraction of maximum g_{syn} arising from activity in neuron i. The quantity $s_\infty(V_i)$ is taken to be a sigmoidal function of the presynaptic membrane potential. The T-type calcium current has been simplified from that of Eqs. (9.118) and (9.119). It does not include a slow recovery variable d but instead only has fast gating. The factor ϕ scales the kinetics and τ_h is the time constant for the h-gate.

The behavior of a pair of mutually inhibitory cells under this dynamics depends on the parameters k_r and V_{syn}. These quantities fix the duration and depth of the hyperpolarizing current. If k_r is small and V_{syn} is sufficiently negative, the two cells will synchronize themselves, undergoing in-phase oscillations. If k_r is large, the pair of neurons will also synchronize, but the oscillations will be out of phase. If V_{syn} is insufficiently negative, the T-type calcium current will not be reactivated by the synaptic current, and the two-cell system will exhibit stable steady state rather than oscillatory behavior.

In a network of all-to-all coupled mutually inhibitory neurons, we encounter a range of behavioral responses as g_{syn} is varied. Assuming that k_r is small, we find that for small g_{syn} cells respond as though they were decoupled. For moderate g_{syn}, the network tends to develop clusters of synchronized cells oscillating out of phase with other clusters. Finally, if g_{syn} is large, an oscillating cluster of cells can silence large numbers of other cells in the population by pegging their membrane potentials at a nearly constant hyperpolarized level. The results just mentioned pertain to cells that are not intrinsically oscillatory. If these cells are endowed with intrinsic oscillatory properties, behavioral responses similar to those encountered in the nonautorhythmic case are observed.

9.10 CALCIUM OSCILLATIONS, EXCITABLE MEDIA, AND CELLULAR AUTOMATA

We began our exploration of adaptive cooperative systems with a discussion of the balance between entropic disordering and energy ordering in equilibrium thermodynamic systems. We then studied the orderly evolution of a number of different physical and biological spatial lattice systems. In the last two chapters, we started to explore dissipative and open systems, observing the manner in which they evolve from disordered to highly ordered steady states. In this last technical section, we will examine an especially interesting instance of spontaneous self-organization, namely that of the intracellular material, in both excitable and nonexcitable cells, responsible for the creation of Ca^{2+} oscillations and waves. We will observe that the internal machinery generating the calcium oscillations can be regarded as an excitable medium, and we will then describe the generation of spiral waves using cellular automata. First introduced some time ago by von Neumann and Ulam, cellular automata may be regarded as simple, dynamic models of excitable media. These models bear great similarity to many of the spatial lattice systems studied in earlier chapters. The several connections and contrasts between cellular automata, and Markov random fields and Ising models, will not be explored in this section. Instead, we will restrict our attention to the problem of spiral wave formation.

9.10.1 Calcium Oscillations and Spiral Waves

In the last chapter we observed that the entry of ionized calcium through NMDA and AMPA receptors initiates a sequence of steps leading to synaptic modification. Calcium is a common signal-transducing device in both non-excitable and excitable cells ranging from bacteria to mammaliam neurons. Calcium functions as a highly localized second messesger and, as we saw in the last section, mediates cooperativity among ionic channels. It diffuses slowly and is maintained at low concentrations in the cytosol by a number of binding and removal strategies. It is pumped out of the cytosol into the extracellular spaces and into the endoplasmic reticulum (ER), and is tightly bound by a variety of molecular species in the cytosol. Calcium is sequestered in the ER by specialized storage molecules that are capable of low-affinity, high-capacity warehousing.[102,103]

In nonexcitable and excitable cells, a series of steps involving IP_3 (inositol 1,4,5 triphosphate) or Ry (ryanodine) cell surface receptors, respectively, leads to the release of calcium from the intracellular stores.[102-194] In many instances complex spatiotemporal patterns are generated. These patterns include pulsate signaling, oscillations, plane waves, and spiral patterns. One possible interpretation of these patterns is that frequency encoding is being used transmit information about the precentage receptor occupancy.[102] Alternatively, frequency encoding may be employed as a means of improving signal-to-noise ratios.[104] Yet another possible function for these differing modes of transmission is their use as a means of discrimination between different signal-generating agents.[105]

9.10.2 Excitable Media

An excitable medium is a spatially distributed system of locally excitable elements. Such a system possesses a stable rest state, is able to both damp out small perturbations, and responds strongly to threshold-exceeding stimuli. A given unit in the medium can be excited by suprathreshold stimuli from one or more adjacent excited units. Once the threshold-exceeding stimulus is received by a unit, it too becomes excited in the same manner as the neighboring unit providing the stimulus. In its excited state a unit becomes refractory and slowly deexcites back to its receptive state where it may again become excited.

In simple excitable media, each element interacts with its nearest neighbors through diffusive couplings. The reaction-diffusion equations used to model excitable media of this type take the form

$$\frac{\partial u}{\partial t} = D_u \nabla^2 u + f(u, v) \tag{9.124}$$

and

$$\frac{\partial v}{\partial t} = D_v \nabla^2 v + g(u, v) \tag{9.125}$$

For a two-dimensional system $u = u(x, y, t)$, $v = v(x, y, t)$, and the Laplacian is

$$\nabla^2 = \frac{\partial}{\partial x^2} + \frac{\partial}{\partial y^2} \tag{9.126}$$

In the above expressions D_u and D_v are diffusion coefficients, and the functions f and g describe the nonlinear kinetics of the medium. The phase plane diagram presented in Fig. 9.22 represents the local response properties for an excitable medium in terms of its excitation and recovery variables. The curves $f(u, v) = 0$ and $g(u, v) = 0$ are our familiar nullclines. As we observed when we considered the generation of action potentials, stimuli that exceed the threshold initiate a looping trajectory in the phase plane. We observe that the increase in the excitation variable u is followed by a rise in the value of the recovery variable v and then a decrease in the excitation variable, leading to a drop in the recovery variable and a restoration of excitability as the system returns to the rest state.

Excitable media comprise a diverse class of dynamic cooperative systems. Organized as spatially distributed systems of identical repeating units, they are capable of propagating pulses and planar, circular, and spiral waves across long distances. Two well-known examples are neural action potentials and waves of neuromuscular activity, and so excitable media include excitable membranes. Contractions, and other rhythmic and pulsatile responses by smooth muscle, cortical, and retinal tissue; oscillations, multiple stable stationary states, and multi-armed spiral waves by reactants in the Belousov-Zhabotinskii (B-Z) chemical process; and spiral waves of cyclic adenosine 3′,5′-monophosphate (cAMP) activity by the social amoeba *Dictyostelium discoideum* are well-studied examples. In one dimension, we may encounter solitary waves and wave trains. In two dimensions, there are two forms of traveling waves—expanding concentric waves and rotating spiral waves. In three dimensions, we find two analogous forms—expanding spherical waves and rotating scroll waves.

The data on intracellular calcuim waves suggest that the internal machinery generating spiral and other patterns in *Xenpous laevis* oocytes form an excitable medium. In this interpretation the calcuim stores serve as repeating

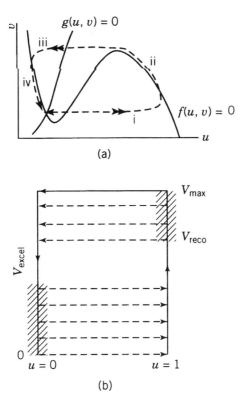

Figure 9.22. (a) Phase plane portrait showing the u and v nullclines intersecting at the unique stable rest state. A threshold-exceeding perturbation initiates the looping trajectory labeled i to iv. (b) Cellular automation containing resting, excited, and refractory states used to model the dynamic behavior of the system. (From Gerhardt, Schuster, and Tyson[106]. Reprinted with permission of the American Association for the Advancement of Science.)

units arranged to form a two-dimensional lattice system. In this last technical section we will examine how this type of cooperativity can be studied using cellular automata functioning as a simple nonequilibrium, dynamic analogue of a Markov random field or Ising system.

9.10.3 Cellular Automata

To motivate our use of cellular automata, we replace the trajectory depicted in Fig. 9.22a with the simplified picture of Fig. 9.22b. In this scheme the excitation variable u may assume one of two values, 0 or 1. The value zero represents the left-hand segment of the $f(u, v) = 0$ nullcline, while the value unity replaces the right-hand portion. The recovery variable v increases whenever $u = 1$, and decreases when $u = 0$. Excitation occurs if v is sufficiently recovered, and deexcitation takes place when v is adequately raised. In its simplest form, this is a description of a system with two states, one in which both u and v are zero, and one in which both variables are unity. In the model we will consider shortly, the recovery variable, as shown in the figure, will be discretized into several values.

Systems composed of regular arrays of units characterized by two variables each of which can assume two or more discrete values and can evolve in discrete time steps according to a set of local rules (i.e., by means of nearest-neighbor interactions) are called *cellular automata*. The rules can be either deterministic or probabilistic, but in either case they are not subject to Boltzmann weighting and detailed balance. As a consequence we may use cellular automata to model the spatiotemporal evolution of dynamic systems far from equilibrium and to explore the emergence of stable or periodic ordered patterns in initially disordered systems. As was the case for spatial lattice systems of coupled oscillators, new order-disorder properties appear. For example, in contrast to equilibrium statistical mechanical systems, one-dimensional cellular automata may undergo continuous phase transitions. More generally, cellular automata have phase transitions, possess critical exponents, and obey scaling laws as do other cooperative lattice systems.

The cellular automata used to describe spiral waves of intracellular calcium obey a set of transition rules[106,107] which are derived from the simplified phase diagram of Fig. 9.22b. As depicted in the diagram, the excitation variable can assume one of two values, 0 and 1, while the revovery variable may take integer values from 0 to v_{max}. The state $u = 0$, $v = 0$ is the resting state. If $u = 1$, the unit is excited, and if $u = 0$, $v > 0$, the cell is in a recovering state. Our first rule

$$u_t = 1 \rightarrow v_{t+1} = \min(v_t + g_{up}, V_{max}) \qquad (9.127)$$

states that if the system is in an excited state at time t, then at time $t + 1$ the recovery variable should increase by an amount g_{up} provided that v is below its maximal permitted level. If the system is in a recovering state, then our next rule

$$u_t = 0 \rightarrow v_{t+1} = \max(v_t - g_{down}, 0) \qquad (9.128)$$

tells us that the recovery variable will decrease by an amount g_{down} as long as v is greater than its minimal value zero. These two rules apply to instances where there are threshold-exceeding interactions. The next three rules cover instances where there are no local interactions. Our third rule states that if the excitation variable is zero at time t, then it will remain at this value at time $t + 1$:

$$u_t = 0 \rightarrow u_{t+1} = 0 \qquad (9.129)$$

The remaining two rules pertain to situations where the excitation variable is unity at time t. If the recovery variable is not at its maximum level, then the excitation variable stays at its current value of unity. However, if the recovery variable is at V_{max}, then there is a transition from the excited state to the recovering state

$$u_t = 1, \quad v_t \neq V_{max} \rightarrow u_{t+1} = 1 \tag{9.130}$$

and

$$u_t = 1, \quad v_t = V_{max} \rightarrow u_{t+1} = 0 \tag{9.131}$$

We are now ready to consider the interaction model. We define the neighborhood for each unit cell, or intracellular calcium store,[108] in terms of a radius r, which is assigned an integer value depending on the spatial resolution required and phenomenon being modeled. The neighborhood system is then defined as all cells lying within a square of linear dimension $2r + 1$ centered on the unit. A resting or recovering cell undergoes a transition to an excited state if the number of excited units within its neighborhood exceeds a threshold value k_{ex}:

$$k_{ex}(v) = k_{ex}^0 + [r(2r+1) - k_{ex}^0]\left(\frac{v}{V_{max}}\right) \tag{9.132}$$

In this expression we see that the threshold is an increasing function of the recovery variable over the range from 0 to V_{ex}. This threshold dependence guarantees that the excitability of the medium will depend on the extent of its recovery. Similarly an excited cell undergoes a transition to a recovering state if the number of resting or recovering units within its neighborhood exceeds a threshold value

$$k_{re}(v) = k_{re}^0 + [r(2r+1) - k_{re}^0]\frac{v - V_{max}}{V_{re} - V_{max}} \tag{9.133}$$

Here we observe that the threshold is a decreasing function of the recovery variable over the range V_{re} to V_{max}.

Experimentally we observe planar waves propagating at speeds in the range 10 to 30μ/s. Also seen are pulsatile foci that generate circular propagating calcium waves following the planar wave transit, and the mutual annihilation of colliding wave fronts. Two important ingredients of any model of these phenomena are (1) dispersion and (2) wave front curvature. The first property refers to the dependence of the speed of wave propagation on the time elapsed since the preceding wave passed through that region; that is, it depends on the extent of the recovery of the medium or its periodicity. The second ingredient pertains to the difference in normal velocities between planar and curved wave motion. In particular, the normal velocity v_N is related to the planar wave speed c, the diffusion coefficient D, and the curvature K through the linear relation $v_N = c + DK$. The first aspect was treated through the above-mentioned rules and interaction model. The second was handled by using r and k_{ex}^0 as adjustable parameters. The model is able to reproduce the linear relation between normal velocity and curvature and also gives results consistent with the dispersion property.

9.11 ADDITIONAL READING

In this chapter we did not discuss pulse-coupled oscillators. This type of model has been used to describe the synchronization of the flashing of fireflies by Mirollo and Strogatz.[109] Summaries of the experimental data on rhythmic activity in the central nervous system have been presented by Gray[110] and by Singer and Gray.[111] The review article by Wolfram[112] provides a good introduction to deterministic cellular automata, and the paper by Kinzel[113] provides an overview of stochastic automata.

9.12 REFERENCES

1. van der Pol, B. (1927). Forced oscillations in a circuit with nonlinear resistance (receptance with reactive triode). Phil. Mag. **3**, 65–80.
2. van der Pol, B., and van der Mark, J. (1928). The heart beat considered as a relaxation oscillation, and an electrical model of the heart. Phil. Mag. **6**, 763–775.
3. Hodgkin, A. L., and Huxley, A. F. (1952). A quantitative description of membrane current and its application to conduction and excitation in nerve. J. Physiol., **117**, 500–544.
4. FitzHugh, R. (1960). Thresholds and plateaus in the Hodgkin-Huxley nerve equations. J. Gen. Physiol., **43**, 867–896.
5. FitzHugh, R. (1961). Impulses and physiological states in theoretical models of nerve membrane. Biophys. J., **1**, 445–466.
6. Nagumo, J., Arimoto, S., and Yoshizawa, S. (1962). An active pulse transmission line simulating nerve axon. Proc. IRE, **50**, 2061–2070.
7. Morris, C., and Lecar, H. (1981). Voltage oscillations in the barnacle giant muscle fiber. Biophys. J., **35**, 193–213.
8. Hodgkin, A. L., and Katz, B. (1949). The effect of sodium on the electrical activity of the giant axon of the squid. J. Physiol., **108**, 37–77.
9. Hodgkin, A. L., and Rushton, W. A. H. (1946). The electrical constants of a crustacean nerve fibre. Proc. Roy. Soc. Lond., **B133**, 444–479.
10. Rall, W. (1962). Electrophysiology of a dendritic neuron model. Biophys. J., **2**, 145–167.
11. Rall, W. (1964). Theoretical significance of dendritic trees for neuronal input-output relations. In P. Reiss (ed.), *Neural Theory and Modeling*. Stanford: Stanford University Press.

12. Rall, W., and Shepherd, G. M. (1968). Theoretical reconstruction of field potentials and dendrodendritic synaptic interactions in olfactory bulb. J. Neurophysiol., **31**, 884–915.
13. Rall, W., and Rinzel, J. (1973). Branch input resistance and steady attenuation for input to one branch of a dendritic neuron model. Biophys. J., **13**, 648–688.
14. Winfree, A. T. (1967). Biological rhythms and the behavior of populations of coupled oscillators. J. Theor. Biol., **16**, 15–42.
15. Kuramoto, Y. (1975). Self-entrainment of a population of coupled non-linear oscillators. In H. Araki (ed.), *International Symposium on Mathematical Problems in Theoretical Physics*. Lecture Notes in Physics, vol. 39. Berlin: Springer-Verlag, pp. 420–422.
16. Kuramoto, Y. (1984). *Chemical Oscillations, Waves and Turbulence*. Berlin: Springer-Verlag.
17. Kuramoto, Y., and Nishikawa, I. (1987). Statistical macrodynamics of large dynamical systems: Case of a phase transition in oscillator communities. J. Statist. Phys., **49**, 569–605.
18. Wilson, H. R., and Cowan, J. D. (1972). Excitatory and inhibitory interactions in localized populations of model neurons. Biophys. J., **12**, 1–24.
19. Wilson, H. R., and Cowan, J. D. (1973). A mathematical theory of the functional dynamics of cortical and thalamic nervous tissue. Kybernetik, **13**, 55–80.
20. Cohen, A. H., Holmes, P. J., and Rand, R. H. (1982). The nature of the coupling between segmented oscillators of the lamprey spinal generator for locomotion: A mathematical model. J. Math. Biol., **13**, 345–369.
21. Ermentrout, G. B., and Cohen, J. D. (1979). Temporal oscillations in neuronal nets. J. Math. Biol., **7**, 265–280.
22. Kopell, N., and Ermentrout, G. B. (1986). Symmetry and phaselocking in chains of weakly coupled oscillators. Commun. Pure Appl. Math., **39**, 623–660.
23. Adrian, E. D. (1950). The electrical activity of the mammalian olfactory bulb. Electroenceph. Clin. Neurophysiol., **2**, 377–388.
24. Freeman, W. J. (1975). *Mass Action in the Nervous System*. New York: Academic Press.
25. Gray, C. M., König, P., Engel, A. K., and Singer, W. (1989). Oscillatory responses in cat visual cortex exhibit inter-columnar synchronization which reflects global stimulus properties. Nature, **338**, 334–337.
26. Gray, C. M., and Singer, W. (1989). Stimulus-specific neuronal oscillations in orientation columns of cat visual cortex. Proc. Nat. Acad. Sci. USA, **86**, 1698–1702.
27. Eckhorn, R., Bauer, R., Jordan, W., Brosch, M., Kruse, W., Munk, M., and Reitbock, H. J. (1988). Coherent oscillations: A mechanism for feature linking in the visual cortex? Biol. Cybern, **60**, 121–130.
28. Jagadeesh, B., Gray, C. M., and Ferster, D. (1992). Visually evoked oscillations of membrane potential in cells of cat visual cortex. Science, **257**, 552–554.
29. Livingston, M. S. (1991). Visually evoked oscillations in monkey striate cortex. Soc. Neurosci. Abstr., **17**, 73.3.

30. Engel, A. K., Kreiter, A. K., König, P., and Singer, W. (1991). Synchronization of oscillatory neuronal responses between striate and extrastriate visual cortical areas of the cat. Proc. Nat. Acad. Sci. USA, **88**, 6048–6052.
31. Engel, A. K., König, P., Kreiter, A. K., and Singer, W. (1991). Interhemispheric synchronization of oscillatory neuronal responses in cat visual cortex. Science, **252**, 1177–1179.
32. Buzsáki, G., Horváth, Z., Urioste, R., Hetke, J., and Wise, K. (1992). High-frequency network oscillation in the hippocampus. Science, **256**, 1025–1027.
33. Ylinen, A., Bragin, A., Nádasdy, Z., Jandó, G., Szábo, I., Sik, A., and Buzsáki, G. (1995). Sharp wave-associated high-frequency oscillation (200 Hz) in the intact hippocampus: Network and cellular mechanisms. J. Neurosci., **15**, 30–46.
34. Gochin, P. M., Gerstein, G. L., and Kaltenbach, J. A. (1990). Dynamic temporal properties of effective connections in rat dorsal cochlear nucleus. Brain. Res., **510**, 195–202.
35. Ahissar, E., Vaadia, E., Ahissar, M., Bergman, H., Arieli, A., and Abeles, M. (1992). Dependence of cortical plasticity on correlated activity of single neurons and on behavioral context. Science, **257**, 1412–1415.
36. Vaadia, E., Haalman, I., Abeles, M., Bergman, H., Prut, Y., Slovin, H., and Aertsen, A. (1995). Dynamics of neuronal interactions in monkey cortex in relation to behavioural events. Nature, **373**, 515–518.
37. Bressler, S. L., Coppola, R., and Nakamura, R. (1992). Episodic multiregional cortical coherence at multiple frequencies during visual task performance. Nature, **366** 153–156.
38. Miles, R., Traub, R. D., and Wong, R. K. S. (1988). Spread of synchronous firing in longitudinal slices from the CA3 region of the hippocampus. J. Neurophysiol., **60**, 1481–1496.
39. Meister, M., Wong, R. O. L., Baylor, D. A., and Shatz, C. J. (1991). Synchronous bursts of action potentials in ganglion cells of the developing mammalian retina. Science, **252**, 939–943.
40. Sillito, A. M., Jones, H. E., Gerstein, G. L., and West, D. C. (1994). Feature-linked synchronization of thalamic relay cell firing induced by feedback from the visual cortex. Nature, **369**, 479–482.
41. Abeles, M. (1991). *Corticonics: Neural Circuits of the Cerebral Cortex*. Cambridge: Cambridge University Press.
42. Malsburg, C. von der (1981). The correlation theory of brain function. Internal report 81-2. Max-Planck-Institute for Biophysical Chemistry, Gottingen.
43. Malsburg, C. von der, and Schneider, W. (1986). A neural cocktail-party processor. Biol. Cybern., **54**, 29–40.
44. Treisman, A. M., and Gelade, G. (1980). A feature integration theory of attention. Cogn. Psychol., **12**, 97–136.
45. Crick, F. (1984). Function of the thalamic reticular complex: The searchlight hypothesis. Proc. Nat. Acad. Sci., **81**, 4586–4590.
46. Damasio, A. R. (1990). Synchronous activation in multiple cortical regions: A mechanism for recall. Semin. Neurosci., **2**, 287–296.
47. Nicolelis, M. A. L., Lin, R. C. S., Woodward, D. J., and Chapin, J. K. (1993). Dynamic and distributed properties of many-neuron ensembles in the ventral

posterior medial thalamus of awake rats. Proc. Nat. Acad. Sci. USA, **90**, 2212–2216.
48. Nicolelis, M. A. L., Baccala, L. A., Lin, R. C. S., and Chapin, J. K. (1995). Sensorimotor encoding by synchronous neural ensemble activity at multiple levels of the somatosensory system. Science, **268**, 1353–1358.
49. Somers, D., and Kopell, N. (1993). Rapid synchronization through fast threshold modulation. Biol. Cybern., **68**, 393–407.
50. Somers, D., and Kopell, N. (1995). Waves and synchrony in networks of oscillators of relaxation and non-relaxation type. Physica, **D89**, 169–183.
51. Strogatz, S. H., and Mirollo, R. E. (1988). Collective synchronization in lattices of non-linear oscillators with randomness. J. Phys., **A21**, L699–L705.
52. Strogatz, S. H., Marcus, C. M., Westervelt, R. M., and Mirollo, R. E. (1989). Collective dynamics of coupled oscillators with random pinning. Physica, **A36**, 23–50.
53. Mirollo, R. E., and Strogatz, S. H. (1990). Jump bifurcation and hysteresis in an infinite-dimensional dynamic system of coupled spins. SIAM J. Appl. Math., **50**, 108–124.
54. Sakaguchi, H., Shinomoto, S., and Kuramoto (1987). Local and global self-entrainment in oscillator lattices. Prog. Theor. Phys., **77**, 1005–1010.
55. Strogatz, S. H., and Mirollo, R. E. (1988). Phase-locking and critical phenomena in lattices of coupled nonlinear oscillators with random intrinsic frequencies. Physica, **D31**, 143–168.
56. Ermentrout, G. B., and Kopell, N. (1984). Frequency plateaus in a chain of weakly coupled oscillators. I. SIAM J. Math. Anal., **15**, 215–237.
57. Ermentrout, G. B., and Kopell, N. (1990). Oscillator death in systems of coupled neural oscillators. SIAM J. Appl. Math., **50**, 125–146.
58. Golomb, D., Hansel, D., Shraiman, B., and Sompolinsky, H. (1992). Clustering in globally coupled phase oscillators. Phys. Rev., **A45**, 3516-3530.
59. Kaneko, K. (1990). Clustering, coding, switching, hierarchical ordering, and control in a network of chaotic elements. Physica, **D41**, 137–172.
60. Kaneko, K. (1991). Partition complexity in a network of chaotic elements. J. Phys, **A24**, 2107–2119.
61. Edelstein-Keshet, L. (1988). *Mathematical Models in Biology*. New York: Random House.
62. Strogatz, S. H. (1994). *Nonlinear Dynamics and Chaos*. Reading: Addison-Wesley.
63. Schuster, H. G., and Wagner, P. (1990). A model for neuronal oscillations in the visual cortex I. Mean-field theory and derivation of the phase equations. Biol. Cybern., **64**, 77–82.
64. Schuster, H. G., and Wagner, P. (1990). A model for neuronal oscillations in the visual cortex II. Phase description of the feature dependent synchronization. Biol. Cyber., **64**, 83–85.
65. König, P., and Schillen, T. B. (1991). Stimulus-dependent assembly formation of oscillatory responses: I. Synchronization. Neur. Comput., **3**, 155–166.
66. Schillen, T. B., and König, P. (1991). Stimulus-dependent assembly formation of oscillatory responses: II. Desynchronization. Neur. Comput., **3**, 167–178.

67. Bush, P. C., and Douglas, R. J. (1991). Synchronization of bursting action potential discharge in a model network of neocortical neurons. Nueral Comput., **3**, 19–30.
68. Koch, C., and Schuster, H. (1992). A simple network showing burst synchronization without frequency locking. Neur. Comput., **4**, 211–233.
69. Wilson, M. A., and Bower, J. M. (1990). Computer simulation of oscillatory behavior in cerebral cortical networks. In D. S. Touuretzky (ed.), *Advances in Neural Information Processing Systems*, vol. 2. San Mateo: Morgan Kaufmann, pp. 84–91.
70. Buhl, E. H., Halasy, K., and Somogyi, P. (1994). Diverse sources of hippocampal unitary inhibitory postsynaptic potentials and the number of synaptic release sites. Nature, **368**, 823–828.
71. Traub, R. D., Miles, R., and Buzsáki, G. (1992). Computer simulation of carbachol-driven rhythmic population oscillations in the CA3 region of the in vitro rat hippocampus. J. Physiol., **451**, 653–672.
72. Whittington, M. A., Traub, R. D., and Jefferys, J. G. R. (1995). Synchronized oscillations in interneuron networks driven by metabotropic glutamate receptor activation. Nature, **373**, 612–615.
73. Llinás, R. R., Grace, A. A., and Yarom, Y. (1991). *In vitro* neurons in mammalian cortical layer 4 exhibit intrinsic oscillatory activity in the 10- to 50-Hz frequency range. Proc. Nat. Acad. Sci. USA, **88**, 897–901.
74. Larson, J., and Lynch, G. (1986). Induction of synaptic potentiation in hippocampus by patterned stimulation involving two events. Science, **232**, 985–988.
75. Huerta, P. T., and Lisman, J. E. (1993). Heightened synaptic plasticity of hippocampal CA1 neurons during a cholinergically induced rhythmic state. Nature, **364**, 723–725.
76. Traub, R. D., Miles, R., Wong, R. K. S., Schulman, L. S., and Schneiderman, J. H. (1987). Models of synchronized hippocampal bursts in the presence of inhibition II. Ongoing population events. J. Neurophysiol., **58**, 752–764.
77. Traub, R. D., Miles, R., and Wong, R. K. S. (1989). Model of the origin of rhythmic population oscillations in the hippocampal slice. Science, **243**, 1319–1325.
78. Barlow, H. B. (1972). Single units and sensation: A neuron doctrine for perceptual psychology? Perception, **1**, 371–394.
79. Getting, P. (1989). Emerging principles governing the operation of neural networks. Ann. Rev. Neurosci., **12**, 185–204.
80. Engel, A. K., König, P., and Singer, W. (1991). Direct physiological evidence for scene segmentation by temporal coding. Proc. Nat. Acad. Sci. USA, **88**, 9136–9140.
81. Finkel, L. H., and Edelman, G. M. (1989). Integration of distributed cortical systems by reentry: A computer simulation of interactive functionally segregated visual areas. J. Neurosci., **9**, 3188–3208.
82. Sporns, O., Gally, J. A., Reeke, G. N., Jr., and Edelman, G. M. (1989). Reentrant signaling among simulated neuronal groups leads to coherency in their oscillatory activity. Proc. Nat. Acad. Sci. USA, **86**, 7265–7269.
83. Sporns, O., Tononi, G., and Edelman, G. M. (1991). Modeling preceptual grouping and figure-ground segregation by means of active reentrant connections. Proc. Nat. Acad. Sci. USA, **88**, 129–133.

84. Sompolinsky, H., Golomb, D., and Kleinfeld, D. (1991). Cooperative dynamics in visual processing. Phys. Rev. **A43**, 6990–7011.

85. Malsburg, C. von der, and Buhmann, J. (1992). Sensory segmentation with coupled neural oscillators. Biol. Cybern., **67**, 233–242.

86. Singer, W. (1993). Synchronization of cortical activity and its putative role in information processing and learning. Ann. Rev. Physiol., **55**, 349–374.

87. Steriade, M., Amzica, F., and Contreras, D. (1996). Synchronization of fast (30–40 Hz) spontaneous cortical rhythms during brain activation. J. Neurosci., **16**, 392–417.

88. Munk, M. H., Roelfsema, P. R., König, P., Engel, A. K., and Singer, W. (1996). Role of reticular activation in the modulation of intracortical synchronization. Science, **272**, 271–274.

89. Kuffler, S. W., Nicholls, J. G., and Martin, A. R. (1984). *From Neuron to Brain*. Sunderland: Sinauer Associates.

90. Llinás, R., and Yarom, Y. (1986). Oscillatory properties of guinea-pig inferior olivary neurones and their pharmacological modulation: An *in vitro* study. J. Physiol., **376**, 163–182.

91. Llinás, R. (1988). The intrinsic electrophysiological properties of mammanlian neurons: Insights into central nervous system function. Science, **242**, 1654–1664.

92. Rinzel, J., and Ermentrout, G. B. (1989). Analysis of neural excitability and oscillations. In C. Koch and I. Segev (eds.), *Methods in Neuronal Modeling*. Cambridge: MIT Press, pp. 135–169.

93. Jahnsen, H., and Llinás, R. (1984a). Electrophysiological properties of guinea-pig thalamic neurones: An in vitro study. J. Physiol., **349**, 205–226.

94. Janhsen, H., and Llinás, R. (1984b). Ionic basis for the electroresponsiveness and oscillatory properties of guinea-pig thalamic neurones in vitro. J. Physiol., **349**, 227–247.

95. Castro-Alamancos, M. A., and Conners, B. W. (1996). Short-term plasticity of a thalamocortical pathway dynamically modulated by behavioral state. Science, **272**, 274–277.

96. Destexhe, A, Babloyantz, A., and Sejnowski, T. J. (1993). Ionic mechanisms for intrinsic slow oscillations in thalamic relay neurons. Biophys. J., **65**, 1538–1552.

97. Wang, X.-J., Rinzel, J., and Rogawski, M. A. (1991). A model for the *T*-type calcuim current and low threshold spike in thalamic neurons. J. Neurophysiol., **66**, 839–850.

98. LeMasson, G., Marder, E., and Abbott, L. F. (1993). Activity-dependent regulation of conductances in model neurons. Science, **259**, 1915–1917.

99. Steriade, M., McCormick, D. A., and Sejnowski, T. J. (1993). Thalamocortical oscillations in the sleeping and aroused brain. Science, **262**, 679–685.

100. Krosigk, M. von, Bal, T., and McCormick, D. A. (1993). Cellular mechanisms of a synchronous oscillation in the thalamus. Science, **261**, 361–364.

101. Wang, X.-J., and Rinzel, J. (1993). Spindle rhythmicity in the reticularis thalami nucleus: Synchronization among mutually inhibitory neurons. Neurosci., **53**, 899–904.

102. Tsien, R. W., and Tsien, R. Y. (1990). Calcium channels, stores and oscillations. Ann. Rev. Cell Biol., **6**, 715–760.
103. Clapham, D. E. (1995). Calcium signaling. Cell, **80**, 259–268.
104. Astri, A., Amundson, J., Clapham, D., and Sneyd, J. (1993). A single-pool model for intracellular calcium oscillations and waves in the *Xenopus laevis* oocyte. Biophys. J., **65**, 1727–1739.
105. Lechleiter, J., Girard, S., Clapham, D., and Peralta, E. (1991). Subcellular patterns of calcium release determined by G protein-specific residues of muscarinic receptors. Nature, **350**, 505–508.
106. Gerhardt, M., Schuster, H., and Tyson, J. J. (1990). A cellular automaton model of excitable media including curvature and dispersion. Science, **247**, 1563–1566.
107. Lechleiter, J., Girard, S., Peralta, E., and Clapham, D. (1991). Spiral calcium wave propagation and annihilation in *Xenopus laevis* oocytes. Science, **252**, 123–126.
108. Markus, M., and Hess, B. (1990). Isotropic cellular automation for modelling excitable media. Nature, **347**, 56–58.
109. Mirollo, R. E., and Strogatz, S. H. (1990). Synchronization of pulse-coupled biological oscillators. SIAM J. Appl. Math., **50**, 1645–1662.
110. Gray, C. M. (1994). Synchronous oscillations in neuronal systems: Mechanisms and functions. J. Comput. Neurosci., **1**, 11–38.
111. Singer, W., and Gray, C. M. (1995). Visual feature integration and the temporal correlation hypothesis Ann. Rev. Neurosci., **18**, 555–586.
112. Wolfram, S. (1983). Statistical mechanics of cellular automata. Rev. Mod. Phys., **55**, 601–644.
113. Kinzel, W. (1985). Phase transitions of cellular automata. Z. Phys., **B58**, 229–244.
114. Traub, R. D., Whittington, M. A., Stanford, I. M., and Jefferys, J. G. R. (1996). A mechanism for generation of long-range synchronous oscillations in the cortex. Nature, **383**, 621–624.
115. Gray, C. M., and McCormick, D. A. (1996). Chattering cells: Superficial pyramidal neurons contributing to the generation of synchronous oscillations in the visual cortex. Science, **274**, 109–113.

GLOSSARY

Allosteric proteins Molecular agents that control and coordinate biochemical events in cells. These proteins possess two or more distinct binding sites, one active and one or more nonactive. When active and nonactive sites couple to one another through conformational changes, the binding of a ligand at the nonactive site will regulate the affinity of the protein for a ligand at the catalytic site.

Assembly coding The dynamic representation of features by the synchronous firing of clusters of cells. Temporal clusters form and reform in response to input signals. Individual neurons belong at various times to different assemblies, and can rapidly associate into a functional group while at the same time disassociating from a different group.

Binding hypothesis The reintegration of different attributes through the synchronous firing of feature encoding cells. Neurons that encode attributes or features that belong together are integrated, or bound together, by their synchronous firing through cooperative processes mediated by the network connectivity. Neurons so bound to form temporal clusters may be distributed across different cortical columns, areas, and hemispheres.

Brownian motion The random, zig-zag motion of dust particles suspended in fluid as first observed by Robert Brown using a light microscope. The two key elements in Einstein's description of brownian motion are its modeling as a Markov process and the notion that only tiny changes in velocity are possible in any small time interval.

Canonical ensemble Maximum entropy distribution for the case where there is one data constraint, usually taken as specifying a mean value for the internal or total energy. These exponential probability distributions are commonly referred to as Gibbs distributions.

Chemoaffinity hypothesis The guidance of axons to their target structures by unique sets of biochemical labels, or markers, that are read by their growth cones. The labels are distributed in a graded fashion, both horizontally and vertically, across the source and target neural structures permitting cell-cell recognition. The markers are acquired during differentiation and uniquely code positional information through their concentrations. These labels, which are supplied by diffusible molecules, are supplemented by local positional cues provided by cell surface molecules placed at decision points and elsewhere along the axonal pathways.

Cliques and clique potentials A generalization of the notion of nearest-neighbor system and nearest-neighbor interactions in the Ising model. A clique is either a single site or a collection of site plets in which every site is a neighbor of every other site. Clique potentials describe the interactions between clique elements.

Convex function A smoothly-varying function possessing a single minimum that can be found starting from any point by means of gradient descent. The curve defined by a convex function always lies below the straight line connecting any two points along that curve.

Critical indices   The exponent of the leading term in power series expansions in $(T - T_C)/T_C$ of the free energy and associated thermodynamic parameters. These quantities characterize the divergences in the thermodynamic parameters of state in the vicinity of a phase transition.

Critical opalescence   Optically visible phenomenon arising as a consequence of density fluctuations in the vicinity of the critical point. As the critical point is approached, regions of a fluid will begin to contain density fluctuation comparable in size to the wavelength of visible light. These density fluctuations generate changes in the index of refraction of the fluid, and as a consequence, the density fluctuations will be optically observable.

Critical point   A point where a system loses stability against fluctuations in its parameters of state. If there are no changes in symmetry properties of a substance as the phases change, the associated first order phase transition will terminate at a critical point. Above a critical point the system can pass continuously between the two phases. In a one component system, second order phase transitions can only occur at a critical point. A critical point occurs at a unique critical pressure P_C and temperature T_C.

Differential adhesion hypothesis   The determination of morphology in systems that cohere while maintaining mobility by the relative strengths of adhesiveness. In such systems the tendency to adhere is related to a preference to minimize the adhesive free energy. The stable states of the system are states of minimal adhesive free energy. The relative strengths of adhesion of the mixed population of cell types or phases will determine the overall morphology of the system, and a hierarchy of positions will occur that reflect the relative cellular adhesiveness.

Entropy (information-theoretic)   A measure of uncertainty or missing information uniquely specified in terms of probabilities by consistency conditions on continuity, monotonicity, and composition. Probabilistic definitions of entropy originate with the work of Boltzmann and Planck.

Entropy (thermodynamic)   A nondecreasing parameter of state of a thermodynamic system. This quantity provides a measure of the amount of thermal disorder present in a system. The term was coined by Clausius from the Greek word for "transformation." For a system not in thermal isolation, the increase in entropy is proportional to the differential amount of heat absorbed by that system.

Equilibrium states   Simple configurations of a thermodynamic system. These configurations are characterized solely by a few parameters of state and are stable against fluctuations in these state variables.

Ergodicity breaking   The restriction of the hypothesis or state space to a particular region of the energy landscape. When ergodicity is broken, the system becomes trapped in one region of configuration space and cannot escape to other valleys of comparable or lower energy.

Excitable membrane   A system capable of propagating pulses and waves of activity over long distances without attenuation. Spatially organized as systems of repeating regenerative units, these structures propagate activity through nearest-neighbor interactions. They have a stable rest state, are capable of damping out small perturbations, and respond strongly to above-threshold stimuli. Action potentials in nerve fibers and pulsatile responses of smooth muscle are two examples of propagation in excitable membranes.

GLOSSARY

Fluctuation-dissipation theorems A class of expressions that relate the relaxation of a system in a nonequilibrium state to the microscopic dynamics in that system at equilibrium. These relations pertain to small perturbations in a linear-response regime. In its simplest form they tell us that the greater the friction, or dissipation, the shorter the period of correlations between velocities of the brownian motion. The mobility, or inverse of the friction constant, is proportional to the diffusion coefficient, and the friction constant is related to the equilibrium fluctuations in the rapidly varying random force. More generally, the fluctuation-dissipation theorems relate the relaxation of a system in a perturbed, nonequilibrium state to the correlations in the spontaneous fluctuations at different times in an equilibrium state.

Frustration The inability to satisfy all constraints at each site of a system in which several sets of interactions or constraints are coactive.

Gibbs-Markov equivalence Assertion that the global character of a Gibbs random field possessing site potentials that are sums of clique potentials is equivalent to the purely local character of a Markov random field.

Gibbs random field A random field whose joint probability distribution is of the form of a Gibbs distribution with a potential, or hamiltonian, that captures the global properties of the system.

Grand canonical ensemble Maximum entropy distributions for cases where there are two or more data constraints, usually taken as specifying a mean values for the total energy and particle number(s). These exponential distributions are commonly referred to as (generalized) Gibbs distributions.

Hebbian synapse A synapse whose state can be modified by use-dependent, spatiotemporally-correlated input in a manner consistent with the Hebb-Stent rule.

Hysteresis A memory effect in which the state of a system depends on its previous history. When hysteresis is present, the state of a system is no longer uniquely determined by the values of its parameters of state. If a process is reversed in a system with hysteresis, the path (sequence of states) traced out will differ from the earlier one. The path formed by repeatedly cycling a system through a set of its parameter values is called a hysteresis loop. Hysteresis is associated with first order phase transitions, where discontinuities in the thermodynamic derivatives occur.

Langevin dynamics A dynamics generated by the Langevin equation. In this dynamics, sometimes called brownian dynamics, a gradient term produces a slow drift that evolves the system towards useful near-optimal low energy states, while rapid random fluctuations dominate the short time behavior, thereby permitting escape from local minima.

Latent heat The molar entropic change, $T\Delta s$, associated with a first order phase transition. This quantity represents the heat absorbed in transforming a given amount of material from one phase to another, and the (molar) entropy difference, Δs, arises from the increase in configuration order that occurs in the more condensed phase.

Limit cycle oscillator A self-sustaining oscillator having a preferred amplitude and waveform to which it will return if disturbed. Limit cycle oscillators describe closed trajectories in phase space that are stable against small perturbations.

Long-term potentiation (LTP) and long-term depression (LTD) Long-lasting (days, weeks, or months), saturation-limited forms of synaptic plasticity triggered by Ca^{2+} influx through NMDA receptors. The threshold and magnitude of LTP and its inverse, LTD, are frequency and prior use dependent. Rapid use produces LTP, slow extended use generates LTD, and the crossover from one to the other depends on prior activity.

Markov chain   A sequence of states $x_1, x_2, \ldots, x_n$, forms in a manner such that the probability that the system is in a particular state x_t at time t depends exclusively upon the probability for the system to be in state x_{t-1} at time $t - 1$. If the sequence of states forms a Markov chain, we do not have to explicitly consider the probabilities for any earlier time steps.

Markovian   A dynamic property of a system that describes how sequences of states are generated. A system is **temporally Markovian** if its state at a particular time depends upon its state at the immediate preceding time, but not on any of its states at earlier times. Similarly, a system is **spatially Markovian** if the states of its constituent elements depend on those of their neighbors, but not on the states of units that are spatially remote.

Markov process   A stochastic process in which the conditional probability that the system is in a particular state at any time is determined by the distribution of states at its immediately preceding time. That is, the conditional distribution of that states of a system given the present and past distributions depends only upon the present.

Markov random field   A random field whose joint probability distribution has associated positive-definite, translational-invariant conditional probabilities that are spatially Markovian.

Maximum entropy principle   Parsimonious inferences using probability distributions that maximize the entropy, subject to the given constraints. We construct probability distributions that are consistent with data, and are otherwise maximally noncommittal with respect to information that is not available.

Mean-field theory   Self-consistent procedures for replacing a global set of interactions with an effective mean-field coupling. The resulting simplified hamiltonians are capable of predicting long-range order. In the Weiss molecular-field theory, interactions between elements of a system are treated in terms of their average effect, and fluctuations are neglected. In other approaches, we treat exactly interactions within a local cluster. All other interactions are treated in terms of their average effect on a given element, thereby neglecting all fluctuations operating length scales larger than a cluster.

Microcanonical ensemble   Maximum entropy distribution for the case where the only constraint is the normalization condition. All states are equally probable, and the probabilities are equal to the inverse of the total number of available states.

Mutual synchronization or entrainment   The synchronization of rhythmic activity brought on in a population of nonlinear dynamic units when each member initially moving with its own native rhythmicity becomes subject to coupling interactions with the others.

NMDA receptors   A subclass of glutamergic excitatory amino acid receptor that is dual voltage and ligand-gated, and allows for a graded entry of Ca^{2+}. Named for the artificial glutamate analog N-methyl-D-aspartate, this class of receptor is thought to serve as a Hebbian molecule. NMDA receptors are activated only when there is a concurrent activation of a sufficient number of adjacent nonNMDA glutamergic receptors to strongly depolarize the postsynaptic membrane.

Order parameter   A fluctuating thermodynamic variable, the average of which provides a measure of the amount of order present in the system.

Parameters of state   Macroscopic quantities that fully characterize the equilibrium states of a thermodynamic systems. These quantities are defined solely by their instantaneous values; small integrated changes in a parameter of state are independent of the paths taken from the initial to the final states.

Partition function Normalization constant for the maximum entropy probabilities. The partition function contains the known information about a system under study.

Phase oscillators or rotators Nonlinear dynamic units having circular or nearly circular limit cycle trajectories and a sinusoidal waveform. Communities of rotators coupled to one another through interactions that depend on the difference of their phases can synchronize their motions by means of a phase pulling mechanism.

Phase transition The loss of stability against fluctuations in the parameters of state of a system. If a system is brought into a condition where it loses its stability against fluctuations in its thermodynamic parameters, that entity will cease to be homogeneous and will split into several portions. In a phase transition, qualitative change in the macroscopic properties of the system is brought about by the cooperative actions of the microscopic fluctuations. In a **first order phase transition**, there are discontinuities in the molar parameters of state such as the molar entropy, energy, and volume. There is a nonvanishing latent heat, and there is hysteresis. Discontinuities in the specific heat may occur as new modes of excitation become accessible. In a **second order, or continuous, phase transition,** the molar parameters of state are continuous. There is no latent heat or hysteresis. Second order phase transitions are characterized by singularities in the specific heat. Phase transitions are signaled by the emergence of order parameters. If the order parameter vanishing discontinuously the transition is first order. Conversely, if the order parameter vanishing continuously the phase transition is second order.

Quenched random variables Random variables that are frozen-in, or fixed, at some value for each element. Random variables do not participate in the kinetic process. Unlike **annealed random variables**, the partition function does not contain a summation over the set of possible values for quenched variables, but instead, the partition function must be averaged over the random variables in order to yield results that are independent of a particular set of frozen-in values.

Random field A set of random variables and an accompanying joint probability measure for all possible values of the random variables over a given lattice or graph.

Random variables Quantities that are of physical interest and for which we define probability distributions. We consider each unique configuration of a system as a point in a configuration space. A random variable, X, is an assignment of a numerical value to each of these points in accordance with a probability distribution. We designate this assignment by the statement $X = x_i$.

Range of correlations or correlation length A measure of the spatial extent of the correlated fluctuations in the thermodynamic variables.

Regression hypothesis The regression or decay of spontaneous fluctuations in a system at equilibrium through the same mechanisms that promote the relaxation of a macroscopic equilibrium-destroying perturbation of that system. The spontaneous fluctuations in a system at equilibrium with its surroundings are indistinguishable from the deviations from equilibrium produced by small perturbations. This theorem provides a physical justification for our use of the same dynamics, as in simulated annealing via the Metropolis algorithm, to describe both equilibrium and near-equilibrium situations.

Relaxation oscillators Nonlinear dynamic units characterized by a phase plane trajectories that have two-time courses, fast and slow, and a sawtooth waveform. Communities of relaxation oscillators coupled to one another can synchronize their motion by fast threshold modulation.

Renormalization group methods Recursive procedures for successively integrating out degrees-of-freedom. The transformations provided by renormalization group methods describe how effective couplings appearing at different length scales relate to one another. Developed initially to study second order phase transitions, RG methods enable us to study the thermodynamic properties of a system in the vicinity of a critical point where fluctuations are important, and they allow us to derive scaling laws.

Simulated annealing An artificial stochastic dynamics that evolves a system through a sequence of equilibrium or near equilibrium states which converges in distribution to a useful near optimal low energy configuration.

Spin glasses Materials in which the interactions between elements are random and frustrated. In dilute magnetic alloys the spin glass phase refers to a low temperature regime in which the local magnetic moments are frozen into spatially random equilibrium orientations.

Spontaneous electrical activity Noise-driven electrical activity in a cell or local circuit. In the retina, this term denotes the prenatal electrical discharges from ganglion cells. Discharges from neighboring cells are correlated and generate traveling waves of activity across the retina.

Spontaneous magnetization The limiting value of the magnetization approached as the external magnetic field goes to zero. The spontaneous magnetization is the order parameter for a ferromagnet.

Spontaneous symmetry breaking The breaking of a symmetry by spontaneous fluctuations in the vicinity of a critical point or by some small asymmetry in the interactions present during the initial preparation of the system. In **broken symmetry,** the ground state of the system does not have the full symmetry of the hamiltonian used to describe the system. In a second order phase transition, the symmetry on the two sides of the transition are related to one another. When the symmetry group that is broken is continuous, as in the breaking of translational or rotational invariance, a new mode of excitation may appear whose frequency goes to zero at long wavelengths (Goldstone's theorem).

Stationary process A Markov process that is independent of the time of measurement. If a system settles down in time to a steady state so that its stochastic properties are independent of the time they are measured, then the process is said to be stationary. A process is stationary if the one-time probabilities are time-independent, and if the two-time conditional probabilities are dependent upon the time differences only, and similarly for the joint probabilities.

Synaptic plasticity The adaptive changes in the efficiency of synaptic transmission triggered by use. The modifications may emerge in the form of anatomical and morphological changes such as sprouting and pruning of axons and dendrites, naturally-occurring cell death, and/or alterations in synaptic membrane properties. The term synaptic plasticity encompasses activity-dependent processes such as long-term potentiation, long-term depression, and use-dependent modifications produced either surgically or by manipulations of afferent input, and the Hebbian mechanisms that promote these forms of synaptic plasticity are thought to underlay learning and memory.

Ultrametric Hierarchical, tree-like organization of the space of states of several prominent NP-hard systems. Below the critical temperature sequences of equilibrium states can be generated by a process of bifurcation. If three states are picked at random, there are only two cases where their mutual overlap is nonvanishing. Either all three pairwise overlaps are equal to one another corresponding to an equilateral triangle in the space of states, or two overlaps are equal and the third is larger than the other two, thereby generating an isosceles triangle with the unequal side smaller than the others.

Universality The notion that the essential properties of a system undergoing a phase transition are independent of the detailed nature of the interactions other than for a dependence upon the dimensionality of the system. Thus all critical phenomena share a common set of features.

Wiener, or brownian motion, process A Markov process formed by a temporal sequence of random variables having conditional probabilities that obey the diffusion equation. In a Wiener process, the random variables are Gaussian-distributed, the mean increment is zero, the mean square increments are proportional to the time intervals, and the increments are independent and stationary.

INDEX

Absorbing states, 53
Acquaintance sets, 202-203
Allosteric proteins, 1, 11
Alpha waves, 338
AMPA receptors, 308, 312, 374
Annealing, in metallurgy, 131, 219
Annealing schedule, see Cooling schedule
Annealing, simulated, see Simulated annealing
Arousal and attention, 338, 340, 381
Arrow model, 155
Assemblies, 17-18, 191-192, 196
Assembly coding, 18-19, 339-340, 378-381
Autobinomial model, 198-199
Auto (linear) models, 188, 196
Autologistic model, 197-198
Autonormal model, 197-198. See also Gauss-Markov models
Autoregression representations, 201, 202-204

Bayesian inferencing:
 in graduated nonconvexity, 238
 in iterated conditional modes, 219-221
 in maximizer of the posterior marginals, 215-216
 in maximum a posteriori estimation, 188, 208-210, 214-215
BCM theory, 9, 279, 281-282, 283-284
 feature selectivity, 298-301
 linear integrator, 290-292
 mean-field network, 303-307
 neurophysiological basis, 307-315
 objective function formulation, 325-329
 receptive fields, 301-303
 rule for synaptic modification, 292-294
 stability properties, 10, 295-296, 306
Besag's auto models, see Auto (linear) models
Beta band oscillations, 338
Bifurcations:
 in coupled oscillator communities, 348

 in spindling, 398
Binary alloys, 67, 111-112, 115-117
Binding and feature integration:
 by assembly coding, 339-340, 378-381
 in random-field models, 187, 226-230
 random fields and assembly coding, 19, 279, 341
Biological membrane, 64
Block circulant matrices, 208
Boltzmann factor, 43, 49-50, 128-129
Boltzman H-theorem, 31
Bragg-Williams method, 87-89, 90-91
Bethe-Peierls-Weiss approximation, 89-91
Broken symmetry, see Spontaneous symmetry breaking
Brownian motion, 7, 254-255, 256-257
 mean-square displacement, 261-263
 velocity-correlation function, 267-269, 271
 velocity increments, 272-276
Burst synchronization, see Synchronized bursting

Calcium oscillations and traveling waves, 400, 401, 405-406
Calcium signaling, intracellular:
 in growth cones, 173-174
 in neurons, 396. See also NMDA receptors
CaMKII, see CaM kinase
CaM kinase, role in synaptic plasticity, 309-314
Canonical ensemble, 31, 36, 43
Carbachol-driven oscillations, 375
Causal random field, 187-188
Cell adhesion molecules, 12, 150-151, 169, 171-172
 dynamic regulation, 172-173
 and synaptic plasticity, 308-309
Cell death:
 naturally occurring, 11, 278-279
 programmed, 278

421

INDEX

Cellular automata, 400–401, 403–406
Chapman-Kolmogorov equation, 255–256, 257, 258
Chattering cells, 381
Chemical potential, 45
Chemoaffinity hypothesis, 151
Chemotropic factors, 12, 150, 151
   chemoattractants, 169–170
   chemorepulsants, 170–171
Circulant matrices, 206–207
Classical rearing, *see* Critical period
Clausius's integral, 41
Clique potentials, 186, 210–212, 223–226
Cliques, 186, 190–191, 194–195
Cluster formation, 9, 339, 340, 381
   and frequency plateaus, 350–352
   for globally coupled phase oscillators, 356
   in hippocampal networks, 376
   in a nearest-neighbor system, 350
   in networks of chaotic elements, 356
   in a reentrant connection model, 380
   in spindling, 399–400
Cluster methods, 139
Combinatorial optimization, 126–128. *See also* Constraints, multiple; Simulated annealing
Common input model, 322–325
Compound Gauss-Markov random fields, 201, 230–232
Concerted transitions, 10–11
Conformational changes, 1
Conservation system, 286–289
Constrained optimization:
   in graph partitioning, 136–137
   in integration, 213–2115, 226–230
Constraints, multiple, 110, 125–126. *See also* Constrained optimization
Continuous simulated annealing, *see* Langevin dynamics
Convexity, 234–235
Cooling schedule, 131–133
Cooperative processes, 1, 5
Correlation length, 70, 96
Counting problem, 62–63
Coupled Markov random fields, 189, 211, 223, 226–230
   and the binding problem, 19, 341, 379
   compound Gauss-Markov model, 230–232
Coupled-phase oscillators, mean-field theory, 10
   a global coupling scheme, 352–356
   population oscillations, 368–371
   rotator models, 344–348
   visual processing and integration, 380
   Winfree's approach, 337

Covariance hypothesis, 16
Cowan-Friedman model, *see* Phenomenological spin model
Critical density, 68
Critical exponents, 64, 69, 101–103
Critical indexes, *see* Critical exponents
Critical opalescence, 66, 96
Critical period, 279, 280, 297–301
Critical points, 66–70, 95–96, 103, 104
Critical slowing down, 139
Critical temperature, 67
   for a ferromagnet, *see* Curie point
   in spin glasses, 124–125
Cross entropy, 57
Curie point, 4, 64, 68, 71, 108

Delay connections, 372–373, 378, 380–381
Delay couplings, 379
Delta waves, 338
Dendritic spines, 307–308
Derin-Elliott model, 197, 211–212
Detailed balance, 53–54, 55
Differential adhesion, 8, 156–157
Diffusible molecules, *see* Chemotropic factors
Diffusion coefficient, 257, 259, 267–269, 275
Diffusion equation, 254, 257, 265–266
Dissipative system, 286, 289
Drift coefficient, 259, 275

Edwards-Anderson model, 112, 118. *See also* Replicas
Edwards-Anderson order parameter, 122
Effective potentials, for reconstruction, 242–243
Elastic string, 238–242
Electroencephalographic recordings, 338, 375, 376
Energy landscapes, 5, 108–111
   in biological systems, 146, 176
   and ergodicity breaking, 125–126
   in the Fraser-Perkel model, 156–157, 160–162
   and NP-hard problems, 129
   ultrametric, 123
Entrainment, mutual, *see* Mutual synchronization
Entropy, 24–25, 34–35, 38
   of a configuration, 28–30
   and disorder, 65, 67, 129
   forms, 30–31
   information theoretic, 24–25
   maximization, 33–34
   and simulated annealing, 133–134
   thermodynamic, 39–42

Equilibrium states, 38–39
  favored, 2, 5
  lowest energy, 42
  ordered, 65
Ergodicity breaking, 125–126, 132
  in graph partitioning, 137
  in traveling salesman problems, 135
Exchange interactions:
  in a ferromagnet, 72
  in a spin glass, 117
  in spin glass models, 118, 119
Excitable media, 401–403
Excitable membranes, 336–337, 381–387
Excitation, recurrent, roles of:
  in hippocampus, 375–377
  in visual cortex, 373–374
Expectation values, 33, 34–35
External parameters, 36–37

Fast threshold modulation, 10, 390–394
Feature selectivity, 280–281
Ferromagnet, 3–4, 5, 70–72
  heat capacity, 67
  spontaneous magnetization, 68–69
Filopodia, 13, 168
Fixed point, defined, 285
Fluctuation-dissipation theorems, 254, 267–269
Fluctuations, 35–36, 70. *See also*
  Fluctuation-dissipation theorem
  critical fluctuations, 4, 65–66, 96, 103–104
  energy fluctuations, 44–45
  regression hypothesis, 253–254
  roles in optimization, 111, 133, 139
Fokker-Planck equation, 256–259
  and Gibbs (Maxwell) distributions, 272–275
  for globally coupled oscillators, 354–355
  and global optimization, 266–267
  and stationary processes, 270
  for a Wiener process, 265–266
Fourier computation, 207–208
Fraser-Perkel model, 8, 156–162
Frequency plateaus, 350–352
Friction constant (coefficient), 256, 259–260, 263
  and the approach to equilibrium, 271, 276
  and the diffusion coefficient, 269
Frustration, 109–110. *See also* Constraints, multiple
  and cooling schedules, 132
  in delay connections, 373
  in the Fraser-Perkel model, 156–157, 162

Gamma band oscillations:
  in the hippocampus, 375, 377
  in the visual cortex, 338, 341, 378, 380–381
Gauss-Markov models, 188, 196. *See also*
  Autonormal model
  compound model, *see* Compound
    Gauss-Markov random fields
  wide-sense autoregressive representation, 201
Geman and Geman, method of, 7, 188–189, 208–215, 223–230
Generalized homogeneous functions, 101
Genetic instruction in epigenesis, 11
Gestalt laws, 379, 380
Gibbs distributions, 24, 36, 43, 46
  in Langevin dynamics, 140–142
  in the Metropolis algorithm, 54–55
  in simulated annealing, 128–131
Gibbs-Markov equivalence, 186–187, 189–191, 192–195
  and simulated annealing, 188
Gibbs random field, 185–186, 194. *See also*
  Gibbs-Markov equivalence
Gibbs sampler, 213–215
Glauber dynamics, 112–114
G-proteins, 169, 174–175
Graduated nonconvexity, 189, 237–242
Grand canonical ensemble, 31, 36, 45–46
Grandfather cells, 378
Graph, 192
Graph partitioning problem, 136–137
  and NP-hard problems, 110, 125, 128, 129
  replicas and a spin glass phase, 137
  ultrametricity, 137
Growth cones, 13, 151, 168–169. *See also*
  Chemotropic factors; Cell adhesion molecules
  signal transduction and integration, 172–175

Hamiltonian, 39
Hamilton's equations, 286–287
Hammersley-Clifford theorem, *see*
  Gibbs-Markov equivalence
Heat, 36, 39–42
Heat bath algorithm, 114
Heat capacity, 45
  for a ferromagnet, 67
  for an Ising chain, 80
  for an Ising lattice, 82–83
  in mean-field theories, 90–91
  in simulated annealing, 133–134
  for a spin glass, 123–125
Hebbian assembly, *see* Assemblies

Hebbian synapses, 15–17, 282, 283
   and NMDA receptors, 279–280, 308
Hebb's rule, 16
Helmholtz free energy, 44, 65
   in mean-field annealing, 236
   in Peierls's argument, 83–85, 103
   in spin glass models, 119–120
   and Widom-Kadanoff scaling, 102
Hemoglobin, 2
Hodgkin-Huxley equations, 10, 336, 341, 382–384
   for the Bush-Douglas model, 385
   for thalamocortical spindling, 395–400
   for Traub's hippocampal network, 385
Hopf bifurcation theorem, 362
Hysteresis, 72
   in coupled oscillator communities, 348
   in spindling, 398
   in the Wilson-Cowan model, 338, 341, 365–367

Inexact differentials, 39
Inhibition, global and mutual, roles of, 341
   in hippocampus, 375–377
   in spindling, 399–400
   in visual cortex, 373–374
Inhibitory circuitry, role of:
   in common input models, 325
   in generating repetitive firing, *see* Inhibition, global and mutual, roles of
   in a mean-field network, 306
   unmasking and plasticity gates, 177, 314–315
Integration, *see* Coupled Markov random fields
Internal energy, 36, 38–39, 46
   in an Ising chain, 79–80
   in an Ising lattice, 81–82
   in a mean-field approximation, 89
Intrinsic images, 226
Irreducible chains, 53
Ising chain, 76–80
Ising lattice, 80–83
Ising model, 7, 63–64, 73–75. *See also* Ising chain; Ising lattice
Iso-orientation patches, 321–322, 325
Iterated conditional modes, 188, 219–223
   for coupled fields, 228–229

Jensen's inequality, 234–236

Kadanoff block spin construction, 97, 244–245

Karhunen-Loéve expansion, *see* Principal component neurons
Kinetic Ising model, 112–115
König-Schillen model, 372–373

Lagrange multipliers, 24, 31–33, 36
   in constrained optimization, 137, 210, 213–214
Lamellipodia, 168
Langevin diffusions, *see* Langevin dynamics
Langevin dynamics, 7, 112, 140–142, 265–267
Langevin equation, 259–264
Latent heat, 66
Lateral geniculate nucleus, 8, 148–150
   laminar structure, 149–151
   multiple interactions, 150–152
   spontaneous electrical activity, 151–152
Lattice gas, 91–96, 104
Learning (spatial) and memory, 283, 310–314, 374
Lee-Malpeli model, 8, 162–168
   nonequilibrium dynamic model, 177
Lennard-Jones potential, 91
Lexicographic notation, 200
Limit cycle, defined, 285, 335
Linear integrator (associator), 281, 290–292
Linear stability and nullcline analysis, 10, 356–361
   in BCM theory, 10, 295–296, 306
   and cellular automata, 401–406
   and fast threshold modulation, 390–394
   in the Morris-Lecar model, 385–390
   in the Wilson-Cowan model, 362–367
Line processes, 209, 223–230
   and the binding problem, 19, 379
   in compound Gauss-Markov models, 231–232
   in graduated nonconvexity, 238–240
Linsker's model, 320–321
Liouville's theorem, 287–289
Long-range order, 68
Long-term depression, *see* LTP and LTD
Long-term potentiation, *see* LTP and LTD
LTP and LTD, 16–17
   and CaM kinase activity, 310–315
   and cell surface molecules, 308–309
   dependence on prior use, 314
   dependence on stimulation frequency, 309–310
   in the hippocampus and visual cortex, 279, 283, 374
   and NMDA receptor activation, 283, 309–310

INDEX   **425**

Magnetic susceptibility, 69, 118
Marker models, 155–156
Markers, 151, 155. *See also* Molecular gradients
Markov chains, 5, 6, 24–28, 50–54
 by Langevin diffusions, 141–142
 by the Metropolis algorithm, 54–56
 in simulated annealing, 131–132
Markovian property, 4–5
Markov mesh random field, 187
Markov P-process, 182–183
Markov process, 5, 7, 255–256
Markov random field, 5, 7–8, 110, 111, 182
 on lattices and graphs, 183–185, 193. *See also* Gibbs-Markov equivalence
 strict and wide sense processes, 199–200
Master equation, 113
Maximizer of the posterior marginals, 188, 215–218
Maximum entropy principle, *see* Entropy, maximization
Mean-field annealing, 189, 233–237
Mean-field theory, 6, 63, 64
 for a cortical network, 303–306
 for coupled oscillators, *see* Coupled-phase oscillators, mean-field theory
 for an Ising ferromagnet, 85–91, 103–104
 in random-field annealing models, 233–237
 for a spin glass, 121
Metropolis algorithm, 6, 25, 54–56, 111
 for binary alloys, 116–117
 for an Ising spin system, 114–115
 for the lateral geniculate nucleus, 168
 for the retinotectal projection, 160–162
 for segmentation, 218
 in simulated annealing, 129–131
 for texture synthesis, 198–199
Microcanonical annealing, 112, 137–139
 and Metropolis algorithms, 139
Microcanonical ensemble, 42–43
 in annealing, 137–139
Möbius inversion, 187, 192, 194
Molecular gradients, 151, 156. *See also* Marker models; Multiple constraint models
Morris-Lecar model, 10, 336, 342, 385–390
 and relaxation oscillations, 390–394
Monte Carlo method, 25, 46–50, 63. *See also* Metropolis algorithm
Multiple constraint models:
 of Fraser and Perkel, *see* Fraser-Perkel model
 of Lee and Malpeli, *see* Lee-Malpeli model

of Whitelaw-Cowan, 155
Mutual synchronization, 3, 342–344, 349–350

Neighborhood system, 183–184, 193
NMDA receptors, 16–17, 155, 312, 339, 374
 Hebbian mechanism, 279–280, 308
Nonlinear dynamic systems, 5–6, 10, 283–290, 335–340
NP-complete, 126–128, 131–132, 135, 137
NP-hard, 110, 112, 126–128, 131–132
Nullclines, *see* Linear stability and nullcline analysis
Nyquist relations, 254, 277

Ocular dominance, 13, 280, 281, 298–303
Ocular dominance columns, 321–322
Oja's rule, 319–320
Order-disorder transitions:
 in cellular automata, 404
 in ferromagnets, 3, 67, 68
 in laser emission, 4, 11
 in the lateral geniculate nucleus, 148–150
 in oscillator communities, *see* Phase transitions in oscillator communities
 in the retinotectal projection, 155
Order parameters, 67, 68–69
 for coupled oscillators, 337, 345–346, 355–356
 for a ferromagnet, *see* Spontaneous magnetization
 for spin glasses, 122–123
Orientation selectivity and binocular interaction, 13, 280, 281, 298–301. *See also* Receptive fields
Oscillator death, 352
Oscillators, phase-coupled, 340, 342–344
 mean-field models, 344–348, 352–356
 nearest-neighbor models, 349–352

Pacemaker cells, 9, 337–338, 342, 352, 381
Pairing gap parameter, 68
Paramagnet, 70–71, 124
Parameters of state, 38
 in the Ising model, 75
Parisi order parameter, 122–123
Particle kinetics, 116–117
Partition function, 34–37. *See also* Helmholtz free energy
Pattern generators, 18, 352, 379
Peierls's argument, 64, 83–85, 103
Pendulum, simple, 289, 335
Phase plane methods, 10, 341, 356–362
Phase-pulling, 10
Phase space dynamics, 284–290

Phase transitions, 45, 65–67, 70. *See also* Order-disorder transitions; Phase transitions in oscillator communities
   avoidance in random field models, 220, 223
   influence on annealing schedules, 132–133
Phase transitions in oscillator communities, 337, 345–348, 355–356
Phenomenological spin model, 321–322
Pickard causal model, 187
Piecewise smoothness, 211, 238
Poincaré-Bendixson theorem, 361–362, 367, 387
Population oscillations, models of:
   mean field, 368–371
   nearest neighbor, 372–373
   Wilson-Cowan, 9, 338, 341, 362–367
Pressure, 41
Prestige-Willshaw model, 155
Principal component neurons, 301, 315–320
Principle of indifference, *see* Microcanonical ensemble
Principle of insufficient reason, *see* Microcanical ensemble
Probabilistic hill climb, 117, 129. *See also* Simulated annealing; Langevin dynamics
Projection pursuit, 326–327
Protein folding, 4, 11

Quenched annealing, 222–223
Quenched random variables, 112, 119, 125
   in coupled oscillator communities, 349
   for the traveling salesman tour, 135
Quenching, in metallurgy, 219

Rana pipiens, 15
Random field, 183, 193
Random variables, 50
   quenched and annealed, 119
Reaction-diffusion equation, 401–402
Real space RG methods, 244–245
Recall, 340
Receptive fields, 8, 13
   in BCM theory, 301–303
   in a common input approximation, 322–325
   dynamic reorganization, 176–177
   dynamic representation, 340
   in Linsker's model, 301, 320–321
   spontaneous symmetry breaking, 295, 301, 321
Reciprocity relations, 254, 276–277
Reconstruction, of images, 204–205, 226
   by constrained least-squares estimation, 205
   using deterministic models, 232–243
   in maximizer of the posterior marginals, 216–217
   using maximum a posteriori estimation, *see* Geman and Geman, method of
   by minimum mean square error estimation, 205–208
   in renormalization group simulated annealing, 243–248
Reentrant signaling, 379–380
Regression hypothesis, of Onsager, 253–254, 256, 276. *See also* Fluctuation-dissipation theorems
Regularization theory, 189
Relaxation oscillations, 10, 342, 360, 367, 390–394
Renormalization group, 63, 64, 96–103, 104
Renormalization group simulated annealing, 243–248
Replicas, 119–121, 135, 137
Restoration, of images, *see* Reconstruction, of images
Retinotectal projection, 146–147, 150
   experimental manipulations, 146, 153–155
   multiple interactions, 150–152
   multiple stable states, 146
   rapid remodeling, 14, 151
Rotators, coupled, *see* Oscillators, phase-coupled

Scale invariance, in natural scenes, 315
Scaling, 64, 96–97, 104, 243–244. *See also* Renormalization group
Schuster-Wagner model, 368–371
Segmentation, 197
   of laser radar data, 228–229
   using maximizer of the posterior marginals, 216–217
   of synthetic aperture radar data, 220–222, 229
Sharp waves, hippocampal, 338, 375
Sherrington-Kirkpatrick model, 112, 118, 121
Short-term plasticity, 394
Simulated annealing, 6, 104, 112, 128–134
Simultaneous autoregressive models, 188, 196–197, 202–204
Smoluchowski equation, 258
Specific heat, *see* Heat capacity
Spindle waves, thalamocortical, 338, 342, 394–400
Spin-flip kinetics, 113–114
Spin glasses, 46, 110, 112, 117–125
   and cooling schedules, 132–133
   and ergodicity breaking, 125–126

Spontaneous electrical activity, 13, 151-152
Spontaneous magnetization, 4, 68-69, 71
  in an Ising chain, 79
  in an Ising lattice, 83, 108, 125
  in a mean-field approximation, 87
Spontaneous symmetry breaking, 4
  in Ising ferromagnets, 69
  in oscillator communities, 337, 346, 356
  in receptive fields, 295, 301, 321
Stationary distributions, 51, 53-54
Stationary processes, 270-271
Statistical mechanics, 33-46
Stent's rule, 16
Stochastic differential equations, 140, 256, 265-267. *See also* Langevin equation
Stochastic matrices, 53
Strict-sense Markov random field, 199
Support sets, 187, 202-203
Suspicious coincidences, 281
Synaptic modification, models of, 9, 280-283, 284, 294
  Bienenstock-Cooper-Munro, *see* BCM theory
  in a common input approximation, 322-325
  Linsker, 301, 320-321
  in phenomenological spin, 321-322
  principal components, 315-320
Synaptic plasticity, 9, 15-17, 278-279
  and adult reorganization, 176-177, 279
  in *Aplysia*, 279, 309
  in the cat, 280, 298-301
  and cell surface molecules, 308-309
  in *Drosophila*, 311, 312-313
  in genetically-altered mice, 311, 313-315
  in the retinotectal projection, 146, 153-155, 176
Synchronized bursting:
  in hippocampal networks, 375-377
  in visual cortex, 373-374
Synchronous activity:
  forms of, 338, 375, 394
  functional roles of, 339-340, 374, 375, 378-381
Synchronous, 7-12 Hz oscillations, somatosensory, 340

TAP equations, 121
Temperature, 38, 40-41
  in Langevin dynamics, 140-142
  in simulated annealing, 128-134
Tetrodotoxin, 16, 150, 279
Textures, analysis and synthesis of, 187, 196, 197, 201

  in the autobinomial model, 198-199
  in the Derin-Elliott model, 211-212
Thalamic relay neurons, 5-6
Thermodynamic limit, 78-79
Thermodynamics, 23-24, 38-42
  first law, 24, 41
  second law, 24, 31, 41
  third law, 42
Theta waves, 338
Topographic maps, 12, 14
  dynamic reorganization, 176-177
  dynamic representation, 340, 394
  in the lateral geniculate nucleus, 163-164
  in the retinotectal projection, 150, 279-280
Transfer matrix method, 63, 76-80
Transition probabilities, 51-53
Traveling salesman problem, 126, 127, 134-135
  and NP-hard problems, 110
  replicas and a spin glass phase, 135
Traveling waves, 9, 337, 340, 352, 376, 394
Triple point, 66, 95-96, 104
Tritonia, 18, 379

Ultrametricity, 122-123
  in graph partitioning, 137
  in networks of chaotic elements, 356
  in traveling salesman tour, 135
University, 69, 97, 104. *See also* Renormalization group

van der Pol oscillator, 336, 359-360, 385, 391-392
van der Waals picture, 63, 64, 94-96, 104
Variances, in parameters of state, *see* Fluctuations
Visual system, epigenesis, 8-9, 11-15
  and classical rearing, 298
  of the lateral geniculate nucleus, 148-150
  multiple interactions, 150-152
  multiple stable states, 146-147
  of the retinotectal projection, 150-151
  and synaptic plasticity, 16-17, 278-280
Voltage-dependent calcium channels, 174

Weak coupling, 337, 342-343
Weak membrane, 240-243
Weiss molecular field, 85-87, 90-91
Wide-sense Markov processes, 197, 199-201
Widom-Kadanoff scaling, 101-103
Wiener processes, 256, 265-266
Wilson-Cowan model, 9, 338, 341, 362-367
Work, 36, 39-42

XY-model, 72-73, 321-322